Roisín Higgins, Manchester 1996.

ENLIGHTENMENT BORDERS

pre- and post-modern discourses

medical, scientific

uniform with this volume

ENLIGHTENMENT CROSSINGS
pre- and post-modern discourses
anthropological

PERILOUS ENLIGHTENMENT
pre- and post-modern discourses
sexual, historical

ENLIGHTENMENT BORDERS

pre- and post-modern discourses

medical, scientific

G. S. Rousseau

MANCHESTER
UNIVERSITY PRESS
Manchester and New York

distributed exclusively in the USA and Canada
by St. Martin's Press, New York

Copyright © Manchester University Press 1991

Published by Manchester University Press
Oxford Road, Manchester M13 9PL, UK
and Room 400, 175 Fifth Avenue,
New York, NY 10010, USA

Distributed exclusively in the USA and Canada
by St. Martin's Press, Inc.,
175 Fifth Avenue, New York, NY 10010, USA

British Library cataloguing in publication data
Rousseau, G. S.
  Enlightenment borders: pre- and post- modern discourses, medical scientific.
  1. Science, to 1979. Social aspects – Philosophical perspectives
  I. Title
  306.4509

Library of Congress cataloging in publication data
Rousseau, G. S. (George Sebastian)
  Pre- and post-modern discourses/G. S. Rousseau.
    p. cm.
  Includes bibliographical references and index.
  Contents: v. 1. Enlightenment crossings — v. 2. Enlightenment borders — v. 3. Perilous enlightenment.
  ISBN 0–7190–3072–2 (v. 1). — ISBN 0–7190–3506–6 (v. 2). — ISBN 0–7190–3301–2 (v. 3)
  1. Social history—18th century. 2. Civilization, Modern—18th century. I. Title.
  HN13.R68 1991
  306'.09'033—dc20   90–19452

ISBN for complete set of 3 volumes 0 7190 3549 x
ISBN for this volume 0 7190 3506 6 *hardback*

Phototypeset by Input Typesetting Ltd., London
Printed in Great Britain
by Bell & Bain Limited, Glasgow

CONTENTS

	Acknowledgements	vi
	Introduction	ix
	PART ONE: MEDICAL	
1	The discourses of literature and medicine: theory and practice (1)	2
2	The discourses of literature and medicine: theory and practice (2)	26
3	Ephebi, epigoni, and fornacalia: towards a historiography of medicine	55
4	Medicine and millenarianism	78
5	Rationalism and empiricism in Enlightenment medicine	118
6	Praxis 1: Bishop Berkeley and tar-water	145
7	Praxis 2: Pineapples, pregnancy, pica, and *Peregrine Pickle*	176
	PART TWO: SCIENTIFIC	
8	The discourses of literature and science (1)	202
9	The discourses of literature and science (2)	213
10	The discourses of literature and science (3)	236
11	Repenser Bachelard	253
12	Science books and their readership in the High Enlightenment	264
13	Wicked Whiston and the English wits	323
	Index	343

ACKNOWLEDGEMENTS

Given the perilous disciplinary borders these essays traverse, it is not surprising that the extended essay (and sometimes not so extended) should have proved my métier. The essay form affords the probing but still unsettled mind, often hankering to make up its mind but plagued by uncertainty, the leisure to ponder alternatives and take roads not trodden. Essays of book length consume so much time and energy on single topics, and are put through such monolithic scrutiny, that they rarely permit their authors the possibility of breaking out of accepted moulds. This is all the more true when disciplinary boundaries are crossed.

For this reason I have relied on the essay form – diverse, dialogical and problematic as it is – to find my own bearings. All of these essays have been published before, but not in this form and not collected in this way. Indeed, the gathering of these essays into this collection and the process of deciding which essays to include has helped me to probe even further.

But I could not have done this alone, or without the sustained dialogues I have held with my friends and colleagues over the years. Chief among these have been ongoing discussions – written and oral – with Richard D. Lehan and Maximillian E. Novak, my colleagues at UCLA, and Roy Porter, my collaborator at the Wellcome Institute for the History of Medicine, whose academic post across an ocean and a continent has not prevented us from debating these matters. To them, as well as to the memory of my teacher and mentor, the late Marjorie Hope Nicolson, these essays are dedicated, partly in the hope that they may eventually serve as a type of record of our own crossing over many perilous borders. The last of these scholars cannot read them, but if the others discover the essays in *Enlightenment Crossings, Perilous Enlightenment* and *Enlightenment Borders* to be a useful reminder of their own important role in activating these debates over two decades, I shall be satisfied.

Others too numerous to name also played a crucial role in the formation of thought represented here by offering encouragement and agreement, as well as disagreement and sharp criticism. They included but are not limited to: Michael Allen, Joanne Trautman Banks, Gillian Beer, P.-G Boucé, Leila Brownfield, Mario Bagioli, Robert Adams Day, Philippa Foot, Clifford Geertz, Ruth Graham, Donald Greene, Roger Hahn, John Heilbron, Gert Hekma, Wallace Jackson, Carolyn Lyle Williams, David Morris, John Neubauer, James Paradis, Ronald Paulson, Richard Popkin, Pat Rogers, Elinor Shaffer, Simon Shaffer, Jean Starobinski, Lawrence Stone, Randolph Trumbach, Richard Zeimacki. I also owe a debt of thanks to my editor John Banks. He has read every one of these chapters with the care of a first-time reader, and promoted me to reconsider their form and content long after I thought I could doctor them no more. These essays could not have made their way through the various stages of publication at a large university press without his experienced editorial eye. My gratitude to him is

Acknowledgements

therefore like that of a son to a father: as time passes the younger knows how much he owes to the older.

These essays appear in their original versions (some with changed titles) except that works originally cited as forthcoming and some of the scholarship have been brought up to date, chapters have been cross-referenced to each other for the reader's convenience, references have been standardised in part, and minor corrections have been made. While assembling the chapters and organising them in the three volumes I thought it particularly important that they be cross-referenced to each other. But there has been no attempt to rewrite the essays in order for them to appear more unified than they are, or to suggest that they now have a common audience. Disguise plays no role in the act of collecting and presenting them here, and the fact is that these essays were written over the span of a quarter of a century for different audiences on different occasions. It insults the reader's intelligence to pretend that this had not been the case, or to glaze over this plain fact by artificially recasting or reshaping them here. The place of all publications is London where not otherwise cited.

The original places of publication of the essays in this volume are as follows: (1) 'Literature and Medicine: the State of the Field', *Isis*, LXXII (September 1981), 406–24; (2) 'Literature and Medicine: Towards a Simultaneity of Theory and Practice', in Joanne Trautman Banks (ed.), *Literature and Medicine: Use and Abuse of Literary Concepts in Medicine* (Baltimore, 1986), 152–81; (3) 'Ephebi, Epigoni, and Fornicalia: Some Meditations on the Contemporary Historiography of the Eighteenth Century', *The Eighteenth Century: Theory and Interpretation*, XX (autumn 1979), 203–26; (4) 'Medicine and Millenarianism: Immortal Doctor Cheyne', in Richard Popkin (ed.), *Millenarianism and Messianism in English Literature and Thought 1650–1800* (Leiden, 1988), 81–126; (5) ' "Sowing the Wind and Reaping the Whirlwind": Aspects of Change in Eighteenth-Century Medicine', in Paul J. Korshin (ed.), *Studies in Change and Revolution: Aspects of English Intellectual History 1640–1800* (London, 1972), 129–59; (6) (written with Marjorie Hope Nicolson) 'Berkeley's *Siris* and English Literature 1750–1800', in G. S. Rousseau and Eric Rothstein (eds.), *The Augustan Milieu: Essays Presented to Louis A. Landa* (Oxford, 1970), 101–37; (7) 'Pineapples, Pregnancy, Pica, and *Peregrine Pickle*', in G. S. Rousseau and P. G. Boucé (eds.), *Tobias Smollett: Bicentennial Essays Presented to Lewis M. Knapp* (New York, 1971), 79–110; (8) 'Literature and Science: the State of the Field', *Isis*, LXIX (1978), 583–91; (9) 'The Discourses of Literature and Science', in *Hartford Studies in Literature and Language*, XIX (winter 1987), 1–19; (10) ' "Till we have built Jerusalem": The Berkeley Symposium and the Future of Literature and Science', in G. S. Rousseau (ed.), a special issue on Science and the Imagination in *Annals of Scholarship*, IV (fall 1986), 1–21; (11) 'Repenser Bachelard', *Annals of Scholarship*, V (autumn 1988), 297–310; (12) 'Science Books and their Readers in the Eighteenth Century', in Isabel Rivers (ed.), *Books and their Readers* (New York, 1982), 197–255; (13) ' "Wicked Whiston" and the Wits: The James L. Clifford Lectures', in John Yolton (ed.), *Studies in Eighteenth-Century Culture* (East Lansing, 1987), 17–44.

UCLA, AUGUST 1990

FOR DICK, MAX AND ROY
with friendship and admiration

and to the memory of
MARJORIE HOPE NICOLSON
who would have applauded the attempt

INTRODUCTION

For as long as I can remember I have been writing essays in the borderlands between academic disciplines and crossing over their perilous boundaries. It has been an act of intellectual will as much as a temperamental proclivity prompting me to break away from the traditional boundaries dictated by university faculties and from the conventional departments academic disciplines designate. There are many implications – personal and profession, theoretical and practical – in crossing boundaries; and the role played by borders in cultural and social differences should not be minimised, either in my own development or in the intellectual life of the times in which I live. These chapters typify those crossings over two decades and constitute the first volumes of my collected essays since I published, almost ten years ago, an anthology of my essays about an eighteenth-century novelist now barely read (Tobias Smollett).

Throughout my writing career – now encompassing the best part of a quarter of a century – I have always looked beyond my immediate academic discipline – English literature of the eighteenth century – to frontiers where different disciplines come together. Some have called me a pioneer in this activity; others, who have felt unduly challenged and discomforted by the resulting dialogue, have claimed that the energy has been misspent and somewhat unintelligible. I would have done better, they have claimed, to remain at home, as it were, and cultivate my own garden.

My response is that I cannot. For many reasons, including ethical ones compelling persons to strive to leave the world better off than they found it, I must always be travelling to another country: to see its territory with an alien, sceptical eye; to taste its exotic and remote realms for myself; to bring my own local knowledge, the knowledge of my own discipline, to its distant borders. This matter of 'borders' penetrates to the essential tension displayed in these essays. The borders between disciplines, like those strange borderlands between countries, have always engendered hybrid cultures (think of the Swiss borders, for example) and bilingual discourses, and it is in these domains that I have elected to study the Enlightenment, about which I care so much.

Nor is the metaphor of countries and borders excessive to describe these disciplinary territories. The humanities and sciences are now practised in such diversity that even those who work in the same field (here is the metaphor under different stress) often discover they do not speak the same language. The theorization of the academy in our time – to echo the phrases of Quentin Skinner and Frederick Crews – has intensified the disparity. Yet, in my own case, I was crossing these boundaries before the advent of poststructuralism, postmodernism, deconstruction, 'difference' – *différance* – and now postdisciplinary discourse. The trope that troubled me throughout the 1960s and early 1970s when some of these essays were written, was that of the 'Renaissance man': a creature with wide-ranging interests in diverse fields who refused to succumb to specialisation.

Enlightenment culture boasted many such figures, of which Franklin and Goethe are perhaps the best known. I did not think of myself as one under any circumstances; nevertheless the 'Renaissance man' represented the ideal for us (then green and young) academics to cultivate in the age of rampant technology and specialisation, when all broad knowledge was threatened with extinction. The Renaissance man did not, of course, have a chance of survival in our specialised universities endorsed by the post-industrial information state. Our research centres mandated just the opposite gaze: narrow, specialised, small pieces of knowledge; isolated practices and specialised discourses performed by technicians who resembled those poor souls toiling away in solitude in Leibniz's imaginary cells.

This ferment of knowledge, as I attempt to document briefly in the head notes to the essays, occurred during the 1970s when various social movements (e.g. the response to the liberalism of the 1960s, the revolt against many forms of authority) and intellectual currents (the first wave of poststructuralism) were palpably felt. It was then, I think, that the academy became theorised and was transformed into the institution we know it to be today. In that long decade extending from about 1968 to the early 1980s, a majestic transformation in the status of knowledge occurred throughout the western world, as it altered from an age of hierarchy and principle to one in which the processes of gathering and classifying information were themselves representative of the highest goal: from modernism to postmodernism; from a structural habit of mind to poststructuralist; from one set of technologies and its attendant practices to a quite different one labouring under the force of the microelectronic revolution. This transformation, about which the historians of the future shall fill many books, was bound to take its toll on the academy – among the most vulnerable of contemporary institutions anyway – and its conventional academic disciplines. Bluntly put, by the end of the 1970s it was clear within the universities that a whole repertoire of buzz words (interdisciplinary, transdisciplinary, crossdisciplinary) had become the signs – the real semiotics – of an oncoming postdisciplinary era in which the traditional disciplines would be hurled to the margins of civilised discourse. Having flown in the face of authority in this way, *post*disciplinary activity institutionalised itself at the centre – in the core – of academic dialogue and discourse.

The transformation posed significant challenges to me personally and professionally. To a certain extent I had been engaged in these encounters among disciplines for some time; crossing the borders, so to speak, before the challenges of the transformation; traversing discourses before the ideology of interdisciplinarity set in; so there was really no question about the novelty of the activity. By the time 'interdisciplinary;' became the favoured code word of administrators and all those seeking funding, I had been (to invoke another neologism) 'interdisciplinating' for some time. The disclosure may seem pungent but it was the methodological consequence of interdisciplinarity in the postdisciplinary age that consumed me rather than any search for novelty: especially the degree to which such a perilous approach could actually produce a more valid historicism by virtue of the broader contexts it created. If the crossing of disciplinary borders could locate previous cultures more accurately, then the risk was worth the effort.

This constellation of categories – culture, context, historicism – is the one this volume explores. It does so through a reconsideration of the ways in which that crucial transformation in twentieth-century cultures was conducted. The fact that the European Enlightenment has always been my preferred chronological

Introduction

period is of little consequence here: it could have been the Middle Ages, the Renaissance, or the Victorian Age, for history is always a dialogue of the present with the past as much as a perception of turning entirely to the past. The matter then is not my own diachronic period, the Enlightenment, but the ways in which that era and its culture have been related to the inescapable present; more explicitly, the ways in which the traditional academic disciplines become the abettors or detractors of that dialogue.

If the borders of these essays, which include critiques and metacritiques and which were written over two decades, are anthropological, scientific, medical, and sexual, their common boundary remains the zone explored in these essays. I make no claim that any *one* method has been invoked, or that an interdisciplinary methodology has emerged after such a long time. On the contrary all my fear and trembling arises from the knowledge that no single method has suggested itself *despite* a constant search for one. This resolution and the process it has entailed represents, I think, my own claim against theory and its grand edifice. Put otherwise, interdisciplinarity itself represents a set of curiosities, a range of possibilities, rather than the beginnings of any method or logic that can be codified and then applied.

The lack of recognised method is, of course, precisely what is held up against it by its detractors who profess that whatever their defects may be, the *traditional* disciplines at least have a methodology (i.e. craft, *kraft*), without which significant inquiry cannot begin, let alone be resolved. But this junction marks the very point complicating these perilous crossings. For we are not merely crossing disciplinary boundaries here but are doing so at a moment when the traditional disciplines themselves have ceased to exist in any organised way. At least this position is the one currently held by almost all those who have pondered these matters, and they themselves come from such a variety of disciplines and diversity of backgrounds that it is inconceivable there could be any collusion or consensus in this position: Kuhn, Weber, Habermas, Latour, Lyotard, Derrida, *et al*.

Something more elaborate must be said, even in a brief space, about this disciplinary-postdisciplinary milieu in which we find ourselves – teachers and students alike – today. The disciplinary still exists – of course – in the institutional sense. Colleges and universities everywhere have faculties, or departments, of history, philosophy, literature, mathematics, physics, and so forth. But – and it is the salient but – the concerns of those faculties and departments have been so remarkably transformed that their designating name signifies little and certainly does not indicate that its current members are asking any of the questions that were being asked a decade or two ago. it is as if these faculties are just the ruins – the shells – of their former selves. They remain vital because they are asking the important new questions, which often are interdisciplinary, not because they each embody a hard won craft or methodology. The semiotics of this state of affairs is that our contemporary institutional arrangements have much to do to catch up with the discourses and practices activated by their members.

These discourses and practices remain at the heart of the matter, and will eventually have to be addressed on a global scale. The essays in these three volumes – *Enlightenment Crossings*, *Perilous Enlightenment* and *Enlightenment Borders* – suggest that such a forum is actually overdue. But there are other concerns perhaps as crucial as the institutional aspect of the various disciplines. Why, some readers will shrewdly ask, are the disciplines so important? Why cannot scholars and critics merely explore wherever their curiosity takes them without concern or regard for disciplinary boundaries? The answer may appear unassailably positive to these questions, but the reality is quite otherwise. Disciplines since the eighteenth century have established their own vocabularies, even code words, and procedures, which function as union cards for admission to the guild. More consequential, disciplines have for two centuries now set the agenda – the kinds of questions – that could be asked by its practitioners. Given that the agenda always determines the conclusions that can be reached in any discussion, whether academic or not, this matter of programmatic agenda is essential if one contemplates traversing disciplinary boundaries in the way I have done here.

But for every reader who interrogates the disciplinary status of these essays, there will be three, I suspect, who want to penetrate to the essays themselves and who will ask questions about the broad range of the essays – will enquire into the background of the author. Here one thinks of Foucault's question, *What is an author?* How can someone whose 'field' is literature range so widely in the borderlands between topics anthropological, scientific, medical, sexual, etc? Is there not some polymorphous perversity here that should be curbed and eradicated? This is the point where postdisciplinary knowledge must be consulted, for – as I have been suggesting – at some point in the 1970s or early 1980s, knowledge ceased being local and disciplinary and became global and interdisciplinary, or – as others called it – discourse thriving on the margins of traditional discourse. Coeval with the movement away from local national structures was a turn to the world economy, world ecology, world environment, world debt crisis – in the vernacular, to one global village. Perhaps this is why Habermas's current project remains the investment in a 'philosophical discourse of modernism' without regard to the disciplinary at all. Others – Feyerabend, Kuhn, Deleuze, Derrida, Bordieu – have never been willing to label themselves historian, philosopher, critic, etc. in the first place. And still others have claimed that the boundaries drawn today are those of class, gender, and race, not of disciplines, institution, or nation.

If this is so, and there is reason to believe it is so, then these essays also serve as the personal chronicle of a scholar who found himself working between the two worlds of the local and the disciplinary versus the global and interdisciplinary. When I was starting out in the 1960s, I had little inkling that the traditional disciplines would soon become transformed as they have been; still less notion that within just a few years, I would soon find myself constricted by the conventional subjects and desperately trying to break out of them. It is as if a physicist were trying to become an interphysicist, or a historian a cultural-anthropological historian. Once she or he starts, the old disciplines appear increasingly constraining in the postdisciplinary age.

By the 1970s, I was self-reflectively searching for a whole series of ways to escape from the tyranny of the disciplines, often without knowing it. And the toll this frustration took is evident in many of the essays, whose genuine subject is the transformation of the traditional agenda of the disciplines in questions, whether one disciplines or another is now irrelevant.

This progression from disciplinary to postdisciplinary organisation also took a toll on my equilibrium. There were long intervals in which I did not feel as if I belonged to *any* disciplines. Some of my opponents called me a defector from the fold who had developed a sharp and polemical style to ward off the skeptics;

Introduction xiii

others claimed that my wide-ranging work in Enlightenment studies was suspicious by dint of not being the product of someone trained in the specific disciplines from which it emerged. But others latched on to these approaches and seemed to applaud the wide gaze. A distinguished historian of literature asked for how long I could expect to get away with not admitting that I had now become a professional historian of science. Yet another claimed that I would have little difficulty if I now called myself a professional historian of medicine. To these and others there is only one response it seems to me: that they are all correct because the traditional disciplines themselves had lost their former vitality. The chaos and confusion lay not in my individual development – as if it had been unique – but rather in the transforming disciplines which had grown miserably out of touch with the reality of the discourses and practices their members were writing and following.

The reader will also see that history remains an *arrière pensée* of these essays. No matter which border I perfidiously cross at this time, I keep returning to history and its vexations. If the purpose of theory is indeed to reinvigorate historical studies, then it has achieved its purpose in my case. From the first of these essays – published late in the 1960s – to the most current at the end of the 1980s – my imagination has always been fixed on the problem of historicism: which historicism, of what variety, how practised, in which rhetorics and tropes and genres? I am thus in no sense a convert: having begun historical – as the head notes detail – I remain historicist. All my doubt has rested on a more proper approach to a deep-layer historicism, especially for Enlightenment studies, than the ones we have at present. But the great challenge of theory during my adult lifetime has been its extraordinary possibilities for historicism. If I did not believe that, I would long ago have abandoned cultural history and returned to the old methods in which I had been trained. Theory, especially feminist theory and the theory that privileges sex, gender and class, has never appeared to me the gremlin it symbolises to some of my university colleagues and fellow scholars because of its inherent ability to re-energise history and challenge conventional wisdom. The challenge of a theorised academy, such as we now have, is a healthy and positive one because theory's ability to ask questions requiring a higher threshold of explanation than prevails in times when theory is not so widely dispersed and ingrained as a natural habit of mind.

Furthermore, during my adult life I have lived through a revolution in knowledge created in large part by the disciplines represented in these essays. In cultural anthropology after Mead and Lévi-Strauss, and the new history of science after Kuhn, Latour, and Bordieu, are found some of the sources of the *post*disciplinary transformations these essays imply. But not all. Much is also owing to Western society at the end of the twentieth century: its global shifts, unprecedented pluralism, run-away technology, and renewed quest for freedom and egalitarianism at any price. This is perhaps why disciplines such as science and medicine have been so important in our time. Without a serious critique of medicine we could not understand how previous men and women had resisted to temptation to cave in under the stresses and strains of the postindustrial age. Yet 'sex' has no academic disciplines, no institutionalised filiation. To be blunt, in which faculty or department would sex and its history be studied? This is the question Foucault often asked, but he supplied no satisfactory answer. Psychology, psychiatry, history, philosophy, sociology, literature? Surely all of these. That is why we cross borders in the first place, to understand evolving discourses like the one of sexuality. But then, why do we still retain disciplines at all?

Finally some readers may wonder why I have been so biographically explicit, and whether it would not have been preferable to write an autobiography or publish memoirs. It is a good question readily responded to. My purpose has not been to document my own development in regard to my *life* – that I will do elsewhere, in another genre altogether – but rather, to suggest the various ways in which that intellectual development was prescribed and determined somewhat beyond my control. Those who contend that I have tried to model myself (to return to the old trope used above) on the Renaissance man have perhaps missed the point, which has been this: do what I could, I seem to have been unable to narrow my concerns and practice in one discipline only; incapable of remaining within the traditional mould and tending to the local agenda. My condition, it seems to me, has been symptomatic of the age.

If this energy appears wildly promiscuous, so it must be. More specifically, if it represents the developing, broad contexts for which many themselves are striving for today, then it will have served to encourage those also on this or a similarly unconventional quest. From my vantage, which is admittedly myopic as well as oblique, it has always appeared that the crossings would eventually open up contexts that had previously remained too narrow. This is one reason the borders are so perilous. Anyway, time will tell.

WORKS CITED

S. Allen (ed.), *Possible Worlds in Humanities, Arts and Sciences: Proceedings of Nobel Symposium 65, 1986* (Stockholm, 1989), section 1: Philosophy: T. S. Kuhn: 'Possible Worlds in History of Science'. Session 3: Literature and Arts: Lubomir Dolezel: 'Possible worlds and literary fictions', with commentaries by Umberto Eco and others. Kuhn's chapter contains some of the ideas of the two worlds of science and language which he developed in his 1984 Thalheimer Lectures at the Johns Hopkins University.

Pierre Bourdieu, *Distinction: a Social Critique of the Judgement of Taste* (1984).

George Dickie, *Art and the Aesthetic in Institutional Analysis* (Ithaca and London, 1974).

Michel Foucault, *Les Mot et les choses: une archéologie des sciences humaines* (Paris, 1966).

Michel Foucault, *Discipline and Punish* (New York, 1977).

Michel Foucault, *The History of Sexuality*, 3 vols. (New York, 1978–86).

E. H. Gombrich, ' "They Were All Human Beings – So Much Is Plain": Reflections on Cultural Relativism in the Humanities', *Critical Inquiry*, XIII (summer 1987), 686–99.

Penelope Gouk, 'The Union of Arts and Sciences in the Eighteenth Century: Lorenz Spengler (1720–1807), Artistic Turner and Natural Studies', *Annals of Science* XL (1983), 411–36.

Loren Graham, Wolf Lepenies, and Peter Weingart (eds.), *Functions and Uses of Disciplinary Histories* (Dordrecht, 1983).

Jürgen Habermas, 'Michel Foucault: the Critique of Reason as an Unmasking of the Human Sciences', in *The Philosophical Discourse of Modernity*, trans. Frederick Lawrence (Cambridge, 1987).

Russell Jacoby, *The Last Intellectuals: American Culture in the Age of Academe* (New York, 1987).

Thomas S. Kuhn, *The Structure of the Scientific Revolution* (Chicago, 1970).

Thomas S. Kuhn, 'Commensurability, Comparability, Communicability', in Peter D. Asquith (ed.), *Philosophy of Science Association*, 1982, 2 vols (East Lansing, 1983), II, 669–716.

Thomas S. Kuhn, 'Scientific Development and Lexical Change' (The Thalheimer Lectures, 1984), unpublished lectures.

Bruno Latour, *Science in Action: how to Follow Scientists and Engineers through Society* (Cambridge, 1987).

Jean-François Lyotard, *The Postmodern Condition: a Report on Knowledge*, Theory and History of Literature (Manchester and Minneapolis: 1984).

Joseph Margolis, 'Theory and Method in the Cultural Sciences', in Robert S. Cohen (ed.), *Boston Colloquium for the Philosophy of Science* (Boston, 1988–9).

M. J. Mulkay, *The Social Process of Innovation: a Study in the Sociology of Science* (1972).

M. J. Mulkay, *Science and the Sociology of Knowledge* (1979).

M. J. Mulkay, *The Word and the World: Explorations in the Form of Sociological Analysis* (1985).

M. J. Mulkay and Will Outhwaite (eds.), *Social Theory and Social Criticism: Essays for Tom Bottomore* (Oxford, 1987).

Alan Sinfield, 'New Historicism and the Primal Scene of US Humanism' (UCLA Department of English talk, spring 1989).

Reba Soffer, 'Why Do Disciplines Fail? The Strange Case of British Sociology', *British Journal of Sociology* XCVII (1982), 767–802.

E. P. Thompson, *Albion's Fatal Tree: Crime and Society in 18th-century England* (New York, 1975).

Weber, Samuel (ed.), *Demarcating the Disciplines: Philosophy, Literature, Art* (Minneapolis, 1986).

Studies by J. T. Klein (*Interdisciplinarity: History, Theory, and Practice*, 1989) and Martin Kreiswirth (*Theory between the Disciplines*, 1990) appeared after this book had gone into proof, but may be useful in these meditations about history and theory in the transition from the disciplines to the postdisciplines.

PART ONE

medical

I

The discourses of literature and medicine: theory and practice (1)

My work in literature and science during the 1970s persuaded me that an even stronger theoretical case remained to be made for literature and medicine. After all, the medical realm eventually touches each of us far more than does the scientific and technological. Even if the freezing of human bodies eventually proves feasible through technological and engineering advances, each of us will still have to die alone. Because physical suffering is an intrinsic part of the human experience, medicine's role in our personal lives is even greater and more significant. The realms of pain – for so long the accepted condition of humanity – teach us why. From the suffering of Achilles to the pain of Gustav von Aschenbach in Thomas Mann's novella, the literary record is replete with examples. It is not unreasonable to claim that medicine exceeds science in its human dimensions.

Yet by 1980, when I wrote this essay, little had been written about the use of literature in medical therapy, and even less about theoretical bridges and connections between the two realms. Here I tried to construct a field called 'Literature and medicine', as I had earlier done for Literature and science (chapter 8).

Now that a decade has elapsed, it seems to me that the essay is too timid and does not extend far enough. I ought to have been bolder than I was in staking out the claim that medicine was of greater relevance to humanity and its discourses than science. I would have made an even stronger case by distinguishing physiological realms – even neurophysiological – from the merely psychological.

> I want to describe, not what it is really like to emigrate to the kingdom of the ill and live there, but the positive or sentimental fantasies concocted about that situation: not real geography, but stereotypes of national character. My subject is not physical illness itself but the uses of illness as a figure or metaphor. My point is that illness is *not* a metaphor, and that the most truthful way of regarding illness – and the healthiest way of being ill – is one most purified of, more resistant to, metaphoric thinking. Yet it is hardly possible to take up one's residence in the kingdom of the ill unprejudiced by the lurid metaphors with which it had been landscaped. It is toward an elucidation of those metaphors, and a liberation from them, that I dedicate this inquiry. (Susan Sontag, *Illness as Metaphor* (1979))

Literature and medicine, unlike literature and science, is not a field that has claimed significant numbers of students, certainly not of historians of science.[1] This is not because medicine and literature interact less frequently or less intensely than science and literature, but because scholars – both literary historians and historians of science – have misunderstood or not been interested in the nature of the interaction. Moreover, medicine has made its appearance at unpredictable times in literature: in Lucretius *De Rerum Natura*, Burton's *Anatomy of Melancholy*, Swift's *Gulliver's Travels*, Johnson's *Rasselas*; in eighteenth-century English novels whose protagonists often suffer from fashionable diseases of the age, Goethe's dramatic poems, nineteenth-century English novels such as George Eliot's *Middlemarch* (in which Lydgate, a physician, epitomises the whole European tradition of the physician); in Proust's *Remembrance of Things Past*, Joyce's *Ulysses*, Thomas Mann's *Magic Mountain;* and in many plays in which some essential aspect of the medical profession is captured (as in the madhouse scenes in Ibsen's *Peer Gynt* or in Alan Dysart's psychiatric ward in Peter Shaffer's *Equus*). The literary historian or historian of science who specialises in a given epoch or period will not observe the entire curve of the interaction.

The irony of this contrast – literature and science versus literature and medicine – is that medicine surely has far more than science to offer literature, and vice versa. Nevertheless, the few students who have studied the interrelation of literature and medicine have either been unable to communicate their enthusiasm to readers or have failed to view the interrelation as so profound, for there has been less scholarship about this subject that about any other area of traditional literary history or conventional history of science.[2] There may be other reasons for a lack of interest: reasons related to professional values, to the traditional separation of subjects into classes (sciences and humanities, and specialisations within these categories, e.g., physics, chemistry, astronomy), or to the considerable knowledge required to write about the interaction.

HISTORIOGRAPHY

The assumptions usually made in the existing scholarship of literature and medicine are these: that literary history, as a branch of general history, is best studied in periods and that meaningful analysis of particular texts requires periodisation (baroque, Romantic, post-Romantic, Impressionistic); that influence construed in the abstract is best understood as chronological influence – a prime tenet of historicists;[3] most crucial of all, that the influence proceeds *from* medicine *to* literature. These assumptions are all reflected in the notion that the writer is 'well grounded' in the medicine of his time. This concept does not necessitate that the writer be trained in medicine (as Keats was), or even that he or she study it privately (as

Sterne did), merely somehow absorb or assimilate the basic medical concepts and assumptions of the age. This irrefutable assumption has been developed in studies of a few principal topics: the influence of ancient Greek medicine on Homeric poems.[4] Renaissance theories of melancholy on Burton's compendious *Anatomy*,[5] physiological theories on the English metaphysical and cavalier poets,[6] sixteenth-century anatomy on Shakespeare's plays,[7] Restoration medicine on the prose satires of Swift,[8] Victorian medicine on Dickens's novels,[9] and Anglo-Irish medicine in Shaw's plays.[10]

Some pronounced tendencies typify this scholarship. First, rarely are there leaps over time: for reasons not usually discussed it is assumed that the writer is influenced by the medicine of his own period, or of the age immediately preceding his. Consequently, one is not likely to discover studies of the influence of Greek medicine on Shakespeare, Renaissance medicine on Dickens, Enlightenment medicine on Proust, although historians of science regularly study the influence of early science on much later science, and literary historians are perpetually studying the influence of early literary techniques on later writers; Greek astronomy on Renaissance astronomy or Galenic medicine on iatrochemical medicine; Chaucer's literary influence on Milton, Shakespeare's on Dickens, Dryden's on T. S. Eliot.[11] Secondly, explanations of the method by which the writer has 'absorbed' the medicine of his age are often lacking. It may be that the omission is necessary, even in the best of possible worlds. Except for cases in which the libraries of writers are extant in sales catalogues – hopefully with their marginalia – or for cases in which the authors have revealed their method of composition, there is no discussion, let alone analysis, of the dynamics of influence. Furthermore, the language used to explain this influence is often vague, dominated by impressionistic words such as 'shaped', 'created', 'impelled', but not 'caused' or 'compelled'. Finally, the same literary works appear again and again as the loci of this variety of influence – perhaps by necessity. To mention a few: Molière's *Le Médecin malgré lui;* Smollett's *Humphry Clinker;* Swift's *Tale of a Tub;* Zola's *Lourdes;* Balzac's *The Country Doctor;* Proust's *Remembrance of Things Past* (for the density of its medical imagery and the number of its medical characters); Tolstoy's *The Death of Ivan Illych* (the literary *locus classicus* of death); Shaw's plays, especially *The Doctor's Dilemma;* Schnitzler's *Professor Bernhardi* (about a professor of medicine); Mann's *The Magic Mountain;* and Solzhenitsyn's *Cancer Ward*. Little of the criticism on these works has been written by trained medical practitioners.[12] Literary studies of the medical components of these books usually adumbrate a few selected categories: delineation of the medical characters and medical episodes, the way such delineation affects the plot, and the author's desire to include social commentary through the use of medical materials.

These interdisciplinary discussions have often avoided crucial questions and produced superficial analyses. Especially lacking is any notion that medicine itself can have formed the writer's primary urge, perhaps because such an idea is too constricting to be entertained seriously by literary critics whose sense of 'the act of creation' is organic rather than mechanistic.[13] Here literary historians have been far more culpable than historians of science. In most such studies, medicine is viewed as a 'secondary cause' of creation. Writers, it is argued, really wishes to probe their own selfhood and imagination; medicine is one of a wide repertoire of subjects enabling them to do so, but within the spectrum of available options it is rarely a *force majeure*. This approach underestimates the creative writer's need for subjects other than the self; it is the result of a post-Romantic view of the writer's self-absorption. But works like Anthony Burgess's *The Doctor is Sick* (1960) suggest that medical content is not simply a metaphor for the writer's sense of self, since medicine has left little mark on Burgess's sentence structure, prose rhythm, and imagery, though it bulks large in the world of the novel. The works of writers who were also trained medical men – Smollett, Goldsmith, Keats, William Carlos Williams, the Portuguese poet Alberto Pessoa, the Czech pathologist Miroslave Holub – perhaps support the case for the importance of external influences on literature more readily then do other types of creative literature, and indeed criticism of such works suffers less from the post-Romantic view than most. But a new attitude must be adopted by literary critics if the state of affairs that prevails today is to change. The internalists – literary historians as well as historians of science who are sceptical of the validity of external influence on the evolution of literary forms and types of science – must compromise their rigidity. Externalists argue that the internalists will eventually be forced to concede that all types of writers enthusiastically read non-literary works and consequently 'import' subject matter, and, furthermore, that the importation permits those materials to be transformed into the stuff of great imaginative literature. Mann's *The Magic Mountain* and Tolstoy's *Death of Ivan Illych*, masterpieces no matter what the aesthetic yardstick of measurement, are no doubt proof enough in themselves: neither Mann nor Tolstoy had any formal medical training, yet each mastered an aspect of social medicine.

FROM MEDICINE TO LITERATURE

The problem of influence in intellectual history is as important to historians of science as it is to literary historians. The arrows of influence in this body of scholarship are always drawn in one direction: *from* medicine *to* literature. This directionalism in literary exegesis – one cannot talk of medical historical scholarship here because there exists very little *medical* writing about *literary* works — has grave consequences. It presumes that an author grows up, as it were, with the medicine he later will use in his imaginative writing; that he need not, as a consequence, read or study it. He is assumed to be capable of imbibing this medicine from the general culture about him: from newspapers, conversation, more recently from the media.

This notion not surprisingly leads critics who hold it to find a deeply felt Freudianism in the writings of Lawrence or Joyce, neither of whom may have read a word of Freud. Moving backward two centuries, they expect to find the appearance of the doctrine of the 'ruling passion' – then the most technical medical and psychological explanation for what we today would call the personality – in the writings of the great eighteenth-century novelists.[14] In these examples, the 'absorption', it is argued, is explicitly found in characters whose basic nature would be different without this superimposed medical richness. The audience discovers the characters to be 'realistic' because they resemble recognisable types in their own society suffering from the same maladies. Literature and medicine, construed in this sense, share a common concern to articulate a culturally conditioned medical perception of general attitudes towards life and death. Thus the humoral types in Ben Jonson's comedies are 'shaped' (this is the suspect word) by humoral theory of about 1600; the love-sick melancholics of seventeenth-century love poetry by medical lore about nervous and bilious diseases; the expiring *belles dames* and pale knights of Romantic literature by medical theories of consumption.

In these studies *from* medicine *to* literature each period is unfortunately associated with a particular type of medicine: the one that is popularised and mythologized and that will influence creative writers.[15] The time lag for influence is not long, sometimes no more than a decade. Rarely is the belief expressed that this popular medicine itself has been determined by non-medical factors: by social necessity (e.g., society's need for a consumptive stereotype on to whom it can project its collective fantasies about wasting away) or psychological need (e.g., the need in our depersonalised age for a condition, cancer, which punishes those who repress their emotions). Least of all is there any notion that creative writing has shaped this medicine. Although literary scholars may probe medical texts for their influence on literature (see below), they do not probe medical texts for the influence of literature; at least they have not as yet. They have not presumed that the source of medical hypotheses may have derived from plays, novels, and poems. So, even today, the arrows continue to move in one direction only: *from* medicine *to* literature.

TYPES OF VALIDITY OF INTERPRETATION: THE INTERNALIST-EXTERNALIST PROBLEM

A survey of the secondary scholarship reveals that literary scholars have detected the medical influences just discussed in five predominant ways. I list them in order of their certainty. First is verbal similarity: when a creative writer uses technical medical language derived from an identifiable source, there can be no doubt about his intentions. He himself may not point to the source, but his source cannot be questioned. An example is Smollett's novel *The Life and Adventures of Sir Launcelot Greaves* (1760–2), the story of two lovers incarcerated in a madhouse to prevent their marriage. The novel contains many pages plagiarised from Dr William Battie's *Treatise on Madness* (1758), but Smollett conceals his source. In this example and others the medicine has shaped the literature, and the influence can be proved beyond any shadow of doubt.[16]

Biographical material affords a second avenue. A writer often leaves diaries or notes identifying his source; investigation may reveal that the writer was reading a particular medical work at the time of composition, and that aspects of his novel or play derive from it, as in the case of Peter Shaffer, who must certainly have been reading or hearing about R. D. Laing while writing *Equus*.[17]

If neither verbal parallels nor biographical materials are available, the scholar reconstructs the writer's reading, using his personal library as a start. Such reconstruction has recently yielded remarkable results, as for the opening pages of *Tristram Shandy*. For a long time critics wondered what Sterne's medical source was for the birth of 'the little gentleman' Tristram. Louis Landa has studied Sterne's library and discovered the source: an eighteenth-century medical controversy about the influence of the 'mother's imagination' on the foetus. The source makes clear why Sterne describes the hero's gestation and birth in embryological language: Tristram is incapable of experiencing the flow of ordinary time because his mother's imagination, impressed on Tristram's mind, was fixated at the moment of conception by her worry that her husband had forgot to wind up the clock.[18]

When these sources are unavailable, the scholar uses his intuition as a reader. Interdisciplinary conjecture also has sometimes produced important analyses that permit the scholar to connect realms otherwise held apart.[19] For the three approaches listed above the source is available; in these two the scholar must rely on wit and learning. But following intuition and arguing that the writer cannot have 'imagined' the disease, the scholar resorts to a kind of blind historicism: the scholar's sense that all knowledge is predetermined by cultural factors. Literary scholars who prefer topics where the evidence of direct influence is slight – who rely on interdisciplinary conjecture – are externalists. They concentrate on cultural movements rather than on single works and authors; they try to show why a type of literature arises when it does, and how it can be indebted to medical developments. Seventeenth-century humoral comedy or the Fabian criticism of medical practice are examples: in each case the scholar shows how medical ideas – humoral pathology and Fabian propaganda – penetrate to the writer's imagination.

FROM LITERATURE TO MEDICINE

This progression has not been studied: here the main difficulty is that historians of science and medicine, who are often internalists historians, distrust literary evidence. No matter how persuasive the material, they consider it defective or insignificant. Their scepticism may be at a nadir now because social history is so fashionable, and literary evidence is considered a significant resource for social history.[20] But before 1800 the literary record is that of the upper classes; by virtue of omission, then, this is a doomed literary record. So the argument goes, and this attitude is as prevalent among social historians of science in the United States as among those in Europe. Another reason for their scepticism is that historians of medicine and medical practitioners feel uncomfortable when discussing literature. Some read it, watch it performed, even profess to cherish it; but they have not been trained to analyse it and are insecure when they do. False piety, even hypocrisy, is evident in this attitude – that literature is invaluable but that it does not count in the development of modern medicine. As a result, these historians and practitioners take the easiest way out: they omit literature altogether, except in epigraphs that adorn their writings.

Yet there are several ways in which the influence of literature has been palpable. Language is a common ground in literature and medicine; metaphors commonly used in both fields require scrutiny: 'wasting away', 'invaded by', 'personality type', words connected to 'consume' and 'consumed by'. These words are cultural signifiers and should be probed in their literal as well as metaphoric dimension, for when used in the latter sense they designate far more than the actual words denote.[21] We also need to learn how the language of cholera in the nineteenth century and leukaemia in the twentieth compares in medical and literary texts. None of this compilation has been performed, and it could yield important results, not least for the social historian.

Literature has also shaped medicine in other ways. Thomas Perceval's *A Narrative of the Treatment Experienced by a Gentleman, During a State of Mental Derangement* requires analysis for his assimilation of fictional techniques of the period. This is not so that new light can be shed on the literature, but to illuminate better Perceval's *Narrative*, the first lengthy self-analysis by the analysand. His account ought to be compared with the novels of Thackeray, Trollope, and Dickens, all written in the same period. It would be similarly productive to compare the text of Richard Madden's *Phantasmata or Illusions . . . : Protean Forms Productive of Great Evils* with Charlotte Brontë's *The Professor*, Ruskin's *Political Economy of Art*, and Trollope's *Barchester Towers*, all published in 1857.[22] Medical authors and patients adopt the values of imaginative writers of their own period. In my examples, it is impossible to say who is the analytic and who the imaginative writer: Brontë probing academic life, Ruskin the politics of the art world, Trollope the church, Perceval the patient, or Madden the nature of insanity. When medical authors explore themselves they invariably reflect the beliefs of their own culture. Other examples of that reflection are the nineteenth-century medical analyses of J.-J. Rousseau's illnesses. His maladies were many; even now there is no agreement about their precise nature. Nevertheless, dozens of doctors wrote treatises in the nineteenth century professing to have discovered the real illness. When this record is compiled and studied it is clear that each physician has 'read into' Rousseau's illnesses the most fashionable disease of his own period.[23]

The 'case history' also deserves investigation. It was invoked by autobiographers long before it was used in medical practice. The Puritans in the Restoration wrote of themselves as 'case histories' two centuries before the concept arises in medicine,[24] yet the similarities have not been noticed. Furthermore, no one to my knowledge has asked how the biographies of literary and medical figures would differ from present biographies if clinical data were taken into account.[25] The doctor should also be compared in his fictional and medical representation for revealed differences. Doctors who consider themselves above such scrutiny need to be viewed in both sets of representation.

But the medical world of yesterday and today is not confined to medical theory. It includes landscapes and settings, instruments and machines, aesthetics and architecture: especially noteworthy in the aesthetic realm are the clinic and the hospital; the interior spaces of operating pavilions, wards, private rooms; the colors and textures of the exteriors and interiors of medical buildings. Research may reveal that imaginative literature (and other arts, especially painting) has provided medicine with some of its ideas for this environment.[26] Certainly there can be no doubt that the scripts performed on television are the source of some of the recent visual changes in these medical spaces.

AUTOBIOGRAPHY, BIOGRAPHY, AND THE CASE HISTORY

This comparison is so potent that it deserves to be explored at greater length. Autobiography and biography thrive on main characters; every medical case history purports to be the whole story of one protagonist. Both varieties employ language that uses vocabulary in certain distinct ways. Each flows in time and space, and each reflects the beliefs and values of the author of the account as opposed to the protagonist. Wish it though we may, there can be no such thing as an 'objective' autobiography, biography, or case history: to delineate a protagonist is to expose one's own beliefs.[27] There are differences, too: in fiction an author writes the imagined history of his protagonist, even though he may be thinking of a real person during composition; in autobiography and biography the history is supposedly less imagined, but inspection demonstrates that this is not always true, as in Boswell's *Life of Johnson*.[28] In medical case histories, doctors write from an intentionally objective point of view, as if they were privileged and stood outside the orbit of ordinary humanity. Yet doctors, as well as writers, are human. They too will grow old and die; they too suffer fears and anxieties as do their patients, and these colour their approach to the whole world, including their profession. Like writers, they possess an imagination and there is no reason to believe they suspends it – or can suspend it – in actual medical practice. They may feel compassion or love for certain of their patients, as writers do for some of their invented protagonists; and these may be the ones they treat differently or whose medical case histories they eventually write. It is impossible for the doctors, finally, not to see themselves reflected in some of their patients, just as writers project themselves on to some of their protagonists.[29]

In autobiography there is another essential difference: the writer *is* the subject; the writer *is* the case history. But this radical conflation renders autobiography all the more germane for the medical case history. If more historians of medicine studied autobiography, they might better understand the case history, especially its limitations. Perceval's *Narrative* is important for practicing psychiatrists today because the patient has interpreted his own case. Translated into concrete terms, the importance of this account derives from the patient's vocabulary in relation to his situation. Viewed as such the patient is an imagined literary character. Historians of science and medicine as well as medical practitioners need not believe that 'imagined literary characters' are alien to their professional concerns. Psychology and psychiatry will naturally profit most from familiarity with these characters, and this may be why literary critics who become lay analysts are often the best of the breed.[30] Every time a patient enters a practitioner's office a literary experience is about to occur: replete with characters, setting, time, place, language, and a scenario that can end in a number of predictable ways. Literature enriches the sense of this daily drama.

MEDICINE AND SCIENCE FICTION

Science fiction has received the least attention of all the literary forms in the study of medicine and literature. Two areas especially – imaginary diseases and medical utopias – require exploration. One need not delve far into imaginative literature to discover writers who conjure imaginary maladies and apply hypothetical causes and methods of cure.[31] Readers have naturally viewed these illnesses within the contexts of the genre, especially as imaginary maladies. They do not read it as criticism of the medical profession when Schratt says to Donovan in *Donovan's Brain*, 'I always thought that protrusion on your head was the purple tumor of health known to have kept our ancients alive for over a century',[32] or when imaginary diseases are found among the natives of Nowhere in Butler's *Erewhon*. None the less, the course of modern medicine ought to cause speculation about diseases whose aetiology cannot be traced to a physical cause.

Science fiction offers further assistance to the interrelation by making distinctions among 'imagined diseases', especially between those with and without somatic symptoms. It is one thing, after all, to repress one's emotions and then develop cancer – a medical condition with clear somatic evidences – and quite another to repress the emotions and then generate a malady in which one feels badly but has no clear somatic signs.[33] Physicians report observing the latter type in increasing quantities today and often send these patients to psychiatrists for treatment.

Science fiction sheds light on this state of affairs because it often transforms a fashionable disease, such as cancer, into a hypothetical one. As in crime, contemporary medicine may be taking cues from the world of fantasy. It may be that some patients are generating diseases they imagine to be novel and therefore of personal advantage. The cultural causes for such generation are ultimately sociological and cannot be discussed here, but this development in itself should not preclude the validity of the connection. In so far as patients are 'imagining' diseases all the time, it may be that their type of conjuring is ultimately not so different from fiction writers who also imagine diseases.

Medical utopias also require more scrutiny by scholars than they have received anywhere. Such utopias are usually imagined worlds in which illness is either non-existent or incapable of complete cure. The suggestion offered by these writers is that people will be healthy only when they are placed in 'healthy' (i.e., utopian) social and political environments, that is, when their economic woes are minimised and they can express themselves as individuals who are not so depersonalised that their egos are crushed. This *direct* link between social milieu and medical health is not often viewed as a topic for serious study by medical scholars or historians with an internalist proclivity. Yet evidence is mounting that certain illnesses (hypertension, coronary disease, even cancer) own their origins to this interaction. Moreover, some of the poor image that psychiatry now enjoys results from its willingness to adumbrate, however primitively, the interaction between the two realms.[34] Such readiness renders psychiatrists vulnerable to charges of nonscientifism and, in extreme cases, even to quackery. But an author such as Samuel Butler, who envisioned medical utopias as a response to the exigencies of his own time, deserves to be read by the medical community.

THE IMAGE OF THE PHYSICIAN AND PATIENT IN LITERATURE

This striking connection seems to have eluded everyone, especially the medical profession. There has been discussion of the obvious physician-poets: Thomas Campion, Albrecht von Haller, Robert Bridges, William Carlos Williams, but little has been written about the image of the physician through the ages. Nor has this matter held much interest for historians of medicine. The image of the physician in specific authors and works and in particular epochs and countries has been studied, but no synthetic history has been written of the physician in Western civilisation.[35]

Literature provides one of the richest archives: it is the lengthiest record, the only resource in which patients and doctors can be viewed from ancient Greece to the present. Surprisingly, it is a consistent record that impugns the physician for specific reasons: his greed, pedantry, and *hubris*. Literature before 1800 does not capture, it is true, the attitudes of the lower classes, but the record of the upper classes is so sceptical of medicine that one may cautiously assume it speaks for the poor as well.[36] Literature also adumbrates the positive side: here the physician is viewed reverentially, particularly by those grateful for his duel function as healer, as 'God's minister of the body'. But on balance the record is unequivocal: the physician, an emblem of greed, has been the target of pungent criticism for almost three millennia. One would think modern medicine should be almost morbidly curious about such a granitic archive.

This material should be compiled and analysed, for propaedeutic reasons (since it is a new subject) as well as for possible reform (for legislators and educators to have a record on which to draw when shaping the social institutions of medicine). Accounts such as those of the eighteenth century in which doctors defend their practices should be studied, as well as the rejoinders.[37] The question why criticism of the medical profession has centred on greed, arrogance, and pendantry needs to be asked, but scholars ought first to compile the record and then ask it. Contemporary doctors may not be impressed by the historical record or collective historical profile (unless the government forces them to be), but their resistance might lessen if they recognise that the study of the literary record can explain the origins – in the sense it is used by certain contemporary philosophers – of negative views of medical men. All three criticism singled out above depend on the notion of certainty in medicine: it is the perception by patients that physicians consider themselves practitioners of a science rather than an art that has fuelled patients' attacks on doctors.[38]

THE PHYSICIAN AS WRITER

This category, more than the others, reveals how the arrows of influence have proceeded *from* literature *to* medicine. When doctors project their 'fantasies of the self' into fictional forms, it is evident that traditional medical history is limited in so far as it can isolate 'the physician' as *constitutive subject*, as Foucault uses the term: a subject viewed in relation to its genesis and to the rhetoric accruing to it. Before we ask how a physician came to think of himself or herself as a writer, we must inquire how he or she developed the capacity (i.e., what imagery he used) to view himself or herself as a doctor.[39] If function preceded image, then he or she must first have constructed a model of his *role* as *physician*, and only later as a social creature with necessary appurtenant activities: as person of general learning, social commentator, community shepherd, writer. Appropriate questions for the subject are various. What kind of literature results when a practising physician such as André Soubiran writes a prolific number of medical novels? How would his best-known novel, *Bedlam*, have been different if Soubiran had not been a physician? What sense of 'the physician' as an 'object' does Soubiran entertain? Likewise for autobiography: how has Erwin Chargaff's fantasy of himself in *Heraclitean Fire* transformed the physician-narrator into a mythical and fictive figure? Even such an astute student of 'mythologies' as Roland Barthes has described the iconography of the priest, but not of the doctor.[40]

The physician-poet type (poet in the sense of imaginative writer) is revelatory at this juncture of function and image for its illumination of the doctor's historicity. Why is it that this type first surfaces in the Renaissance?[41] The ancients boasted of seers whose sphere of competence embraced medicine and literature (the shaman, the prophet, Tiresias), but the type as a recognisably medical-professional one is virtually unknown before the sixteenth and seventeenth centuries, when physician-poets such as Burton and Campion begin to appear in numbers. Yet the remarkable feature of the literary expressions of these men (female physician-poets such as Sarah Mapp do not rise until the eighteenth century and the advent of the bluestockings) is transformation of their recognisable professional self (as physician) into something quite mysterious and exotic: lover, poet, hero, conqueror.[42] The alteration is apparent whether viewed in its Renaissance flesh – Thomas Campion – or in more modern forms – Smollett as eighteenth-century poet or Robert Bridges in this century. A context is needed to interpret this protean movement from professional self to reflected literary image. Otherwise there is no means for explaining the rise of this body of literature (Campion to Bridges).[43] (The patient as writer has no historicity; his identity has hardly reached the stage of 'subject', let alone as transformed subject.)

The clue to these mysteries lies in literature, not, as has often been erroneously assumed, in the medical writings of physicians. For the medical treatise conceived as text already suppresses the very image of the physician for which we are searching; it raises up instead an impersonal discourse, no matter what the grammatical form (whether first person or not), in which role and function are assumed rather than explained. Even more damaging to the search, the medical treatise erects 'the medical history' as its constitutive subject; this, unfortunately, further suppresses the identity of the physician's image and selfhood. Only in the eighteenth century does a tension begin to develop between two alleged subjects: 'the medical history' and 'the physician'. Even in medical treatises, though more patently in the imaginative literature, they begin to contend with one another for attention, and one sees as a consequence the emergence of a new form: 'the physician' as the newly constituted subject of the old 'medical history'. This dialectic of competing subjects was bound to change the destiny of medical writing: by the early nineteenth century the historiography of medicine appears radically altered from its earlier forms.[44] But the key, the clue, exists in imaginative literature – poems, plays, novels, the writings of physician-poets – not in pre-1800 medical treatises. The imagery of these non-medical works – these very same imageries about which literary historians have written so much in recent times – is the source of 'the physician' as object. But practically nothing has been written about this terrain, and certainly not about patients, although there are indications of a start.[45]

Here my historiography naturally glances at the internalist-externalist debates discussed everywhere today, among historians of science and other historians.[46] Only a few externalist historians have appeared to recognise that perception of the field of medicine is culturally conditioned and related to attitudes towards life, death, birth, sex, madness, food, and health. The source of such perception is necessarily located in the vocabulary common to literature and science. To decipher and then decode this common vocabulary requires analysis, as a minimum, of the way it was used in social relations and intellectual discourse. Yet even the most catholic externalists have (intentionally?) excluded imaginative literature from the works gathered under the umbrella of intellectual discourse. In the nineteenth century general historians demonstrated that imaginative literature was a rich archive for understanding the forms of social relation in any era; but their works, despised as monstrosities by the twentieth-century New Critics and devotees of practical criticism, was suppressed at the very moment – the 1950s – when out bravest contemporary Anglo-American externalists were being schooled and forming their deepest beliefs. No such suppression occurred in France, and French externalists, especially those on the fringes of Marxism, have looked at imaginative literature as a constitutive subject whose vocabulary contains the sought-for commonality.

I am referring primarily to books by Foucault, Deleuze, Bachelard, and Barthes – in this order because only Foucault has isolated 'medicine' and 'medical history' as subjects worthy of special attention.[47] Foucault's reasons are unimportant here, but the responses of his colleagues to his isolation of that field are suggestive and ought to be compared with the rejoinders his critics are making today, both internalists and externalists:

> ... if one poses, for a science such as theoretical physics or organic chemistry, the problem of its relations with the political and economic structure of society, doesn't one pose a problem which is too complicated? Isn't the threshold of possible explanation placed too high? If on the other hand, one takes a [field of] knowledge [*savoir*] such as psychiatry, won't the question be much easier to resolve since psychiatry has a low epistemological profile, and since psychiatric practice is tied to a whole series of institutions, immediate economic exigencies, and urgent political pressures for social regulation? Cannot the interrelation of effects of knowledge and power be more securely grasped in the case of a science as 'doubtful' as psychiatry? ... What 'threw me off' a bit at the time was the fact that the question which I posed did not at all interest those to whom I proposed it. They considered it a problem without political importance and without epistemological nobility.[48]

Such externalism and contextuality is of another order than that found in Bachelard and Barthes; and this, ultimately, is why Foucault, who has never isolated 'the physician as writer', has nevertheless written around the margins of the topic. Central to Foucault's best-known texts – *Madness and Civilization: a History of Insanity in the Age of Reason* (1961; trans. 1965), *The Order of Things* (1966; 1970), *The Archaeology of Knowledge* (1969; 1979), *The Birth of the Clinic* (1963; 1973), *Discipline and Punish* (1975; 1977), *The History of Sexuality* (1976; 1978) – is the death of the traditional subject and the birth of the constitutive subject. Yet in affirmation of such discontinuity Foucault has posited, *volens nolens*, the manufacture of just the type of subject needed: 'the physician as type' and then 'the physician as writer'. He did not assist literature and medicine intentionally: his concern was to show the production of constitutive subjects – author, patient, doctor, madman – in relation to a specific historical conjecture at the end of the Enlightenment. Scholars have commented that Foucault's corpus hereby provides the fiercest critique we have of the traditional 'history of ideas' approach.[49] This may be true, but the critique itself ought not to blunt the innovation: by dwelling again and again on the death of traditional subjects, Foucault has continued to imply their birth. Birth when? At some unnamed historical moment. The application of these theories remains to be completed.

THE PHYSICIAN AS CULTURAL HERO AND ANTI-HERO

Does this Foucaultian approach not have rich possibilities for literature and medicine? If there was a 'birth' of traditional subjects (e.g. authors, texts, university departments), was there not also a birth of medical subjects: doctor, patient, invalid, nurse, and now the medical subtype, the paramedic? Was there not also the birth of imageries in the spaces now designated as explicitly medical: the asylum, hospital, clinic, each with its own subset of imageries? Foucault has maintained that there was, and *The Birth of the Clinic* purports to demonstrate the rise of these imageries as constitutive subjects: the asylum, clinic, doctor, and patient.

Yet in all Foucault's discourse the physician himself has somehow disappeared. He was not intentionally obliterated but blurred into a landscape – the night-side of Foucault's kingdom – that refused to romanticise him. In this negation lay an embedded anxiety that is *au fond* political and ideological: the possibility of the birth of 'the physician as hero' as a manufactured constitutive subject. Foucault himself set the stage: a chronology of 1750 to 1820, a scenario of French 'actors' in the Revolutionary period who worshipped heroes and then romanticised them. For Foucault there was danger here: although the moment (1750–1820) was ripe for the birth of 'the physician as hero', the politics of such a category was not. The physician had been anything but heroic: an oppressor, an emblem of regime and authority, the physician had administered misery and institutionalised suffering. The old Addisonian quip – one of dozens coined in the eighteenth century – about doctors inflicting pain more than cure[50] was but another strain, for Foucault, in the chorus of unrelenting disapprobation. As a consequence, the physician could never be culturally heroic. At best, the new constitutive subject in the moment of discontinuity was 'the physician'. The other subject, 'the physician as hero', a sentimental and romantic category, was politically unsound for Foucault.

Within this context, then, Foucault has suggested the anonymity and cultural nefariousness of Establishment figures. In a recent essay on the 'death of the author' he has explained 'the anonymity of the producers of truth' in every society.[51] One wonders if the physician did not become anonymous at the very moment when he lost his heroism, at the end of the eighteenth century. Before then the metaphors that attach to the doctor are culturally approbatory. He is often likened to religious figures, at least in works of literature: 'the minister of health', 'steward of the body', 'shepherd of the soul'. After the early nineteenth century he is secularised and loses his privileged place in the cultural cosmology. Now, almost two centuries later, the physician is once again being demythologised, for in the course of two hundred years much algae has clung to the bedrock of his then newly constituted identity. Calling medical men 'physicians of no value' and 'forgers of lies', Job anticipated an attitude that has prevailed with only minor discontinuity since about 1800. This point of view may be retrieved in historical chronicles: it is more lucidly gathered in the imaginative literature and art (painting, drawing, sculpture) of generations of Western civilisation.[52]

If the patient has never been viewed as culturally heroic in the march of centuries, this is for different reasons.[53] Unlike the physician, he has enjoyed only low status, and in our own time he runs the risk of becoming an endangered species altogether. The patient has another historicity and iconography, one that changes most drastically in the early nineteenth century. Before then, his sufferings and writings are ignored, his pain muted by those – especially physicians – authorised to suppress them. By the mid nineteenth century all this begins to change. The patient and physician reverse roles: the afflictions of ordinary valetudinarians such as widows and servants – in literature the sick common folk of Dickens's novels – are elevated and romanticised, and the patient is now believed to enjoy a heightened state of consciousness, while the pretensions of a Dr Lydgate or the dozens of medical doctors in the fictions of Balzac and others are chastised. These are but the barest outlines of the picture; nothing as yet has been coloured in. Even the size of the canvas remains to be determined.

WHAT THE FUTURE HOLDS

Hopes for literature and medicine as a new province of learning with a high level of explanatory power should not be too unrealistic at this early stage. When the valid question, what can this interdisciplinary field hope to teach, is placed before the various academic disciplines involved, an excessively ambitious answer ('much' or 'everything') ought not to be given. Naturally, a great deal can and will be learned about medical practitioners who were also writers (e.g., Sir Thomas Browne, Smollett, Chekhov, William Carlos Williams) and about writers who were not medically trained and who did not practise but who had serious medical interests (Molière, Defoe, George Eliot, Melville, Mann, Solzhenitsyn, Peter Shaffer), as well as information gathered about the history of medicine that could not have been gained otherwise than from literature. Indeed, the large and pivotal question, why certain writers, such as Proust and Sylvia Plath, are intrigued by medical imagery (the physician, the patient, the hospital, the asylum), while other writers are not, will be seriously asked, perhaps for the first time. Furthermore, the field will also yield much information about the history of literature that thus far has remained undetected because medical concerns, no matter how apparently germane to a post-Darwinian humanism such as ours, nevertheless continue to be considered tangential to the proper domain of literary studies.[54] But these new insights will arise within the context of language, that is, within a belief that all ideas – even medical ideas – are generated in linguistic categories. We should expect to discover, therefore, more about the sensibility of physician-authors like William Carlos Williams.[55] And we may expect to comprehend why the whole corpus of Williams's writings would have been different if he had never been a practising physician. But we should not hope – and this is the quintessential point – at least not for some time, to gain practical wisdom in the form of medical application or therapeutic remedy from a province of knowledge so patently abstract. That is, we probably ought not to encourage medical practitioners to believe that the study of medicine and literature will be of use to physicians in treating their patients, or to ailing patients in understanding their illness. The sympathetic physician or curious patient who reads Lael T. Wertenbaker's *Death of a Man* – a beautifully written novel about a courageous husband and wife who plan for a rational and ethical death during the final year when the husband suffers from stomach cancer – may discover useful clues for his or her own death, but neither physician nor patient should expect to plunder literature primarily to gain insight into the treatment of cancer.[56] Moreover, literature whose thematic content is medical tends also to be romantic – generally so preponderantly romantic that it is dangerous to derive practical applications from it.[57] But so long as present and future students are sanguine about the limitations of literature and medicine as a theoretical domain, and are satisfied to view the interrelation as a branch of literary history – especially by exploring language for its levels of embedded meaning – then new frontiers of knowledge will be crossed.[58] But both groups involved – literary and medical – ought to be deeply curious, about the domain of the other if this interdisciplinary field is ever to advance. Thus far there has been the anticipated resistance in both camps, but whereas literary critics and theorists have recently been willing to study major medical *texts* as if they were masterpieces of literature (e.g., those by Ramón y Cajal, Jung, and

Freud),[59] the medical profession and medical historians have been reluctant to concede that language, even when applied to situations in which patients require sympathy or when surveyed in relation to the placebo concept, plays a significant role in medical theory or practice.

Literature and medicine cannot advance very far until the medical world is persuaded that the territoriality of language extends far beyond the borders imagined by most medical practitioners and researchers. Stated otherwise, literature and medicine, unlike literature and science, which profits from the scholarly endeavours of a new army of historians of science interested in language, will never become an important activity to groups other than academic literary scholars if it excludes or dares to neglect the very medico-historical community whose expertise and angle of vision it so urgently requires.[60] Precisely how to awaken this community to the exigencies of language – its tropes and its rhetoric as well as its infinite complexity – must be studied on another occasion.

CONCLUSION

In the brief space allotted me here, I have necessarily omitted crucial terrains. One contains the perfidious medical metaphors that touch on our daily lives. As Susan Sontag has written: 'Illness is the night-side of life, a more onerous citizenship. Everyone who is born holds duel citizenship, in the kingdom of the well as in the kingdom of the sick. Although we all prefer to use only the good passport, sooner or later each of us is obliged, at least for a spell, to identify ourselves as citizens of that other place.'[61] These are not merely poetic words in prose form: they possess substance. For no study of the interaction that neglects language, even construed in its crudest rhetorical tropes, can hope to make the necessary connections. Of this last point, perhaps more than any other, the historian of science needs to be persuaded. Most culpably, I have said nothing about medicine and the sociology of literature, a conjuncture that has not produced results as yet but that appears to be on the verge of doing so. And I have been silent on the fiscal and institutional realities: where ideas and information may be taught and the prospects for funding courses and institutions. I have omitted all pedagogical considerations on the grounds that readers would want to know *what* literature and medicine is, rather than *where* it is.

At this moment the most crucial aspect for medicine and literature is the destiny of three of its objects – the *physician*, the *patient*, and the *medical history* – in relation to the manufacture of these objects themselves. Throughout I have been suggesting that literature is the supplier to a certain extent – I would certainly not wish to maintain that it is the exclusive agent – and my ten topics offer hints about the nature and source of that literary supply. But it is one thing to suggest arrows of influence and another to demonstrate evidence or prove them. On this proof everything stands or falls, for without some reciprocity – *from* literature *to* medicine as well as *from* medicine *to* literature – there is neither a field nor its state to survey.

NOTES

1   See *Isis*, LXIX (1978), 583–91 (chap. 8 below).
2   The student must search long and hard to find a book such as Adolf Braun, *Medizinisches aus der Weltliteratur von der Antike bis zur Gegenwart* (Stuttgart, 1937), written in the German philological tradition.
3   I use this loaded term as does Karl Popper in *The Poverty of Historicism* (1957), a book that does not trouble traditionalist historicist literary critics.
4   Homer and medicine has been a field in itself during this century; see, e.g., the recent W. S. Morgan, 'Mimyeskomai in Homer' (Ph.D. diss., University of Michigan, 1971).
5   See B. Evans, *The Psychiatry of Robert Burton* (New York, 1944) and Lawrence Babb, *The Elizabethan Malady* (East Lansing, 1951).
6   See R. R. McFarland, 'Poems of Green-Sickness', *Journal of the History of Medicine*, XXX (1975), 250–8.
7   This field, like the Homeric, enjoys a vast secondary literature. A recent example is I. I. Edgar, *Shakespeare, Medicine and Psychiatry* (New York, 1970). For an extensive bibliography of the subject, see E. B. Rajadurai, 'Shakespeare and the Renaissance Sciences' (Ph.D. diss., Kent State University, 1970).
8   See Sherman Hawkins, 'Swift's Physical Imagery: the Medical Background and the Theological Tradition' (Ph.D. diss., Princeton University, 1960). Montaigne attracts similar treatment: see J. S. Taylor, 'Montaigne and Medicine', *Annals of Medical History* (1921), 97–121, 268–83, 327–48.
9   See J. Doggart, 'Dickens and the Doctors;, *Practitioner*, CCIV (1970), 449–53. The Victorian era has attracted some sensitive literary critics who view the influence as deriving *from* medicine *to* literature: see Barbara Hardy, 'Dickens and the Passions', *Nineteenth-Century Fiction*, XXIV (1970), 449–66; Ian Jack, 'Physiognomy, Phrenology and Characterisation in the Novels of Charlotte Brontë', *Transactions of the Brontë Society*, XV (1970), 377–91; David Daiches, 'George Eliot's Dr. Lydgate', *Proceedings of the Royal Society of Medicine*, LXIV (1971), 723–4.
10 This topic also claims many recent titles, e.g., R. Boxhill, *Shaw and the Doctors* (New York, 1969).
11 Even a few works illustrate the point: see M. Kay, 'Quixotic Medicine', *Medical Life*, XXXIII (1926), 170–94, a study of the influence of the medicine of 16th-century Spain on Cervantes, and R. Shryock, 'Before and After: a Medical Drama of 1849 and 1949', *Bulletin of the History of Medicine*, XXIII (1949), 554–76, who demonstrates that the medicine of each period (1849 and 1949) influenced the play the author wrote.
12 An exception is the insightful essay by Heinrich Ulrici, M.D., on Thoman Mann's *Magic Mountain* (*Klinisches Wochenschrift*, XXXI (1925), 78–95).
13 Three of Arthur Koestler's books – *The Act of Creation* (1964), *The Ghost in the Machine* (1976 with new postscript), and *Janus: a Summing Up* (1978) – have done much to encourage this view.
14 I have intentionally selected a problematic example: some of these writers, especially

Sterne, read contemporary technical medicine. Richardson was not only intrigued by it, but corresponded with England's most fashionable and popular physician, George Cheyne, F.R.S., and relied upon Cheyne's medical expertise when planning the 'nervous constitution' of his great heroine in *Clarissa* (1748) and depicting the disease (melancholy) by which Clarissa should eventually die. The question in these examples is how a particular body of medical knowledge has changed an essential aspect of the resulting novel.

15 See, for example, H. Rolleston, 'Poetry and Physic', *Annals of Medical History*, VIII (1926), 1–15.

16 The discovery was made by Ida Macalpine and Richard Hunter, and published in two articles: 'Smollett, M.D., and William Battie, N.D.', *Journal of the History of Medicine*, XI (1956), 102–3, and 'Smollett's Reading in Psychiatry', *Modern Language Review*, LI (1956), 409–11. Smollett, himself a practising physician in London for a number of years, read Battie's treatise in 1758 (the year before he began his novel) in the course of his editorship of the *Critical Review*. He was keen to review himself the medical books received.

17 I have not actually seen this point made in print, but it was assumed to be common knowledge in London literary circles when the play first appeared.

18 See Louis A. Landa, 'The Shandean Homunculus', in C. Camden (ed.), *Restoration and Eighteenth-Century Literature: Essay in Honor of A. D. McKillop* (Chicago, 1963), 49–48. Virtually all of the controversial works were written by medical men whose books were in Sterne's library, some marked with Sterne's marginalia.

19 Foucault's quasi-Marxist history of madness, *Histoire de la folie* (Paris, 1961) is a classic example; but there are others, including several books that span medicine and literature by Jean Starobinski, J. C. Sournia's *Mythologies de la médecine moderne* (Paris, 1969), and R. Mandrou's *Des humanistes aux hommes de science* (Paris, 1973).

20 More persuasive evidence in support of this point than the books published in the formation of societies in the last decade, the start of new journals, and appointments made to university posts. No one would want to argue that the 'history of ideas' is completely out of favour – it is not among the generation of established scholars – but it is not fashionable among young historians on either side of the Atlantic.

21 Susan Sontag (see epigraph) has taken this on a small scale but without any of the historian's instincts. What is needed is a new literary form that combines the intuition of Sontag with the resources of Pelling and Merton. See Margaret Pelling, *Cholera, Fever and English Medicine, 1825–1865* (Oxford, 1978) and Robert Merton, *Science, Technology and Society in Seventeenth-Century England* (1938; New York, 1970). Some preliminary work has been done by Leonard Barkan in *Nature's Work of Art: the Human Body as Image of the World* (New Haven, 1975).

22 Thomas Perceval, *A Narrative of the Treatment Experienced by a Gentleman, During a State of Mental Derangement: Designed to Explain the Causes and the Nature of Insanity* (1838, 1840). There is no modern edition nor post-Victorian study of this intriguing case. Richard Robert Madden (1798–1886), F.R.C.S., was not a distinguished surgeon but was important as a historian and sociologist. His *The Infirmities of Genius*, 2 vols. (1833) combines literature and medicine in an unusual way: by referring the anomalies of characters in literary work to the habits and constitutional peculiarities of the men of genius writing them.

23 I am grateful to Ralph Leigh, Professor of French at Cambridge University and editor of the definitive edition of Rousseau's correspondence, for enlightening me on this point.

24 See G. A. Starr, *Defoe and Spiritual Autobiography* (Princeton, 1965).

25 Several 'ideal figures' suggest themselves here: 'Bloody Mary' of England, Christopher Marlowe, Richard Burton the writer, Pope and Swift, Darwin, Proust,

Monet, Aubrey Beardsley, Virginia Woolf, and several members of the Churchill dynasty, especially Lord Randolph.

26   But see Grace Goldin, 'A Painting in Gheel', *Journal of the History of Medicine*, XXI (1971), 12–23, which studies the way a painting influenced those designing a Dutch hospital.

27   Lawrence Stone in 'Gentlemanly Sexual Behaviour: Case Histories', *The Family, Sex and Marriage in England 1500–1800* (1977), 546–602, may aim at an objective stand, but even in this treatment external considerations enter, beginning with his selection of five or six 'representative gentlemen'.

28   D. J. Greene has argued that most of the alleged facts are fictions in "Tis a pretty book, Mr. Boswell, but . . .' *The Georgia Review*, XXXII (1978), 17–43.

29   The point has been developed by Foucault in 'The Life of Infamous Men', in M. Morris and P. Patton (eds.), *Michel Foucault: Power, Truth, Strategy* (Sydney, 1979), 76–91. A writer's first novel also often displays this trait. A perfect example is *Roderick Random* (1748), Smollett's first novel, a first-person narrative of a young ship's-surgeon. Approximately a fourth of the characters in the novel are medical figures, into whom Smollett has projected aspects of himself.

30   In American there is also a tradition of the clinical psychologist or practising psychiatrist who maintains serious literary interests: examples in this century are Henry Arthur Murray, Professor of Clinical Psychology at Harvard, and Edwin Shneidman, Professor of Psychiatry at UCLA. Both are distinguished Melville scholars acclaimed by others in that field.

31   A few recent examples include L. S. Mercier, *Memoirs of the Year 2500* (New York, 1977); M. Moorcock, *Dying for Tomorrow* (New York, 1977); S. Robinson, *Telempath* (New York, 1977); C. Strete, *The Bleeding Man* (New York, 1977); A. Bester, *The Demolished Man* (New York, 1978); B. Shaw, *Vertigo* (1978); S. G. Spruill, *The Psychopath Plague* (1978); H. M. Harrison, *Plague from Space* (1978); W. T. Webb, *Poisoned Planet* (1978); F. Hoyle, *Diseases from Space* (1980). The secondary literature is not extensive: since World War II see *The Ciba Symposium on Medical Utopia* (Summit, 1945); René Dubos, 'Medical Utopias', *Daedalus*, LXXXVIII (1959), 410–24; F. R. Freemon, 'The Utopian Medicine of H. G. Wells', *Journal of the American Medical Association*, CCXII (1970), 101–2; M. Vida, 'Utopias and the Sociology of Medicine', *Orvostort Kzol* VI (1972), 11–28; Lionel Trilling, 'Aggression and Utopia', *Psychoanalytical Quarterly*, XLII (1973), 214–33; M. Philmus, *Into the Unknown: the Evolution of Science Fiction from Francis Godwin to H. G. Wells* (Berkeley, Los Angeles, 1970); and F. E. and F. P. Manuel, *Utopian Thought in the Western World* (Cambridge, 1979). Medical utopias are nowhere discussed in the recent works of literary criticism on the subject, not even in Darko Suvin, *Metamorphoses of Science Fiction: on the Poetics and History of a Literary Genre* (New Haven, 1979). See also Robert Scholes, *Structural Fabulation: an Essay on Fiction of the Future* (Notre Dame, 1975); S. J. Lundwall, *Science Fiction* (New York, 1977); F. Pohl, *The Way the Future Was* (New York, 1978); J. Brosnan, *Future Tense* (New York, 1979).

32   C. Siodmak, *Donovan's Brain* (New York, 1969), 142.

33   I have searched without success for a character in science fiction who suffers from an 'imaginary cancer' or 'cancer of the mind'; that is, for a fictional patient whose physician uses the language of cancer to describe his condition; but who presents no bodily evidence of pathological abnormality.

34   By so doing psychiatry has earned the reputation of being a defective science, for some not a science at all. During the 1970s the number of medical students who designated psychiatry as the field of their residency reached an all-time low figure. Fearing a shortage of psychiatrists at the end of the 20th century, the American Psychiatric Association hosted a national symposium in March 1980 on 'The Career Choice of Psychiatry'. The poor image of psychiatry is reflected in the earnings of

The discourses of literature and medicine (1)   23

35   its practitioners: according to the *Profile of Medical Practice* (Chicago, 1977), psychiatrists are the worst paid M.D.s.

35   Studies of the physician-poet often make the plea that the figure has been neglected far too long; see e.g. J. B. Hamilton, 'Robert Montgomery Bird, Physician and Novelist: a Case for Long Overdue Recognition', *Bulletin of the History of Medicine*, XLIV (1970), 315–31. For the little on the image of the physician, see H. W. Haggard, *The Doctor in History* (New Haven, 1934); B. Wachsmuth, *Der Arzt in der Dichtung unserer Zeit* (Stuttgart, 1939); P. Findley, 'The Doctor in Literature', *Medical Life*, XLI (1934), 566–84; G. Webb, 'The Role of the Physician in Literature', *Medical Life*, XXXVI (1929), 192–218; J. D. Gordon, *Doctors as Men of Letters* (New York, 1964); G. S. Rousseau, 'Doctors and Medicine in the Novels of Tobias Smollett' (Ph.D. diss., Princeton University, 1966); C. B. Norris, 'The Image of the Physician in Modern American Literature' (Ph.D. diss., University of Maryland, 1969); H. Silvette, *The Doctor on the Stage* (Knoxville, 1967); W. B. Ober, M.D., *Boswell's Clap and other Essays: Analyses of Literary Men's Afflictions* (Champaign, 1979).

36   This is as much a matter of common sense as of historical documentation: it is possible that the lower classes (who were no doubt more religious than the upper) held the doctor in greater esteem, but the record of the last six centuries does not bear out the suspicion.

37   See, e.g., Gideon Harvey (a physician himself), *The Conclave of Physicians, Detecting their Intrigues, Frauds, and Plots against their Patients* (1683), and the denial of his charges by Thomas Guidott (a fashionable Bath physician) in *Gideon's Fleece: or, the Sieur de Frisk . . . Written on the Cursory Perusal of a Late Book Call'd The Conclave of Physicians* (1684). No one has studied the dozens of these attacks and their replies made from 1600 to 1800. Another gauge of this view of doctors is the statistical extent of satire of them, but that indicator may be deceptive in that literary satire does not flourish in all periods.

38   The medical profession could help correct its image if its practitioners would enlighten the public on the extent to which medicine is a science as opposed to an art. Some valuable discussion of the subject is found in Lester S. King, *The Growth of Medical Thought* (Chicago, 1963), 38–40, and King, *The Philosophy of Medicine* (Cambridge, 1978), 237–8. On origins, see Edward Said, *Beginnings* (New York: Pantheon Books, 1975).

39   'Imagery' in the narrow and broad sense, as used by David Bloor, *Knowledge and Social Imagery* (1976).

40   André Soubiran, *Bedlam*, trans. O. Coburn (1956); Erwin Chargaff, *Heraclitean Fire: Sketches from a Life before Nature* (New York, 1978); Roland Barthes, 'Iconographie de l'abbe Pierre', *Mythologies* (Paris, 1957), 54–6. A partial list of such works – but no analysis – appears in J. Trautmann, *Literature and Medicine: a Checklist of Works* (Hershey, 1975).

41   The short, narrowly conceived, biographical essays that exist in this field do not address this question. See, e.g., P. J. Davis, 'Thomas Campion: Lyrick-Doctor in Physicke', *Journal of the American Medical Association*, CCVII (1969), 115–19, or the few remarks in R. B. Hinman's *Abraham Cowley's World of Order* (Cambridge 1960), 58–61.

42   The physician-author hopes that the transformation will ensure a heightened state of his own consciousness – a state usually reserved for his sickest patients.

43   The crucial period for the shift is 1550–1800. Susan F. Cannon says very little about the professionalisation of the physician before 1800 in her otherwise provocative book *Science in Culture* (Folkestone, 1978), and Jeanne Peterson omits imaginative literature in *The Medical Profession in Mid-Victorian London* (Berkeley and Los Angeles, 1978).

44   Thus from the eighteenth century dozens of dissertations such as John Freind's *The*

45   *History of Physick from the Time of Galen, to the Beginning of the Sixteenth Century*, 2 vols. (1725–6), in which individual 'physicians' usurp the territory of the 'medical history'. The closest to a study of medical historiography is Michel Foucault, *The Birth of the Clinic* (1963; Eng. trans. 1973).

45   E.g., in medical sociology; see N. D. Jewson, 'The Disappearance of the Sick-man from Medical Cosmology'. *Sociology*, X (1976), 225–44.

46   See, e.g.: (history of science) M. Hesse, 'Hermeticism and Historiography' an Apology for the Internal History of Science', in R. H. Stuewer (ed.), *Historical and Philosphical Perspectives of Science* (Minneapolis, 1970); (general history) Hayden White, *Metahistory* (Baltimore, 1975); (literary criticism) E. D. Hirsch, Jr., *Validity in Interpretation* (New Haven, 1967); (sociology) Barry Barnes, ' "Internal" and "External" Factors', *Scientific Knowledge and Sociological Theory* (1974), 99–125.

47   Foucault's interest in medicine appears in his first book: *Maladie mentale et psychologie* (Paris, 1954), and continues to his most recent, the projected 6-vol. history of sexuality. See also A. Sheridan, *Michel Foucault* (1980), 36–45.

48   Foucault, 'Truth and Power', in *Michel Foucault* (n. 29), 29–30.

49   See Elinor Schaffer, 'The Archaeology of Michel Foucault', *Studies in the History and Philosophy of Science*, XIV (1976), 269–75 and G. S. Rousseau, 'Ephebi, Epigoni, and Fornacalia: Some Meditations of the Contemporary Historiography of the Eighteenth Century', *The Eighteenth Century: Theory and Interpretation*, XX (1979), 203–26 (chap. 3 below).

50   Joseph Addison, in *The Spectator*, ed. D. F. Bond, 5 vols. (Oxford, 1965), I, 90 (24 Mar 1711). The passage is discussed by Fielding H. Garrison in 'Medicine in the Tatler, Spectator, and Guardian', *Bulletin of the History of Medicine*, II (1934), 481.

51   Trans. as 'What is an Author?' in D. F. Bouchard (ed.), *Language, Counter-Memory, Practice: Selected Essays and Interviews* (Ithaca, 1977), 113–38.

52   Job 13:4. Art and medicine are rarely connected, and when they are the arrows are from medicine to art, causing art to be underestimated in the same way as literature (but see Grace Goldin, 'Painting in Gheel', n. 26). There is no study of medical iconography (e.g., of paintings such as Daumier's 'Sad Physician').

53   See however E. Lain Entralgo, *Doctor and Patient* (1969).

54   Historians of science should recollect that the study of literature as a university subject remains no more than a century old. During this relatively brief period literary study has developed in primarily two divergent directions: either as a pseudo-science which attempts to outlaw subjective considerations and to prove certain hypotheses (see I. A. Richards, *Science and Poetry* (1926); the Chicago Critics; the New Critics; and now the structuralists and their American disciples); or as a non-scientific activity that draws upon the critic's intuitive powers and asks him to evoke impressions and memories that are anything but scientific (for this type see S. Sontag, *Against Interpretation* (New York, 1961), and P. Goodman, *Speaking and Language: Defence of Poetry* (New York, 1972)). Because the second of these approaches considers the first a pseudo-science, and never attempts, moreover, to prove hypotheses, it has paradoxically been able to maintain some interest in medicine, or, at least, in medical metaphors: see S. Sontag, *Illness as Metaphor* (New York, 1979). For the divergence of these two approaches, see Harry Levin, *Why Literary Criticism Cannot be a Science* (Cambridge, 1968).

55   In 'The Man Who Loved Women: the Medical Fictions of William Carlos Williams', *Georgia Review*, XXXVI (1980), 840–53, Marjorie Perloff designates several of Williams's novels as 'medical fictions' but does not delve into the connections that exist between these medical fictions and Williams's pervasive erotic impulses.

56   Lael T. Wertenbaker, *Death of a Man* (New York, 1967). Some day there may be practical application, e.g., the inculcation of moral values learned from imaginative literature, but literature and medicine is still a theoretical domain and should probably

not establish itself as an applied humanity. This position is supported by L. Binet and P. Vallery-Radot in *Médecine et littérature* (Paris, 1963) and *Verlaine à Aix-les-Bains* (Paris, 1958), a medico-literary study.

57 Many literary scholars are still reluctant to concede that the portrait of patients in modern novels is romantic and sentimental; that romanticism leads many physicians not to consider these literary works as verisimilitudinous. Lists of such novels appear in J. Trautmann, *Literature in Medicine* (n. 40) and M. A. Simpson. *Dying, Death, and Grief* (New York, 1979), 31–49.

58 Especially in the domain of biography and autobiography and in ways never conceptualised by J. Collins in *The Doctor Looks at Literature* (New York, 1923). For a fine example of integration of medicine and philosophy, see K. Dewhurst, *John Locke Physician and Philosopher: a Medical Biography* (1963).

59 See R. Benitez, 'La novela científica en España: Ramón y Cajal', *Revista de Estudios Hispánicos* (1979), 25–39 and H. Tzitsikas, *Santiago Ramón y Cajal: Obras literarias* (Mexico City, 1965); E. Maier, *The Psychology of C. G. Jung in the Works of Hermann Hesse* (New York, 1953); M. Serrano, *El círculo hermético* (Santiago, 1965); E. Faas, *Offene Formen in der modernen Kunst und Literatur* (Munich, 1975); T. R. O'Neill, *The Individuated Hobbit: Jung, Tolkien and the Archetypes* . . . (Boston, 1979). The literature on Freud now forms a vast industry; recent important works include W. Schonau, *Sigmund Freuds Prosa: Literarische Elemente seines Stils* (Stuttgart, 1968); J. E. Godo and G. H. Pollack, *Freud: the Fusion of Science and Humanism* (Chapel Hill, 1968); S. Timpanauro, *The Freudian Slip: Psychoanalysis and Textual Criticism* (1976); C. Lévesque, *L'étrangeté de texte: Essais sur . . . Freud . . . Derrida* (Montreal, 1976); L. Bersani, *Baudelaire and Freud* (Berkeley and Los Angeles, 1977); and Joseph H. Smith, (ed.) *The Literary Freud: Mechanisms of Defense and the Poetic Will* (Psychiatry and Humanities Series, 4) (New Haven, 1980).

60 Psychoanalysis is the only branch of medicine with which literary scholars are familiar and in which they maintain interest, on the grounds that psychoanalysis alone concerns itself with 'the word'. See S. Felman (ed.), *Literature and Psychoanalysis, the Question of Reading: Otherwise, Yale French Studies*, LV/LVI (1977), entire issue. Recently in America many disciples of J. Lacan have applied his concepts to literary criticism; see also J. Lacan, *The Four Fundamental Concepts of Psychoanalysis*, ed. J. A. Miller, trans. A. Sheridan (New York, 1978).

61 Susan Sontag, *Illness as Metaphor*, 3.

2

The discourses of literature and medicine: theory and practice (2)

I had a second chance to reconfigure the field in 1984–5 when Joanne Trautman Banks, the editor of The Letters of Virginia Woolf *and the author of a bibliography of the intersections of literature and medicine* Literature and Medicine: an Annotated Bibliography, rev. ed. *(Pittsburgh, 1982), invited me to write for a new journal called* Literature and Medicine. *I published this essay in that journal in 1986.*

This time I emphasised discourse theory and practical therapy, the two most crucial areas on my agenda then. The first was there because so many health-care professionals did not then (and still do not today) realise the healing power of the word in the therapies they dispense; the second arose out of my belief then (and now) that the humanities aggressively search out the practical applications of their ideas, especially in the post-disciplinary theoretical era in which the academy is fighting for survival.

In this and other chapters my references to 'literature and medicine' (whereas 'literature and science') are usually to be taken as referring to both subjects together.

> ... theory cannot be developed or tested without *critique*, and critique must involve the direct identification of alternative positions in a polemical manner. If one cares about ideas, it is difficult to write about error (or imputed error) without a certain sharpness of tone. I hope I have always argued *with reasons* (E. P. Thompson, *The Poverty of Theory*, 403)

INTRODUCTORY PREMISES

My reason for beginning with a brilliant quotation about the essential nature of theoretical discourse is not to exalt sharpness or point to Marxist affiliation, but to show that I have been attentive to the tropes in which my theoretical discussion is necessarily generated. In fact, I have been self-reflective about language throughout this essay.

My thesis here – unlike the more primitive thesis of an earlier paper[1] – is that the theory and practice of literature and medicine cannot be split at this stage. The field is too young; we (I mean those who cultivate it) need to have a better sense of how it can develop.

This attachment to practice should not imply that theory is not embedded in my 'practical' statements: theory is always present in research even when the researcher remains silent about it, or when it appears in confused fashion. The more pressing matter for literature and medicine is not a dichotomy between theory and practice but the sense of the field harboured by those who work in it today.

This is the matter to be addressed, as a consequence of which I abjure any artificial split between theory and practice and invite my colleagues to discuss the methods they would like to see used. My six sections are generated in the name of *method* rather than theory for this reason. To some, this position may seem evasive; some may prefer a statement at the outset of my own relation to contemporary theory. My attitude remains, first, that there is not room here and, secondly, that contemporary literary theory is less urgent than the approach I call method. The reasons have to do with a potential for utility built into literature and medicine. I hope this attitude will not imply that a developing subject can afford to overlook theory – unequivocally it cannot. But the matter is rather that the fundamental issues about theory in relation to literature and medicine are *specific* in the way they (the issues) relate to the developing traditions of medicine; and they (the issues) cannot be summarily reduced to 'laws' described in a few paragraphs. The result is that my discussion of method implies a theoretical dimension; and by adopting this approach I hope I am not evading theory at all. My emphasis on a method of inter-relationship as a fundamental procedure is directed toward this theoretical end.

There remains ideology, which is also always embedded in theory and practice. The question I ask is primitive: what is the ideological content of the two terms – 'literature' and 'medicine' – used throughout this essay to identify a 'field' as yet theoretically incoherent? My belief is that literature and medicine ought not to continue without self-awareness of the theoretical status of the basic terms used to designate the field. But both the terms and their practices have altered over time; their ideological content remains in a terrific state of flux today. Contemporary discussion involves the use and application of the terms 'literature' and 'medicine' to earlier historical periods when the (then) contemporary meanings neither coincided with, or even approximated, the reality. Theory insures that we continue to know *what* we are discussing when exploring the reciprocities of these domains. Theory forces upon us the making of basic, working definitions.

Throughout my discussion I am troubled by a utilitarianism I fail to disguise. My hunger is that literature should prove itself *useful* to the medical profession in the healing process. But even this craving embeds 'practical' issues containing 'theoretical' underpinnings; and I would need another essay to explain how I arrived at this belief. But the methodological consequence of my utilitarianism is that literature and medicine will not be any more coherent – theoretically or philosophically – whether it is useful or not. On the other hand, utility and theoretical coherence are not parties: something practical and clearly profitable from the viewpoint of utility need not justify itself in the name of logical unity or rational coherence. Thus, the irrelevance of my utilitarianism to theoretical coherence is not a phenomenon to which I am blind.

Yet the disparity troubles me. On the one hand, I cannot easily endorse a literature and medicine altogether devoid of utilitarian good in healing. On the other, I want a field – a domain – methodologically sophisticated enough not to be the theoretical sham it remains today. And I take no comfort in having to *select* between a utilitarian literature and medicine that runs the risk of remaining incoherent, and a nonutilitarian 'discourse of literature and medicine' capable of coherence. Each alternative seems perfidious. What is more, despite my enthusiasm for an eventually *useful* subject, I rather suspect, however naively, that no matter how useful literature may be to a relatively few doctors, it will not change what happens in the office or hospital. But when I reflect on the current methodological incoherence of the field, I am so troubled by its magnitude, that I want to shed my ill-fated utilitarianism and devote my attention exclusively to method – this within a plea that the field must soon become more coherent if it hopes to survive in the realms of respectable critical discourse.

Furthermore, any discussion of theory today requires consideration of privilege and ideology in relation to tradition and method. Tradition itself is a privileged concept, as well as a theoretical/practical matter. Much recent commentary has entailed the search for, or recovery of, tradition in one group of another (blacks, women, gays) in order that the group can reconstitute itself along the lines of the privileged. Reconstitution then absorbs the very tradition for which its devotees are searching; and validation of the new, now acquired, tradition becomes a primary focus of the programme.

But what is the tradition in, or of, literature *and* medicine, and how could it reconstitute itself as one of the privileged discourses? This is the matter of this essay, but it would be dishonest to claim it has vexed me to the degree that the utility question has.

Finally, my discussion is guilty of privileging the physician in many ways, not least at the expense of other, perhaps more spiritually inclined, healers, and certainly as a type of Renaissance humanist. The riposte that I have done so in the name of historical *fact* may not persuade those whose sense of a 'medical tradition' places greater weight than I give here to homespun therapies (shamanism, demonism, witchcraft, superstition, folk medicine, homeopathy, etc.). Yet it seems to me that history will eventually have the last word here, and that such privileging of the physician as a type of humanist hardly falsifies the historical record. The calamity would be *not* to acknowledge the ideological basis of this position of privilege:to pretend that physicians had been humanists only, and nothing else. I make no such claim, although I emphasise their humanism at the expense of their demonism.

Because I believe it would be a mistake to split theory and practice at this stage, I title the discussion a 'simultaneity', hoping that its future practitioners will concurrently work for both.

TOWARDS A METHODOLOGY OF INTER-RELATIONSHIP

At least since the eighteenth century, the heuristic value of theories has been evident. The doctor and poet Erasmus Darwin, a primary candidate for literature and medicine, stated the case *for* theory this way:

> Extravagant theories . . . in those parts of philosophy where our knowledge is yet imperfect, are not without their use; as they encourage the execution of laborious experiments, or the investigation of ingenious deductions, to confirm or refute them. And, since natural objects are allied to each other by many affinities, every kind of theoretic distribution of them adds to our knowledge by developing some of their analogies.[2]

New affinities as refined discrimination: this was the chief value of theory for Darwin. It remains primary among theorists today. But when a field itself is new, double advantage obtains: not merely analogies and affinities but a powerful heuristic tool capable of revitalising old subjects and permitting new insights into existing relationships.

There is a crucial, third consideration in this case because medicine has been omitted from the cultural debate for so long.[3] An example from bibliotherapy clarifies the point. Imagine a thousand cancer patients voluntarily reading Solzhenitsyn's *Cancer Ward*. The patients are divided into random and control groups; each is followed for an identical period. If a large number of the control group improves, we can assume that bibliotherapy is a treatment of possible value whose research should be further funded. Assuming a positive response, funding can be sought, eventually creating the need for experts in the field. But suppose the experiment cannot be undertaken; do we then renounce bibliotherapy, or retain it as a *possible* practice until such time as the experiment(s) can be performed? I hope the latter – and a 'tool' such as literature and medicine permits speculation about results.

The relation of printed discourses does not altogether differ: imaginative primary literature, like medical writing, is culture-bound. This seems the simplest of points yet has profound implications because few scholars gaze simultaneously on both types of texts. Although each type – literary and medical – relates to previous texts in a continuum governed in part by formal rules of genré, as classical critics stressed (i.e., glancing backwards, Dante to Virgil, Dickens to Fielding, Shaw to Foote; in medicine, Vesalius to Galen, Freud to Sydenham), each type also relies, to a certain extent, on the symbols and values of the time, no matter to what rhetoric resorts.

In literature and medicine a single critic pronounces on both sets of discourse; or there may be two critics. The crucial activity is *parallel* scrutiny of two 'types' of texts' but there is as yet no agreement about the dialectic by which to relate these works. Both discourses are culture-bound, as already indicated, suggesting that each specimen is firmly rooted in the cultural differences that gave rise to it; and the critic's utility is this: by relating them he or she greatly illuminates the 'maker' and 'nature' of the text, as well as the culture in which they were generated. The critic's procedure is inexorably eclectic: he or she may focus on accepted master-pieces of literature and medical classics or minor works, but the critic does not function here as the custodian of a 'great tradition' or as the preserver of a canon. The critic's task is broader – it involves ordinary human achievement as well as peaks of distinction. Furthermore, the critic is committed to include medicine. That is, he or she recognises that some fundamental aspect of the culture – of difference – has been misunderstood, even misrepresented, without this inclusion. As George Cabanis claimed in his important *Coup d'oeil sur les revolutions de la médecine*, nothing tells us so much about a culture as its systems of medicine.[4]

Such parallel method offers an alternative to the current, if unfortunately limited, practice of studying *either* imaginative literature *or* medical classics. Instead of isolating, for example, Thomas Campion, Sir Thomas Browne, Foote, Mandeville, Defoe, Smollett, Goldsmith, Keats, Flaubert, George Eliot, Proust, Chekhov, Bridges, and so forth; or, alternatively, treaties on Elizabethan blood, Commonwealth writings on the 'religio medici', Restoration books about plague, medical texts on masturbation and insanity, Victorian theories of delusion, the Russian cholera epidemics of 1892–3, etc.; instead, these can be diachronically paired,[5] certainly not out of known influence but toward heuristic advantage:

Physician-authors or writers	Examples of the type of synchronic text
Thomas Campion	Thomas Wright, *The Passions of the minde*
Thomas Browne	Gideon Harvey, *Vanities of Physick and Philosophy*
Mandeville and Defoe	John Hancock (treaties on plague) Thomas Lodge, *A Treatise on the Plague*
Smollett	medical textbooks on 'nervous sensibility'
Goldsmith	Dr William Battie, *Treatise on Madness*

Keats	Thomas Burgess, *The Physiology of Blushing*
Schiller	Schiller's massive medical-physiological writings
Flaubert	Felix Pouchet, *Essai sur l'historie naturelle*
Beddoes	Johann Schoenlein, *Natural History of the Diseases of the Europeans*
Robert Bird	Samuel Warren, *Passages from the Diary of a Late Physician* (1832)
George Eliot	Richard Madden, *Phantasmata, or Illusions*
Proust	Paul Dubois, *Psychic Treatment of Nervous Disorders* (Eng, trans, 1905), the medical writings of Philippe Ricord, Joseph Rollet of Lyon, Brissaud's medical works on asthma
Chekhov	P. A. Arkhangelsky, *The Treatment of Cholera*
Robert Bridges	writings of the Victorian 'gentlemen physicians,' especially of the 1870s, and Bridge's own medical poem *Carmen Elegiacum . . . de Nosocomio Sti. Bartolomae Londinensi* (1878)
Schnitzler	Freud on hysteria
William Carlos Williams	Sir Charles Scott Sherrington, *Man on His Nature* (1940)
André Soubiran	Max Theiler, *The Virus* (1951)

Diachronic analyses, performed horizontally or synchronically, as above, yield constellations of metaphors. Eventually, if enough pairs are compared, it becomes evident that although the verbal craft of each set of authors varies, both sets are culture-bound. The sets are not literally *determined* by the culture – this goes too far in denying individuality. But both sets respond to social exigencies weighing upon their formal (internal) arrangements, and this similarity is represented in the precise nature of their metaphoricity. By isolating common metaphors – say metaphors of delusion in Eliot and Madden – by deciphering their patterns and contexts, we pinpoint a 'moment' in the life of the metaphor, as well as better comprehend the cultures' differences. The approach rests on an assumption that metaphors have organic lifespans, as do human beings; that they are born, mature, develop into adulthood, grow senscent, decay, and die; equally significant, that they arise at particular moments for good reason and are not merely the creations of chance or random play.[6] Literature and medicine thus rests on a conception of metaphor that is moored to literary history and cultural anthropology and, as I have been attempting to suggest, to medical history as well. Without such broad perspective and analysis, whole sets of metaphors would remain neglected, archaeologically buried, as it were, concealing a substratum of the culture under consideration.

In all these examples, the medical component within literature and medicine is construed as a *printed* text. As in the above columns, again proceeding horizontally, the critic's texts are dual: literary *and* medical – and both sets fall into approximately the same chronological period. The implications are various: for one thing, metaphors are assigned a primary role in the search for affinity, an activity placing monumental trust on the power of analogy. For another, it assumes a rhetorical threshold: it keeps the discussion of narrative on a high plane, using the best tools of semiotics and narratology, and does not indulge an unacceptably simple-minded notion of 'literature'. Finally, it assumes, *faute de mieux*, that we want to know *more* about the culture (differences) being isolated. All this acknowledges that the inferior writing of certain medical texts, notwithstanding, the time has come to treat these texts as 'literary' artefacts. The understandable objection that most medical texts are unworthy of the critic's attention (the Harveys and Willises, Freuds and Jungs, constituting exceptions), is unacceptable on grounds that the selection of texts is wide open. To be sure, some critics will have to learn something about medicine; but better this education than the pretense that medical texts are not worth scrutiny – or immune from scrupulous textual investigation. The point of the activity is not historical fussiness (authorship, sources, influences) but cultural completeness: learning what was imagined, as well as discovering possible therapies embedded in the wisdom of literature. If, indeed, medicine has been omitted from the cultural history debate, as I have been suggesting, then it should be plain that any analysis of a society that omits its medicine must be incomplete. The method outlined here does not eliminate those who care little about the historical past; it merely sets up a minimum set of conditions without which serious work cannot begin.

THE PHYSICIAN AS HUMANIST: EMPATHY AS CRITICISM

The image of the physician and other health-care professionals in our time is probably too various and complex for sweeping generalisation. On the one hand, the physician is maligned as a technician; on the other, he or she is not diminished in personal relations or non-medical contexts: the physician as reader, musician, patron-appreciator of the arts, everyday philosopher, etc. But the lay public holds on to such ambivalent views of the physician's type owing to the costs of health-care, that its (i.e., the view's) monolithic negativity has been heard to interpret with any confidence.[7] The public will not alter an image that has endured for decades; but the *self-image* of these medical professionals is another matter.

In our century nothing has influenced the physician's profile more profoundly than the loss of his or her identity as the last of the humanists. Until recently, physicians in Western European countries received broad, liberal educations, read languages and literatures, studied the arts, were good musicians and amateur painters; by virtue of their financial privilege and class prominence they interacted with statesmen and high-ranking professionals, and continued in these activities throughout their careers. It was not uncommon for Victorian and Edwardian doctors, for example, to write prolifically throughout their careers: medical memoirs and autobiographies, biographies of other doctors, social analyses of their own times, imaginative literature of all types. In twentieth-century America, the pattern has changed; only the most imaginative physicians can hope for this artistic lifestyle as a consequence of the economic constraints and housekeeping demands placed upon the doctor in a world where servants have disappeared.[8] Also, the diminution of 'humanist' content in the training of physicians has lent an impression – perhaps falsely so but nevertheless pervasively – that medics are technicians, anything but humanists. As a by-product, it has nurtured a myth (already old by the eighteenth-century Enlightenment) that medicine is predominantly a science rather than an art. Both notions require adjustment if physicians hope to return to their earlier enriched, and probably healthier, role.

The issue of privilege is embedded here, especially the notion of doctor as humanist. Without definition 'humanist' is a meaningless word, emptier now than it was twenty years ago in view of the attacks on 'humanism' by various Churches and conservative movements. At stake, more than historical traditions of Christian humanism and their recent fate, are the physician's interpretative skills. These are hardly empty or meaningless abilities. No one doubts that contemporary physicians interpret signs, diagnose symptoms, read clues – are semioticians of a type. To be sure, not all physicians are equally sensitive, or vigilant, in these activities; yet *all* doctors, like *all* artists, are *au fond* interpreters and doctors' interpretations of the chaos about them survive by empathy. Artists also sympathise with the natural chaos surrounding them; this act typifies their adoration of Nature and is a type of Keatism 'negative capability'. (I realise, of course, that empathy and sympathy are not synonymous concepts in the domain of medical care.) The physician's especial gift is that through a type of compassion – as much as through education or intellect – he or she can envision an imagined world. In this sense too, physicians are a type of artist. Imagination, as Romantic artists knew, survives on sympathy and empathy; remove these and imagination shrivels: it deceases on an inadequate diet of memory. Whether the artistic 'maker' composes, paints, writes, dances, or sculpts, empathy with the things of this world nurtures his or her imagination. Artists empathise with living creatures and the natural things of this world external to themselves – plants, trees, mountains, oceans, clouds, skies, rocks – as well as with its timeless universals, the so-called human condition. These sustain them by continuous interchange – it is not enough for artists to 'remember' them. Such a view may rely too heavily on romantic sensibility – not enough on medical reality — but there is much truth to the affinities between the doctor's imagination and the artist's.

I would proceed further and suggest that doctors *must* imagine a fictive world, in addition to a real one, if they are to perform their work. Novelists imagine characters they will invent by empathising with them; the resulting degree of verisimilitude depends upon his psychological leap more than on stylistic bravura or technical craft. Writers who possess craft alone do not get far. Is the same not true of doctors whose Hippocratic oaths extend to the waiting room only? Some physicians both sympathise and empathize because an early illness, as in the cases of physician-authors Robert Montgomery Bird and Robert Bridges, forced them to abandon their medical practices (Bird grew mentally ill after he stopped practising). Samuel Johnson, whose writing is often a discourse of literature and medicine, would have agreed; in his *Life of Boerhaave*, the great Dutch physician of the Enlightenment, Johnson claimed that Boerhaave's 'own pain taught him to compassionate others'. If these similarities exist, then literature and medicine share another domain which has rarely been explored. The fact that computers may permit already overworked doctors to rely increasingly less as humanists, is ancillary to the point. The greater matter is that we need to ponder the images and roles of the doctor, for new images may give rise to new therapies.

Even so, it will not do to link everything with everything. In what *precise* sense, we ask, is medical diagnosis based on imagination? To what extent is empathy implied in the clinical situation? There is convincing evidence that the doctor's compassion (his or her clinical version of empathy) causes patients to improve, but it exists in a number of forms ranging from the doctor's role of hope and confidence in the healing process to his or ability to 'read the patient' correctly.

The affinities are not abstruse. Literature helps the doctor to read, explicate and interpret, as well as to control language; literature holds up the widest assortment of varieties of human behaviour. By expanding the physician's universe literature permits the doctor greater awareness of, and — by extension — sympathy for, his or her patients. Proof of these processes must be sought elsewhere; my point here is that literature and medicine should play a part in conceptualising the matter in the first place. Every doctor knows that a large part of his or her clinical practice is devoted to interpretation of the psychological problems of patients. Not all these require psychotherapy or analysis; most patients want reassurance in the form of sensitivity, understanding. A few sentences, uttered in the right tone, will do. For the multitudes of doctors who are not so linguistically gifted as are most writers, who have not mastered the terse but perfect sentence, more words are necessary. A laconic doctor can offend patients, even lose them; their conditions worsen. But the more words doctors use, the greater the likelihood of using them inappropriately and creating problems for themselves and their patients. Doctors must therefore learn to control their own language. Here literature and medicine is a rich resource that holds up the mirror and lamp of the medico-linguistic drama.

The doctor who cannot empathise with his or her patients is no better off than the poet who bears no compassion, as Coleridge might have said, for the slimy creature of the deep. The analogy needs no labouring, yet analogies to the doctor's specific use of language in clinical situations are harder to pinpoint. Medical schools have devoted some attention to language but it is insufficient. The specific issue regards the practise doctor *after* he or she leaves medical school: in adulthood and maturity; when, in middle age, doctors grow disenchanted with their self-image as a mechanical technician; when they may be searching for something higher and greater; when they realise that they cannot conceptualise these matters, let alone settle them, until they think about 'words' as the precursors of concepts. Then, an already *developed* subject, literature and medicine, may be useful – as useful as any other source of education; and a journal like this one may be near the top of their lists, ranked with the specialised ones in their fields.

CATHARSIS IN LITERATURE AND MEDICINE: THE ANALOGOUS METHOD

Our sense of the Greek word *catharsis* and its extended concept derive, of course, from Aristotle in the *Poetics*: neither can be discussed without recourse to the realms of literature and medicine, as classicists have known for a long time.[9] Aristotle seems to have believed that spectators, who observed doomed figures fall into misery from high stations of power, somehow grow emotionally more stable. The audience's pity and fear. Aristotle thought, causes emotional rearrangements leading to new psychological strength. It is off hypothesis, alien to many of our beliefs in a post-Pavlovian, post-Freudian, age. Yet it is not remote, mechanical, or foreign to our beliefs about learning. It has appeal: especially the notion that we the spectators grow strong through the suffering of others, and without harming ourselves. The idea may be defective: profit is usually made at a price – at least takes a toll. But the great virtue of the Aristotelian premise is that we get something for practically nothing. We need only watch and 'suffer' ourselves.

As classicists have commented, the original Aristotelian notion is both an aesthetic *and* medical theory. Aristotle himself seems to have sensed no discrepancy between them: this strengthened his belief in it psychological validity. Samuel Johnson has captured the Aristotelian notion in more accessible terms:

> The passions are the great movers of human action; but they are mixed with such impurities, that it is necessary they should be purged or refined by means of terror and pity. For instance, ambition is a noble passion; but by seeing upon the stage, that a man who is so excessively ambitious as to raise himself by injustice, is punished, we are terrified at the fatal consequences of such a passion. In the same manner a certain degree of resentment is necessary; but if we see that a man carries it too far, we pity the object of it, and are taught to moderate that passion.[10]

Empathy remains at the heart of the matter. For one thing, Aristotle and Johnson both agree that the audience's weeping is part of the process of purgation; for another, they concur that these are physiological experiences for the spectators. Tragedy – they imply – permits us to experience pity, illness fear: the two are analogous experiences. By viewing the terrific fear of these tragic figures, we – the spectators – resurrect the self-pity we would have in our own death. Yet we survive: the observation, and the passing of time, combine to regenerate our insight. The combination cleanses us through physiological and psychological catharsis. Thus the spectator in the theatre is not so alien to the physician who observes his or her patient. The analogy is imperfect but strong, and creative writers have shown why, as the lists of works in Joanne Trautmann's bibliography demonstrate.[11] Modern literature glitters with characters who are regenerated to a new awareness of life, and a heightened perception of reality, through illness. The notion has had a powerful grip on the creative psyche. Perhaps the point is as relevant to our discussion of method as to any perceived medical reality: the analogous method of literature and medicine beckons the play of mind, and sustains its students by encouraging them to consult the record of humankind's imagination in literature. In the case at hand – catharsis – we wonder under what circumstances emotional purging promotes regeneration and heightened body awareness. Novelists have written stories, poets poems, on this theme, but the question has never been seriously addressed. There seems to be no natural forum as yet, although indications are that there soon will be.

These are *both* literary and medical matters. The ideas involved – catharsis, purging, regeneration, renewed emotional stability, heightened states of awareness – are concepts that make sense only when we recognise their expressive forms. If spectators recollect fear and pity, ought not physicians? I believe that our young physicians lack this acquired skill to be 'witnesses' or 'spectators' in suffering. But the critic's mind should be unfettered here: after all, what is the difference between a spectator who sees Antigone or Oedipus fall, and a physician who observes his or her young patient succumb to myocarditis, leukaemia, AIDS? Aren't both groups transformed by the experience?

Privilege and tradition enter again. It is not physicians – as traditional histories claim – but nurses who have witnessed loss, dying, and grief. To nurses, terminally ill patients disclose private thoughts they wouldn't dream of telling anyone else; and many physicians insulate themselves from these charged transactions by a protective wall. Yet, nurses did not write much in comparison to doctors. So we necessarily rely upon physicians, but not without realising that we have privileged them. Over three hundred years ago, Thomas Browne, the prolific physician-author of *Religio Medici*, grasped the doctor's privileged position in an epoch when clergymen and physicians were professional rivals: 'to preserve the living, and make the dead to live . . . is not impertinent unto our profession; whose study is life and death, who daily behold examples of mortality, and of all men least need artificial mementos, or coffins by our bed side, to minde us of our graves'.[12] Likewise, contemporary psychiatrists are privy to secrets withheld from the rest of the society. By virtue of this greater wisdom, the shameful-physician has possessed a different angle of vision, as Chekhov continued to stress to his correspondents.[13] The embarrassment is that we have paid him – the shaman in earlier societies, more recently the physician – so little critical attention. In the pages of *The New England Journal of Medicine*, Lewis Thomas has pleaded for the shaman's version of the tradition but has claimed few followers among medics.

Some patients and observers may be broken by the spectacle of catharsis. Others may go mad, but the benefits outweigh the defects. Though suffering taxes some beyond the point of reason, as the Greeks knew, it instructs those of a rational and imaginative cast. Yet suffering must be embedded in language to be conceptualised. So the questions remaining is whether this heightened angle has expressed itself in written language distinctive enough to be called 'the physician-writer's'. The answer depends upon the rigidity of the question. It is tempting to respond negatively because literary history supplies so many contradictions. But negative response entails a profound consequence – a disjuncture between experience and formal representation (mimesis) that remains a fundamental tenet of traditional literary criticism. The disparity begs for other questions to be put: are form and content wedded in the writings of physician-authors in a readily identifiable way? If Selzer's essays and stories are not instantly recognisable as having been written in a physician-writer's mode – however 'Brownian' or baroque his style – and non-physician Sylvia Plath's poems not recognisably *dis*similar (although not medically trained she was steeped in medical reading), then what does it mean to invoke 'physician-author's language'? Medical training and experience, no matter how superior, is an insufficient index; some linguistic features also ought to appear. Gazing beyond, if no such language exists within the conventional epochs of literary history, is there at least a sign, or signature, as it were, of the author's hand in the work? How does it matter to the literature of Thomas Browne, Goldsmith, Keats, Chekhov and dozens of others, that they attended medical school, to William Carlos Williams's version of Modernism that he practised medicine? Selzer's language is relevant here. A gifted stylist, he could find no language *common* to literature and medicine in modern literature; he returned to the baroque cadences of Thomas Browne for a model, and in the process has become something of a baroque stylist himself. But what *are* the continuities and discontinuities of this 'common language'? Chekhov seems to have had some notion, although he rarely verbalised it in a correspondence which teems with medical references: 'I don't doubt that the study of the medical sciences seriously affected my literary work; they significantly enlarged the field of my observations, enriched me with knowledge, the true value of which for me as a writer can be understood only by one who is himself a physician: they also had a directive influence and probably because I was close to medicine I avoided many mistakes.'[14]

Is this delusion: 'can be understood only by . . . a physician'? Presumably the 'mistakes' and 'influence' extend beyond medical verisimilitude: i.e., faithfully depicting this illness or that symptoms, getting the medicine right. But do they extend to language? How much is Chekhov claiming by insisting that he is not 'one of those artists who think that they can arrive at everything by the intellect alone'?[15] Chekhov is silent throughout these correspondences on the physician-author's language, compelling the critic to supply what he, the perceptive Chekhov, had omitted. It is curious that Chekhov was not troubled by a common language in the way Selzer has been.

The analogous method asks other questions, suggesting that literature and science, however useful, offers inadequate analogues to literature and medicine; not least because few writers have performed serious laboratory research or conducted experiments, whereas everyone has some familiarity with illness and healing. Literature and medicine is a different field of inquiry – though 'difference' does not readily yield itself. What difference is there between a physician writing from experience about illness and suffering, and an imaginative-empathetic Kafka or Mann conjuring wards and sanatoriums? Miroslav Holub, the Czech poet-pathologist, suggests while in a quasi-Platonic mood in 'The Root of the Matters': '*There is poetry in everything. That / is the biggest argument / against poetry*' (section 2 19–21).[16] Perhaps true: yet the poetry of medicine goes largely unsung; whatever literature and medicine will become, it is a field that dares to ask these questions. That it has found few answers on which to build is unsurprising. It remains a fledgling in its critical awareness and sense of ideological boundaries.

THE IMMENSE PROBLEM OF DEMARCATION: MEDICINE BREAKS APART FROM LITERATURE

The historical 'break' – a discontinuity of discourse – occurred in the eighteenth century. Before then, huge globs of printed matter – not merely prose – passed as *either* 'literature' or 'medicine', Distinctions barely existed in most countries, and the prose of the 'doctor' was often as refined as could be found. If names and titles are concealed, it is impossible now to distinguish passages from the 'literary' forms that contain them, as could be demonstrated if there were space. This accounts, in part, for the predictable reception during the Restoration and early eighteenth-century of Browne's *Religio Medici*. Furthermore, physicians were pre-eminent among those who published, especially in England, France, and Holland. While today it is the rare doctor who writes creatively, it was then the norm, and any statistical count would show how prolific doctors were.[17]

About 1700 or 1750, 'literature' had not yet separated out into the divisions we take for granted today. Authors were not yet identified with the categories their social or professional types would produce today: i.e., full-time poets writing poems, novelists novels, and so forth. Pope's physician, William Cheselden – a noted surgeon who operated on the famous and rich – helped Pope edit Shakespeare;[18] today it is a rare physician who knows the folios well enough to assist any professional Shakespearean editor. Physicians before 1650 or 1750 – convenient beacons, not firm beginnings or endings – often wrote religious tracts, scientific treatises (natural philosophy), forensic, matter, memoirs and diaries: all were then adjudged 'literature'. The Warton brothers, influential critics writing literary history in England in the middle of the eighteenth century, formulated quasi-scientific criteria for 'imaginative literature' that eventually set the category 'literature' apart. Erasmus Darwin, already mentioned in connection with theory as a heuristic tool, was no Samuel Johnson, but he lived in the same town (Lichfield) and shared the splendors or horrors – depending on the case – of Grub Street. Today, any critic would decide whether Darwin was *primarily* doctor or writer, (ditto for Sir Thomas Browne). But the question would have appeared foolish to the age of Johnson,[19] and today we ask whether the late Geoffrey Keynes, surgeon and brother of the economist, was primarily doctor or literary figure? Clearly, he was both. Keynes amassed one of the finest book collections anywhere, and was a bibliographer and writer about English literature. He also practised surgery for six decades and wrote in his memoirs that, given the choice he preferred medicine.[20] Keynes was alert to the distinction in a way Darwin would not have been because the break had been relatively recent. Indeed, 'imaginative literature' is a category developed by Romantic poets and English universities in the nineteenth century; before then, the sense of great writers existed of course: Johnson writing about Shakespeare, Dryden, Pope. But the notion that *professional* types produced this great literature had not yet come into being. The development has impact on literature and medicine.

Before the age of professional specialisation – at the end of the eighteenth century – physicians wrote for a number of reasons: to refute false knowledge, for fame or financial gain, for social status. It was *fashionable* to write, one prominent Georgian doctor in Bath then commented; most successful physicians did.[21] Yet this is hardly the reason given by a practising GP writing poetry today: 'I have had to write poetry or go bonkers.'[22] In the transition from fashionability to psychic regeneration, a facet of the break or separation is found. Before the eighteenth century writers are not self-conscious of the methods of composition they have used in relation to their own professional status. The same is true of physician-authors. It is therefore futile to expect them to use metaphor with this disparity in mind, and their writing may therefore be good test-cases for determining whether their use of language is different.

A limited sense of literature and medicine develops if literature after 1800 alone is consulted – especially if the Enlightenment is overlooked. A provincial literature and medicine then arises: historically naive, narrow-minded, 'Whiggish' in the sense that the late Cambridge historian, Sir Herbert Buttefield, contended that we have distorted earlier versions' originality by making them conform to our patterns.[23] But what was its archaeology, in Foucault's sense, before 1800? Retrieval grows complicated when we recall the institution of 'the physician' differed from his modern counterparts before (approximately) 1800: his manufacture, constitution, being, as well as the social arrangements that overlapped into his self-image.[24] The routine life of a physician in William Harvey's England differed considerably from that in Newton's or Darwin's. Further transformations in the last century cause the Victorian physician to appear now as an anachronism – an extinct species, if we are literal-minded enough to reconstruct his type from daily schedules (including rounds) and personal diaries (hundreds of which survive).

The synchronic method outlined above (horizontal scrutiny of printed texts) necessitates the exploration of works of *non*-physician-authors of the same period for understanding of similarity and difference in the use of language. If the evidence were produced here, it would suggest that the author's profession – doctor, statesman, novelist, etc. – is ultimately less consequential for his use of language than are his social class and intellectual milieu. The exigencies of his culture and mentality count for more. It matters, naturally, that he is a doctor rather than a lawyer or statesman – this status permits him ranges of knowledge and types of perspective. But ten texts by ten Enlightenment doctors will vary in composition, language and metaphor, suggesting how treacherous it is to diminish the originality of writers by attributing too much to social determination. These ten, compared with another ten of the *same* period by *non*-doctors, vary less – this is the crucial aspect to grasp for any theory of literature and medicine. The link has limits; we can attribute just so much to it; after that, textual discrimination prevails, and the honest critic is compelled to say what is *not* determined by literature and medicine.

This last matter may seem elusive to readers wondering what literature and medicine have in common; even to attentive readers uninitiated into theoretical thinking. But it addresses questions much debated in our pluralistic times by textual critics, such as, to what degree does an author's personality contribute to the resulting text, and what do we need to know about his or her life and times? Can we interpret the text meaningfully if we pretend it is an anonymous document? The answers to these questions vary according to the philosophical beliefs of those answering them. But whatever the critic's persuasion, the fact remains that literature and medicine possesses as yet little of its own theory, so little that 'schools' such as phenomenology and deconstruction seem remote. Deconstruction actually poses no serious threat to literature and medicine, and it is preposterous to claim that deconstructionism has prevented literature and medicine from developing a theoretical basis. Debates about personalities and privileged tropes will remain unresolved whether or not deconstruction is applied. Nevertheless, recent pluralism shows to what extent a *theory* of literature and medicine depends upon an antecedent theory in literary criticism at large; and this is why it isn't possible to be serious about a theory of literature and medicine unless one is willing to assume the burdens of a certain amount of general theory.

Concomitantly I have been suggesting that the medical component of literature and medicine also reveals an embedded *theory* of medicine; of cause and effect in illness, of the theory of disease, and, of course, as always, of the type of language in which these relationships are expressed. I cannot imagine any theory of reciprocity being useful, let alone widely accepted, if carte blanche it abjures history: a history of real medicine, just as a history of real literature.

PATIENTS AS AUTHORS: ARCHETYPES OF SUFFERING

If the theoretical approaches I have been outlining have any validity, then the patient-author ought to be even more intriguing than the physician.[25] Their differences are not insignificant. Unlike the physician, the patient has actually suffered, unless one pinpoints a handful of physician-authors who, like Bird and Bridges, were chronic patients as well as physician-authors. If empathy is the noblest of the passions, suffering instructs as well as affects and products a new, abstract, intellectual awareness with values and aesthetics of its own. The transitions are barely understood. That 'suffering alone teaches' – *pathei mathein* – is an ancient Greek piety (Aeschylus) still relevant. Doctors' alleged objectivity is no doubt valuable here, may even permit them to 'gaze into' the patient more deeply than they can gaze into themselves. Yet suffering is unique among types of experience and is, moreover, meta-linguistic in that many patients – even the most articulate – insist that they cannot verbalise what 'they have been through'.

While patients, as a group, have been less able to write than doctors, many patients have been educated, literate, even erudite: why then have they not written as much, and as diversely, as physicians? More curiously, why have authors who themselves were chronic patients – Pope and Keats, Samuel Johnson, Proust – written little about their illnesses? Can it be that writing on *non*-medical subjects somehow alleviated, or purified – catharsis again – their pain?[26] There is enough of an historical record from which to reconstruct answers to these enduring questions about the patient-author as a distinct type writing in recognisable styles; questions pertaining to the uses of analogy and metaphor; and procedures arising from diachronic comparisons within synchronic approaches to texts. If catharsis-as-recollection is as effective as catharsis-as-observation/participation, as Chekhov continued to iterate, it must be doubly so for the ailing patient. The doctor 'suffers' because he knows too much, the patient because he 'experiences.' Both 'suffer', yet the patient needs to purge himself or herself (symbolically as well as literally) even more than the physician. The reason is that nothing subdues patients so effectively as conceptualising themselves – as Foucault intimated and Sontag substantiated – 'conquering the disease'.[27] The type of language used remains the issue in this conquest: specifically, comparison of the patient's language with the physician's, and the way patient and physician reflect and parody the language of the 'other' one. Patients who invent tropes – i.e., the language of some seventeenth-century schizophrenics, diaries of madmen, the patient-authors cited by Sontag – conceptualise and internalise the disease in ways that must eventually prove of great interest to literature and medicine.

What are the writings of patients – especially those who have not been medically trained? Much depends on the patient's education, of course, and on the rhetorical conventions to which he or she has subscribed. Patients in Shakespeare's England will not describe themselves like those in Proust's Paris or Schnitzler's Vienna. Similarities do appear:

categories of thought so consistent in their metaphoric coherence (the patient as hero, disease as predatory invader-villain, health as conquest, the patient's courage as derived from sources external to himself or herself, the physician's moral weakness and potential corruption, the act of dying as the predator's investiture of himself or herself, regeneration as overthrow and ejection of the invader-villain) – so coherent that one wonders if there could be a heritage of myths about suffering in the Jungian archetypal sense. Over many centuries patients have continued to refer to states of minds experienced during illness that portend a set of quasi-universal structures of the mind: radical discontinuity from health to sickness; invasion of the body by the predatory disease; a seeming interminability of suffering; the beauty embedded in pain – this despite apparently lethal disease; alteration of physical space about the patient; new moral perceptions and heightened mental states. All these suggest structures and permeate hundreds of written accounts. There is no death of material: from seventeenth-century diaries, like those of Dr Napier and Samuel Pepys, to those of Charles Darwin and Alice James, plus novels, plays, poems, letters. The archive is large and remains an untapped repository capable of unlocking the door to secrets about pain and suffering.

Our myopia in the late twentieth century results from the way we gaze at medicine – the way we refuse to view its ramifications in the corners of ordinary life. It is naive to think we harbour any objective view of doctors, hospitals, patients, ailments; we still respond to their stereotypes, symbols, spaces, not their literal reality.[28] We still view patients from the *doctor's* point of view – for complex reasons, and not entirely because the doctor's records over time have given us a specific image. Even in the seventeenth century, when doctors were fewer per capita, and when their relation to patients was different from what it is today, the patient had already been mythologised. Ralph Josselin, the now obscure seventeenth-century vicar in Essex whose diaries contain hundreds of references to his illnesses and suffering, rarely consulted doctors (they weren't available in north Essex). Yet he saw the implication of disease throughout human history and recorded some of the elusive links. Others – Pepys, Pope, Boswell – constantly refer to pain and suffering, to the melancholy illness produces; but do not extemporise: they record their symptoms as faithful chroniclers and rarely interpret the signs or analogise them.

Today it is hard to know what to make of these disjunctions between patients and healers. The demarcation is exacerbated by the difficulty we have in generalising about these quasi-archaeological, quasi-paleolithic, types. Each chronicler – Napier, Josselin, Pepys, Pope, Boswell, Darwin, Alice James – must be evaluated ultimately on his or her own terms; syntheses are hard to make.

Yet it is tempting to search for a synthesis about 'the patient's style' if it can be supported. It is true, a new history of medicine would pave the way.[29] but despite the solid achievements of the social history of medicine generated in the last few decades, no coherent model has developed. Nor can any develop until the history of medicine, anthropology, and social history combine their forces with imaginative literature – until we return to the gaze of the whole. Even if the model existed, the disparities would not magically disappear. The mental life of patients over time – the way suffering is experienced in relation to the *expectation* of suffering – is an elusive realm in the history of consciousness.[30] No wonder Lain Entralgo thought much of it was meta-linguistic.[31]

The signs of literature and medicine need to be codified into a grammatology. Suffering and pain are retrievable, it seems, precisely because their tropes have thus far been shaped by medical savants: doctors, surgeons, nurses whose monolithic codes can be cracked, demythologised, and interpreted. Moreover, the patient's codes help, and if there is a latency period between the self-detection of symptoms and the patient's attribution of 'dis-ease' into a medical category, the patient only becomes 'sick' when he or she submits to medical tropes indicating infirm types. It is not perverse to notice that the *classifications* of illness by patients remain taxonomies of the medical profession, nor does the trend abate in our time. Crudely speaking, patients present a formidable hurdle for those who wish to retrieve a history of suffering and pain, just as historians of gender have been foiled by the disappearance of whole classes of evidence. Yet the annals of suffering are Brobdignagian in contrast to the remnants of pornography and erotica. The record of suffering constitutes a gold mine waiting to be quarried, if we will only learn to decode its signs and languages.

Linguistic analysis is the best start. An example culled from the diary of an erudite patient demonstrates the method:

> I have swallowed the weight of an Apothecary in medicine, and what I am the better for it, except more patient and less credulous I know not. I have learnt to bear my infirmities and not to trust to the skill of physicians for curing them. I endeavour to drink deeply of Philosophy, and to be wise when I cannot be merry, easy when I cannot be glad, content with what cannot be mended, and patient where there can be no redress. The mighty can do no more, and the wise seldom do as much.[32]

This is Elizabeth Montagu in 1739, the non-compliant English Bluestocking wracked by choler (sore throat and cold) and disobedient of her doctor's orders despite having 'swallowed the weight of an Apothecary'. What should we make of these words? How comprehend the ambivalences (captured in the symmetry of the syntax) and mental state ('drink deeply of Philosophy')? Montagu's disgust with doctors, to which she devotes much energy, silently sings a litany from the Middle Ages to the Enlightenment: not all memoirists have been so disobedient, or realigned their pain by a *decision* not to obey. The consequences of such disobedience make fascinating reading if studied over time. Montagu represents a middle point betrayed by her syntax of 'when' and 'with', but some patient-writers have been more oxlike in their outright refusal to comply. Montagu seems to have concluded – as a type of eighteenth-century Norman Cousins – that philosophy, merriment, and laughter do as much good as pills, potions, and doctors. The tropes of 'realignment' are as significant for literature and medicine as the images used by physician-authors. The desideratum is not excessive: we need a theory of literature and medicine, as well as practical field work, which places value on the *patient* as well as the physician, which grasps suffering in *specific* contents, i.e., whether or not to obey the doctor, and, which contemplates pain.[33]

The language attaching to self-diagnosis and dosing constitutes another approach. Annals of imaginative literature are permeated with patients who have acted as their own doctors and whose diagnoses appear in letters and diaries, as well as in fictionalised forms. As various as are the literary forms, so are the tones. Wesley, the Methodist healer and great Lutheran symbolic figure for the Romantics, records his diagnoses and doses: 'pounded garlick, applied to the soles of my feet . . . I broke through the doctor's order not to write, and began transcribing a Journal for the press'.[34] How swift the disobedience! In 'pounded' and 'broke' a shorthand code emerges that offers a clue to Wesley's anxiety when dosing. In the same epoch Dr Smollett, the Scottish physician-author of *Humphry Clinker* (1771), performed a somber self-diagnosis of his psychological condition which he conveniently placed into the mouth of one of his fictional figures:

> Know, then, I can despise your pride, while I honour your integrity; and applaud your taste, while I am shocked at your ostentation. – I have known your trifling, superficial and obstinate in dispute; meanly jealous and awkwardly reserved; rash and haughty in your resentments; and coarse and lowly in your connections. – I have blushed at the weakness of your conversation, and trembled at the errors of your conduct. – Yet, as I own you possess certain good qualities, which over-balance these defects, and distinguish you . . . as a person for whom I have the most perfect attachment and esteem, you have no cause to complain . . . and as they [your faults] are chiefly the excesses of a sanguine disposition and looseness of thought, impatient of caution or control; you may, thus stimulated, watch over your own intemperance and infirmity, with redoubled vigilance and consideration, and for the future profit by the severity of my reproof.[35]

The harsh language of this passage differs from contemporary self-diagnoses by Montagu, Wesley, and Lord Hervey (Pope's 'Sporus').[36] Close reading reveals with what particularity the patient dramatizes his conflict. The larger point of literature and medicine is that the language of patients, monitored like EKGs over time, reveals much about the web of health, especially its precariousness.

Those who suffer learn how to cope by observing fellow-suffers; this is why the physician is ultimately limited, as patients in history have intimated. Yet the record of self-diagnosis includes more healing of *others* than of one's self. Even today, we continually play surrogate doctor, however unconsciously; the psychological processes involved are mirrored in literature, a remarkable lamp illuminating this profile of pain: letters, diaries, novels, stories. Chekhov was *both* physician and patient, observer and observed, self and other, in just this sense: his mediation between the so-called 'objective' gaze of the doctor and the anguish of the patient. Chekhov told Plescheyev, who had suffered gruelling illness as a political exile in Siberia, what it meant for him (Chekhov) to be 'subject and object' at the same time: furthermore, 'as a medico . . . I [Chekhov] described the psychic pain correctly, according to the rules of psychiatry'.[37]

Patients who imagine themselves as healers – whether in a hospital bed or as visitors – use signs and language. Yet, as the long tradition of the English novel exemplifies, those who serve as lay healers also attempt to convert spiritually. Until the late nineteenth century, the altar was the focus of the hospital ward – in its centre or at its head – and 'conversion' was a principal function of the nun-nurse. Even Wesley wandered about England preaching to the sick and poor, proselytising sinners, indulging their sexual fantasies as he doled out the radical medicines, barks, and potions, the electric 'shocks' described in *Primitive Physick* (1747). The result was not always conversation to a *medically* democratic religion – Methodism – but to even more primitive forms of Christianity. In American literature (Hawthorne, Emerson, Melville) the conversion is often meta-religious: to a new form of awareness of self; in our own times, to a new life style that endows health on the victim as the reward of physical regeneration. In extreme cases in the literature of Hawthorne and Melville, conversion utterly destroys the character. Such destruction is also found in Shaw's *The Docotor's Dilemma*, Whitman's regeneration 'of the body', and in the commentary on suicide Flaubert implied in *Madame Bovary*.

But if the patient is a separate self he or she is also a member of society, in sickness and health. The revelations of patients – even the poet as patient – constitute a prolific genre that has only recently been explored in this way; group illness, especially epidemic illness, represents another genre, as the literature of plagues and famines, miasmas and medical holocausts reveal. It is one thing to imagine oneself suffering from a *unique* set of physical circumstances that coalesce into a condition (phrensy, mania, catarrh, more recently, asthma, flu, pneumonia), and another to be one of hundreds, or thousands, consumed by epidemic disease (plague, cholera, leprosy, diphtheria, mass hysteria, now AIDS and the fevers of nuclear winters; in science fiction and fantasy, the imaginary diseases of other planets). The annals of illness emerging from the *patient's* point of view capture distinctive emphases: to be sick *alone* differs from crowd affliction to such degree that 'illness' is an inadequate label to describe both experiences. What is an 'illness' anyway? – first we need to know how, when, and where the sufferer suffers. As ploughed over a 'fiction' as Defoe's *Journal of the Plague Year*(1722) has not been explored in these terms, replete though it is with descriptions of dying.[38] No romantic ironist, as were Dostouevsky, Tolstoy and Mann – who were also obsessed by the relations of art and disease – Defoe's aim was different: his 'journal' displays an aesthetics of epidemic illness gilded by surrealistic silhouettes of corpses littering a depopulated city. If Defoe's genius permitted him to capture London's pulse as an Ortega y Gasset of the masses' sufferings, it also enabled him to extract the poignant beauty of group miasma.

But time takes a toll even in the history of pain. The categories invoked in our post-Romantic frame of mind centre on disease as a prime mover: as the inspiration of art – its muse; in Wordsworth the 'word wounds' as much as the bow. Yet there are other potent analogies, as in the writings of Samuel Richardson, who knew what he was doing in *Clarissa Harlowe* when he designed Clarissa's illness as the last word on the moral worth of her shattered life. 'The art of dying' ethic culminated in the nineteenth century; it was hardly born then. Joseph Addison, England's leading literary critic during the Enlightenment, pronounced what it was like 'to die as a Christian'. But the tradition has now been lost except as a curiosity among specialists.

Missing in this discussion is a sense of the patient-artist, patient-writer, in relation to a culturally distinct society. We have tended to think about Defoe and Flaubert, Proust and Céline, Mann and Gide, Williams and Selzer, as if each were an island apart – had existed in universal worlds without real hospital walls, real doctors, real medical emergencies. No one ought to diminish the individuality of the physician-writer's imagination, or set up a priori barriers to the possibility of an archetypal mental 'set'. If a distinct community affords patients support systems for their illnesses, this radically alters the patient's self-perception and affects those writers who fictionalize him or her.[39] There are moral, as well as aesthetic, implications here that touch on the eventual 'literary artefact'. Au fond the matter is not ethical: e.g., whether a given society does enough – for example, to sustain those dying of AIDS . Sufficiency apart, the diaries of terminally ill AIDS patients would be of the greatest interest to literature and medicine if compared with diaries of those suffering from politically more neutral diseases.[40] As the laity suspects, the progress of no two cancers is identical; but plagues and epidemics – AIDS – are viewed differently. The literary consequences of the disparity are considerable, not only the effect on the narrating voice and its tropes but the sense of time and space, and the metaphors chosen to describe the categories of invasion and combat, deterrence or submission.

When does the sufferer become a patient? The question appears primitive but is not stupid. Medical sociologists have asked it under one condition (Parsons's 'sick role') and Foucault under another (the manufacture of categories). Foucault says, when the sufferer 'enters the medical zone' and comes under the sphere of influence of 'medical control'.[41] This may be so: the manufacturer of 'the patient' differs, after all, from the onset of the conferred status. Today status is conferred when the white tag with computer number is placed around the wrist. Yet suffering and pain have hardly been the exclusive privilege of manufactured patients: those who have placed themselves under the care of medical attendants. Suffering – surely for providential reasons – extends its dominions more widely. The Baudelaire who suffers the ecstasy of agony and the 'flowers of evil' in 'sweet affliction' while composing *Les Fleurs du mal*, whose suffering so captured the attention of the young Marxist critic Walter Benjamin, suffers as poignantly as those confined to sanatoriums.[42] The historical record – even of great writers – is permeated with examples of persons enduring mental and physical pain who were *not* patients. Imaginative literature demonstrates what intellectual history has been unable to: that until recent times illness has been surcharged with *meaning*. The whole history of morality has been tied to it. Literature, better than other art forms because denotative and explicit, reveals how the ethic 'no man is an island' has applied to suffers who happened *not* to be patients. Sickness, indeed, has been a condition of the greatest interest to the social historian, as it has been a private pilgrimage for those with a particular destiny. But literature has also offered an alternative model to the medical one: even disease need not be limited to the medical model that removes pain from the human condition and reduces it to malpractice suits. Literature permits pain. In literature we see, as it were, that our historical ancestors were not merely compulsive neurotics, obsessed with sickness and death, but sufferers whose involuntary melancholy sometimes permitted them to extract and preserve health.[43]

Pain rather than suffering has been the great teacher and moral leveller until relatively recent times. Pain, not cure, is the condition described over and over again in Renaissance diaries and eighteenth- and nineteenth-century European novels. From Bunyan, Defoe, and Richardson down through the Gothic and major Victorian novelists; yet these writers have been largely neglected by historians of medicine on grounds that literary evidence is untrustworthy. Untrustworthy of *what*? – perhaps of statistical trends, but not untrustworthy of the mentalities literature and medicine should seek to recover. Many of the rituals described in poems and stories of the last three centuries have been attempts to avert pain of one type or another. Whether consolation has come from child or lover, minister or doctor, the aim has been negation, or at least regulation, of pain: bodily pain arising from social conditions in cities and towns, as well as private mental anguish. If the medical demarcations sketched above teach us anything in the domain we conveniently called literature and medicine, it is that a segment of the past has been obliterated by our twentieth-century attempt to discover cure at all costs. In the shift, pain has been wiped off the slates of previous mentalities. It is as if cure had always signified an avoidance of pain. Nothing could be further from the truth; for centuries pain was the known condition of human life.[44]

THE DISCOURSE OF LITERATURE AND MEDICINE

An adequate method of literature and medicine would not stop here but continue to treat more of the people: their sociopolitical institutions, broadly interpreted, sense of life, manners, customs, needs, as well as the patient's languages and signs: in short 'the patient as text'. This last concept involves nothing less than a semiotics over historical time, and has been so frequently invoked by now that it begins to sound like a cliché. Yet 'patient as text' remains a vital concern: an end in itself, as crucial as suffering and pain.

As must also be evident, a main concern of my program is the education of doctors in the interpretation of 'texts' so they can 'read' their obligatory ones: their patients. If literature and medicine ever does anything useful, it will be here: in educating the doctors to 'read his or her text' (patient) more subtly.

But immediate utility is not the only goal of literature and medicine. Utility remains a primary goal among several, but the others cannot be discounted. Among them is a critical discourse of literature and medicine whose texts have hardly been reconstituted, let alone retrieved. This retrieval involves the record of their overlaps and reciprocities. Because no text ever owes its genesis to a single 'Ur-text' or original there exist no primary texts in paleolithic days (pre-Homeric) that can be isolated. But the discourse of literature and medicine was old by the Renaissance, and it flourished in the seventeenth- and eighteenth-century Enlightenments.

It must be retrieved and reconstructed as part of the subject's future. Retrieval will involve, as a minimum, the discussion of logical relations (i.e., the inherent logic of medical theories, differences between mere overlaps and casual reciprocities), as well as the 'manufacturer' of the categories that obviously constitute the dramatis personae of its narratives: necessarily doctors, patients, nurses, healers; simultaneously texts, genres, rhetorics, tropes, the treasure-house of metaphor and metonymy.

To affirm then, when asked what literature and medicine is, by an analogue of the sort that it is a subject wherein 'the patient is the text', diminishes its potential too much, through reductionism. The cliché succeeds by its half-truth (i.e., there are sense in which the patient clearly is *not* the text), in part as a charitable gesture (i.e., patients have been overlooked for so long), as well as through metaphoric vividness: new affinities appear when 'the patient as text' is explored. Metaphor entices in just this way, and 'the patient as text' is no doubt worthy of more attention than the analogy has received.

But for all its good intentions, this approach narrows the domain of literature and medicine too rigidly by privileging the utilitarian goal over others. And whereas I disclosed at the outset my own ambivalence about utility versus theoretical coherence, I certainly would not maintain that the theoretical aspect is *un*important.

This philosophical or metaphysical dimension will, I suspect, determine whether literature and medicine ever enlarges to anything more than a medical humanist's pipe dream. For two decades a handful of medical schools have been trying to educate students in the medical humanities, a group of subjects deemed by many to be theoretically incoherent. Yet these subjects continue to be taught, in part, because there has been a consensus that doctors need more humanistic training than they have been receiving. But this approach to literature and medicine exclusively treats of its utilitarian dimension, and implicitly addresses its political and ideological status since funding remains at stake in developing academic programmes. The *discourse* of literature and medicine is, however, entirely and exclusively linguistic: found in texts, buried in genres, locally tied to rhetorics and tropes. It can only be unraveled by linguistic analysis: the same explication de texte critics apply to other texts in other domains. The fate of the discourse of literature and medicine is therefore necessarily the same as that of other critical discourses awaiting – like departing jets on the runway – exegesis. Its medium remains linguistic, its content mimetic (as are all other contents). Its greatest challenge remains the relation of the text to its various 'con-texts', its milieu created by signs and acts of speech no less than in other discourses. The fact that doctors and patients, rather than tragic heroes and comic heroines, dominate its pages is, from the theoretical vantage, arbitrary. What counts is recognition of the ultimate linguistic status of the discourse.

Not to recognise this other, equally urgent, if less obviously utilitarian, activity of literature and medicine diminishes the whole enterprise.

It is well to exploit such clichés as 'the patient as text', and to recognise how poetry and health, poetry and reason, were united in the primitive symbolic imagination in a single mythical figure: Apollo.

But new subjects do not validate themselves merely by half-truths and approximations. Sooner or later it will be patent even to those with a casual interest in literature and medicine, that its discourse forms an essential part of its (the subject's) accumulated heritage. If 'the patient as text' still requires exploration, so does the discourse. However rudimentary, the first step must ensure the recognition that this discourse is first, and foremost, a *language*: a written record, with Burton and Bacon, Browne, and Defoe, Smollett and Keats, Shaw and Chekhov, among its many compilers. Much has also been contributed to the discourse by the obscure doctors and writers, nurses and patients, shamans and votaries, whose names are no more known to us today than the signatures on unvisited tombstones in country churchyards.

The discourse of literature and medicine needs to be interpreted just as any other discourse is: according to a priori assumptions about language and its signifiers, and the relation of this language, or sets of languages, to the things represented. Eventually, the student of the discourse of literature and medicine will need to declare his or her allegiance to one set of assumptions no less than the student of any other discourse, be it psychiatric, biologic, or economic.

If the day arrives when the phrase 'literature and medicine' universally designates a convenient, shorthhand label for 'interpretation of the discourse of literature and medicine,' it will not be excessive to learn if the interpreter is – for example – a traditional or radical historicist, vestigial New Critic, Marxist, Lacanian, deconstructionist, feminist, etc., or some combination of these. It will indeed be crucial, and all along I have been suggesting that we have not yet arrived at that point.

NOTES

1 See G. S. Rousseau, 'Medicine and Literature: the State of the Field,' *Isis*, LXXII (Sept. 1981), 406–24 (chap. 1 above). I am grateful to Drs Laurel Brodsley, Gloria Gross, and David Morris for commenting on various versions of this essay, and to Professor Joanne Trautmann Banks whose expertise as an editor I have utilised far beyond the call of her own duty; I myself am responsible, of course, for the essay's imperfections.

2 Erasmus Darwin, *The Botanic Garden, a Poem, in Two Parts* (New York, 1807), xxii; the poem was first published in 1789.

3 The few exceptions – e.g., Peter Gay's two-volume study of *The Enlightenment* (New York, 1966–9) and Keith Baker's and Roy Porter's books – should be applauded.

4 Pierre Jean George Cabanis, *Coup d'oeil sur les révolutions et sur la reforme de la médecine* (Paris, 1804), 9–10.

5 I.e., diachronic analysis of synchronic texts. By advocating this method I mean to

privilege metaphor above – for example – synecdoche, and to place greater importance on it than on tropes of space and time as a phenomenologist would.

6   I.e., literary history rather than contemporary metaphor theory: e.g., the debates between unitarians and dualists, the writings of M. Black, D. Davidson, M. B. Hesse, M. Johnson, R. Rorty, J. Searle *et al.* on metaphor. See also the writings of Clifford Geertz and other anthropologists.

7   The view, which includes fees, pedantry, and the certainty of knowledge in medicine, has a long history extending at least as far back as the Renaissance: see especially the writings of Keith Baker and Foucault.

8   The British Victorian 'gentleman physician' and the current American 'city doctor' would make a fascinating study in contrasts with the differences *vis-à-vis* writing outdistancing the smiliarities.

9   On the medical dimension of catharsis within the context of Greek tragedy and literary criticism see: J. Tate, 'Tragedy and The Black Bile', *Hermathena* xxv (1937), 1–25; Hellmut Flashar, *Melancholie und Melancholiker* (Berlin, 1966); Donald William Lucas, *Aristotle: Poetics* (Oxford, 1968), appendix ii, 273–90.

10 *Boswell's Life of Johnson* ed. George Birkbeck Hill, rev. and enl. L. F. Powell, 6 vols. (Oxford, 1934), III, 39.

11 Joanne Trautmann and Carol Pollard, *Literature and Medicine: an Annotated Bibliography*, rev. ed. (Pittsburgh, 1982; orig. pub. 1975).

12 Thomas Browne, *Hydrotrophia*, in *Selected Writings of Sir Thomas Browne* ed. Geoffrey Keynes (Chicago, 1968), 116.

13 See, for example, Chekhov's letter of November 1888 to Souvorin, his publisher and literary correspondent: 'Those who have mastered the wisdom of the scientific method and are able to think scientifically experience many charming temptations . . . Those who possess the scientific method feel with their souls that a musical composition and a tree have something in common, that both are created in accordance with equally regular and simple laws. Hence the question: What are those laws? Hence the temptation to work out a physiology of creative art . . .' This long letter, merely dated November, continues in this fascinating speculative way and denies that literature and science have opposite methods and opposite goals; the letter appears in S.S. Koteliansky and Philip Tomlinson, *The Life and Letters of Anton Chekhov* (eds.) (1925), 129–30, but is not found in Avrham Yarmolinsky's more recent anthology, *Letters of Anton Chekhov* (New York, 1973). It is curious that Tolstoy should have thought Chekhov would have been an even greater writer than he was if he had *not* been a doctor, this on the assumption that medicine cramped him and confined his vision; for Tolstoy and Chekhov and medicine, see Sophie Laffitte, *Chekhov 1860–1904* (New York, 1973), 74.

14 See Chekhov's 'case history' of himself dated 11 Oct. 1899 and written to G. I. Rossolimo, a neuropathologist who had been his classmate in the Moscow University medical school; quoted in Yarmolinsky, 352.

15 Yarmolinsky, 353. For Chekhov art and reason were not superior to science and observation; indeed the empirico-scientific strain ran deep in Chekhov's anti-rationalist temperament, as is even clear from his treatment of Professor Nikolai Stepanovich in 'A Dull Story' (1889).

16 Miroslav Holub, 'The Root of the Matter.' in *Selected Poems*, trans. Ian Milner and George Theiner (Baltimore, 1967), 95. Holub's poetry is saturated with medical imagery, especially in such poems as 'Pathology' as well as 'On the Immortality of Poets'.

17 For the diverse types of writing, fictional and non-fictional, of medical doctors in the Renaissance, Enlightenment, and afterwards see: A. Barbeau, *Life and Letters at Bath in the Eighteenth Century* (1908); Norman Moore, *The Physician in English History* (Cambridge, 1913); A. S. Collins, *Authorship in the Days of Johnson* (1927); Thomas

The discourses of literature and medicine (2)

Kirkpatrick Monro, *The Physician as Man of Letters*, 2nd ed. (Edinburgh, 1951); J. W. Saunders, *The Profession of English Letters* (Toronto, 1964); John D. Gordon, *Doctors as men of Letters* (New York, 1964).

18 See Marjorie Hope Nicolson and G. S. Rousseau, *'This Long Disease, My Life': Alexander Pope and the Sciences* (Princeton, 1968), 7–129.

19 Because the professionalisation of medicine then was so different from what it has become, i.e., the social status of the doctor.

20 Geoffrey Keynes, *The Gates of Memory* (Oxford, 1981), 307.

21 Thomas Mathias, *The Shade of Alexander Pope on the Banks of the Thames*, 3rd ed. (1799), 29.

22 The British GP-poet Henderson Smith's confessional description of himself in Howard Sergeant, *Poems from the Medical World* (Lancaster, 1980), 178; also see my essay about this very anthology: G. S. Rousseau, 'White Aesthetics', *Literary Review*, XIX (1980), 25–6.

23 I.e., the Whig versus Tory view of the past described by the late Cambridge historian Herbert Butterfield in *The Whig Interpretation of History* (1931).

24 The definitive work on this development remains to be written, but for an idea of the primary resources see: John Mason Good, *The History of Medicine, so far as it relates to the Profession* (1795).

25 For the patient in cultural history see: G. Williams, *The Age of Agony* (1975) and *The Age of Miracles* (1981); Arthur Kleinman, *Patients and Healers in the Context of Culture* (Berkeley, 1980); Anthony R. Moore, *The Missing Medical Text: Humane Patient Care* (Carlton, 1978); Roy Porter (ed.), *Patients and Practitioneers: Lay Perceptions of Medicine in Pre-Industrial Society* (Cambridge, 1986); rather little has been written about the patient's *language*.

26 The point here is not now to retrieve the 'patient' historically but linguistically.

27 Both Foucault and Sontag have stressed this point, however idiosyncratically, Foucault by focusing on the varieties of subjugation of the patient, Sontag by isolating and deploring the metaphors that attach to specific patients when afflicted with specific diseases. See Michel Foucault, *The Birth of the Clinic* (New York, 1972); and Susan Sontag, *Illness as Metaphor* (New York, 1978).

28 See Foucault's *The Birth of the Clinic*; Ian Kennedy, *The Unmasking of Medicine* (1981); David Armstrong, *The Political Anatomy of the Body* (Cambridge, 1983).

29 One that takes account of imaginative literature as a primary concern of its programme, as well as concern itself with the tropes of medicial discourse.

30 See the Iowa Symposium on 'The Language of Pain and Fear', *Iowa Review*, XI (spring/summer 1980), 115–40.

31 See P. Lain Entralgo, *The Therapy of the Word in Classical Antiquity*, ed. and trans. L. J. Rather and John M. Sharp (New Haven, 1970).

32 E. J. Climenson (ed.), *Elizabeth Montagu*, 2 vols. (1906), 1;36.

33 If Susan Sontag has lambasted the metaphors that attach to such major illnesses as tuberculosis and cancer, Ivan Illich has taught in various of his books the limits of medicine and sceptical attitudes towards the professional doctors who practise it; but Illich did not substitute imaginative literature dealing with medical topics.

34 Nehemiah Curnock (ed.), *The Journal of the Reverend John Wesley*, 8 vols. (1909–16), VI, 65 and IV, 91.

35 Tobias Smollett, *The Adventures of Ferdinand Count Fathom*, ed. Damian Grant (1971), 2, and my analysis of the passage in G. S. Rousseau, *Tobias Smollett, Essays of Two Decades* (Edinburgh, 1982), 7–8.

36 See John Hervey's manuscript, 'Account of My Own Constitution and Illness; (1731),

in Romney Sedgewick (ed.), *Some Materials Towards Memoirs*, 3 vols. (1931), III, 961–87.
37 See Yarmolinsky, 92.
38 Dying, grieving, and suffering continue to be the most neglected topics within literature and medicine; for example, L. A. Landa's fine edition of Defoe's *Journal of the Plague Year* (Oxford, 1968) does not even mention this crucial component of the work.
39 Medical historians (Webster, MacDonald, Holmes) have shown this pattern to be true in early modern Europe.
40 The retrieval of published and unpublished diaries of the seventeenth to nineteenth centuries would better demonstrate the point.
41 Foucault, chap. 1, 'Spaces and Classes'.
42 I would like to insist emphatically that suffering, grieving, and dying are the *primary* substantive concerns of literature and medicine, though I recognise that there are others who find gender and sexuality central subjects.
43 This is why the history of medicine and social history naturally attach themselves to literature and medicine.
44 See Kenneth Keele, *Anatomies of Pain* (Oxford, 1957); and Brian Inglis, *The Diseases of Civilisation* (1981).

3

Ephebi, epigoni, and fornacalia: towards a historiography of medicine

My work on literature and medicine – viewed as an emerging discourse about overlaps and differences, in a complex world of reciprocities – was not generated as the exclusive province of theory, nor with a view to therapy only. Throughout the 1960s and 1970s I continued to find myself drawn to the history of medicine as a discipline that had never been fully appreciated.

As the 1970s drew to their conclusion various strains began to coalesce for me, even if imperfectly. I was thinking of Foucault and also of the developing discourse of postmodernism, of the way Foucault's work had spawned a new history of madness; of the possibility that cultural anthropology would eventually revolutionise methodologies in the humanities and sciences; and finally of the ways in which the traditional history of medicine was evolving as a discipline in a post-disciplinary era.

In that decade I had been so overwhelmed by Foucault's brilliant leaps of imagination that it was difficult to disagree with him about anything. His veiled rhetoric and poetic prose sweeps readers away and sways them into his orbit whether or not they agree. Foucault seemed to utter the best of one's most abstract conceptualisations but which, upon further reflection, one had never been able to fructify. This essay was generated in a mood of scepticism within veneration. It was gratifying to learn that the astute scholar J. G. Merquior, among others, agreed with me and wrote more critically of Foucault than I have here (J. G. Merquior, Foucault (Berkeley 1985), 64).

I was also captured by the Grecian, neo-Hellenic revivalism of the Enlightenment – itself a potential sub-discipline of Enlightenment studies – at a time when specialisation was being encouraged by a wide variety of projectors and Encyclopedists whose ideas were lucidly expressed in such remote ephemera as Steele's periodical The Plebeian, in which the young Spartan ephebes became a metaphor – a mere trope – for commentary about sex and madness in England in the early eighteenth century.

The essay now seems crude in many ways, not least in its ability to announce a thesis commingling the various strains. It is very much the product of its moment. But it captures, I think, the degree to which these post-disciplinary discourses, so prevalent today in ideologies such as the New Historicism, continued to haunt me. I published this essay in The Eighteenth Century *in 1979*.

So they made the oven into a goddess of that name (*Fornax*); delighted with her, the farmers prayed that she would temper the heat to the corn committed to her charge. At the present day the Prime Warden (*Curio Maximus*) proclaims in a set form of words the time for holding the Feast of Ovens (*Fornacalia*), and he celebrates the rites at no fixed date; and round about the Forum hang many tablets, on which every ward has its own particular mark. The foolish part of the people know not which is their own ward, but hold the feast on the last day to which it can be postponed. (Ovid, *Fasti* ii, 525–2, Loeb Library translation)

The first half of my title indicates the three strains I want to amplify: notions about arrival and the coming of age, as in the postpubescent ephebi of ancient Greece; concepts about lineage and succession, especially succession in the official sense, as in the notorious case of the seven sons of Adrastus, the so-called epigoni, whose attempt to revenge their father against his enemies failed; and concerns about ambiguity in ritual experience, as in 'the moveable feasts' – the Roman fornacalia – in which some cultists were too stupid to know to which *curia* or sect they belonged and as a consequence celebrated the wrong feast on the wrong day.[1] I want to probe these strains within the context of writing about the history of medicine.

It seems to me that all three strains pertain to the current state of higher criticism and literary sensibility as much as to my chosen subject, medical historiography, not to be confused with medical history. If I have exaggerated the connection, it is by design rather than chance; for I am positively certain that humanistic disciplines interconnect in a given epoch (certainly they do in the eighteenth century) by means of languages, which if attentively studied reveal the connecting links.[2] This interconnection notwithstanding, I want to explore why it is that certain contemporary developments in literary theory are taking a toll on the writing of the history of medicine. In a sense I presuppose my exploration can amount to no more than a footnote on Foucault's Emersonian self-confession, 'je suis pluraliste' (in part it is a delusion) – the same Foucault whose rhetorical tropes have captured the attention of small segments of the literary and historical community of scholars, although these same writings have not mesmerised the historians of science and medicine to anything approximating the same degree. But I also wish to probe contemporary writing about the history of medicine in our period and to comment upon its remarkable lack of diversity.

There is nothing extraordinary about the yoking of ephebi, epigoni, and fornacalia, all three historic developments. The yoking itself indulges a narcissistic fantasy: namely that the connection possesses greater meaning for the yoker, in this case, the author, than for the reader. Nevertheless, in the three historical developments so well known to historians of the ancient world, one sees a metaphor developing for American scholarship, for the direction it has taken since the Second World War, and probably as a result of professional and sociological reasons having to do with matters announced in the opening paragraph. Construed as a metaphor of modern scholarly life, the yoking is endlessly fascinating.

Everyone knows that the ephebi entered formal military life shortly after puberty; that they served an apprenticeship, usually stretching for a few years not 'from the appearance of the first down', as Xenophon says, but about five years later; and that they served military masters who taught them the art of war and the love of Greece.[3] Historians studying the institution, especially its longevity, have been amazed by the success of the institution, especially its vigour and length of success.[4] It endured for centuries, fading only with he decline of Greek civilisation itself. Even the Romans kept up the institution, although they altered its name. The crucial aspect in both the Greek and Roman world was the moment of entrance: the law of Greece was adamant that it must occur at the onset of the eighteenth year, with no provisions made for precocious ephebi.

The mythological epigoni are both more and less real. The seven sons of the wretched Adrastus, whose name means the unescapable, blindly joined their father in an attack on Thebes. But they were miserably beaten. Only Adrastus returned home, owing to his wonderful horse Arion. Ten years later Adrastus and the seven spigoni tried again; and although they vanquished their foe, one of the epigoni was slain in battle. Still, they blindly followed their father, and for their blind subservience paid dearly.

The fornacalia, about which Ovid has written so vividly in the *Fasti* (part of which passage is cited in the epigraph above), is less easily grasped on either the literal or imaginative level. Perhaps most striking is its vocabulary of *curiosae*,[5] the foolish part of the people who could not discover their religious identity. The feast of ovens (*fornax*), from which the festival ritual takes its name, was celebrated on different dates, according to the sect which one belonged. Neither Ovid nor any other ancient writer tells us *why* the *curiosae* were too stupid to know which sect they had joined. All doubt concerns the source of their stupidity, not the reasons for 'moving the feast'. Besides, it is one thing to be foolish and quite another to be *curiosae*, a word capable of so many interpretations as to render the *curiosae* almost anything one wishes: curious, diligent, officious, spylike, pompous, foolish, sectlike, caring.

If there is a moral in this yoking, it applies to the most contemporary of scholarly situations, at least for students of Enlightenment culture. We American scholars are being told over and again that we are ephebi who have 'come of age', that our scholarship is now surely the best in the world. At the same time and perhaps paradoxically, we are derisively called *epigoni*, not only by the French but by the British who view us as servile sons – epigoni – of the so-called unescapable father, certain 'higher' French critics. Essay upon essay, review after review, now opens with commentary about 'the American epigoni', and one need not read extensively to learn that Derrida and his French intellectual brethren – Deleuze, Ricoeur, even Foucault – are the new Adrastoi, the new unescapable fathers, and we Americans their sons, although very few among us will concede it in such stark and bald terms.

An even more dire circumstance may exist. A version of the decadent Roman confusion just described has set in among even the most staid type of veteran scholars. American scholars who previously thought they knew their academic identity have now grown so thoroughly confused by the new pluralism that one wonders what name to give them. It is a development perhaps parallel to that of Ovid's erring *curiosae*. The case is not merely that certain 'critical types' – as the sociologists might call them – are winning the only few vacant posts but rather that scholarly kudos is also going to these types,[6] that is, to the model epigoni who faithfully follow the father. The establishment is placing great valve on the type of financial reward. In literary criticism, the rewarded species is not of the traditional brand of solid historical scholarship but a newer brand of derivative French structuralism practised by professions in New Haven, Irvine, Santa Cruz, and elsewhere. In some American quarters the debate about the merits of one brand over another and the ensuing confusion about just rewards for the deserving are so prevalent that entire departments are embroiled in endless logomachy. Even in less war-stricken departments the confusion about pluralism takes stringent forms, particularly in cases related to tenure and promotion.[7] The whole development adumbrates a new randomness in humanistic studies.

None of these observations would seem pertinent to the historiography of medicine in the Enlightenment, a subject generally thought to have nothing to do with the life of contemporary writing. Yet it has everything to do with it, for during the last few decades two distinct varieties of the history of medicine have been written: the general survey of study based on research and the thesis-oriented 'history' which some 'discovery' has prompted the author to write. Ultimately the rhetoric invoked by each is of the first importance, as I will demonstrate. But before delving into that controversial area, it is essential to chart out the two territories just mentioned.

Whereas both types are often diachronically arranged, the former usually aims at surveillance, thoroughness, even completeness and exhaustibility, while the latter rarely relies on these qualities. The latter is usually work prompted by the discovery of some material which in turn leads to a thesis, such as discovery of the existence of a secret society, or the significance of a hitherto neglected figure about whom something radical has been learned, or the serendipitous encounter of a cache of manuscripts now compelling the scholar to reread and reconstruct his field or subject. The former hardly maintains such grand aims: conceived in the womb of incompleteness, it bases all its validity on the grounds that its field is uncharted; that a great mass of scholarship has accumulated over the years; and that all sorts of disparate sources urgently need to be brought together under one cover. Moreover, whereas the former type claims thoroughness in exhausting the scholarly field (the history of medicine is particularly vulnerable here), the latter admits to being controversial and in its best forms even to being dangerously wrongheaded. The latter wants to compel other scholars to put down their work and review 'the new discovery'; the former engages the critics and scholars only when they bask in 'the blessed isles of leisure' (the comparison is especially appropriate for recent literary criticism). Who now reads new 'histories' of this or that? Nor does it seem to matter how distinguished the historian is. But disciples are made out of the stuff of the latter type, and discipleship in traditional and untraditional forms is what is debated in the universities today. Disciples, always eager to please their fathers, soon learn how to gain power in the universities and create standards for value.

All this reverts to the original point about ephebi and epigoni arriving and obeying, and about the kinds of books the new epigoni are intent on writing. The diachronic survey or history, no matter how arcane or erudite its subject, cannot disorient and puzzle the reader about 'value' in scholarship to the degree that thesis-oriented books do. This is why it pays to consider a category called 'history as discovery'. To continue in this line of thought, the general historian or historian of medicine who ventures into the former field (i.e., to chart the entire history of some development) normally does not presuppose that his history will be an exercise in obscurity. How could it be? To be so would jeopardise its most intrinsic nature, as I will shortly demonstrate. But in this realisation one also sees precisely why contemporary higher French critics so widely idolised by their epigoni in our country have no truck with history.[8] Nor do their disciples: history in any form except as rhetorical deconstruction is anathema. Both fathers and sons have carried the solipsistic reserves about history to the limit.

But it is less than clear where the French and Americans, father and sons, stand on the second variety: theme-oriented history as I have been describing it; and I doubt that the medical historians or those who generalise about the writing of medical history take any stand at all. A number of critics have been affirming that 'it all depends' upon the principles invoked. Such scepticism may be adequate as a prolegomenon to critical practice, but in practice, the doubt acts to permit author-critic to destroy with seeming impunity traditional historical assumptions about the validity of periodisation, the role of biography, and the value of facts themselves. It is not so simple a matter as implying that the higher French critics and their already mentioned American epigoni are out to destroy traditional history.[9] More accurately, their intent is further to free the text from shackles of any sort, so that it can cease from being a text and become a psyche or consciousness. Only this remarkable degree of freedom will allow texts to approximate the randomness and vagary of human consciousness.

The application of these ideas to the history of medicine in the Enlightenment is as intriguing as the French-American connection I have been describing. If we pause to observe, the facts will speak for themselves: there are almost no modern histories of medicine in the eighteenth century.[10] Furthermore, there are few histories written by Americans. Granted, there is Fielding Garrison's *History of Medicine*, originally published in 1913, revised six times by 1929 (the date of the reset and revised edition cited here) and reprinted once again after that. Elsewhere I have written that it typifies the 'telephone directory approach to the history of medicine' because it organises itself by listing names, dates, and works, almost in the manner of Greek epigraphy,[11] the study of the inscriptions on tombstones. Yet the more germane aspect of Garrison's history is that it superficially surveys a field rather than develops a thesis. Like several other recent works – particularly Oswei Temkin's *The Falling Sickness* (1971), Ida Macalpine and Richard Hunter's *Three Hundred Years of Psychiatry* (1963), Lindeboom's various books about Boerhaave, and W. Parry-Jones's *The Trade in Lunacy: a Study of Private Madhouses in England in the Eighteenth and Nineteenth Centuries* (1972) – it surveys and describes rather than analyses. It chronicles in the name of exhaustibility. Furthermore, it was not written, as I have attempted to suggest, because the author stumbled upon manuscripts or books waiting to be discovered, materials that altered the author's sense of an entire field or area, medical, historical, scientific. The works just mentioned were undertaken to patch holes and fill ditches: because previous histories and biographies were inadequate. Lindeboom's work on Boerhaave presents a slightly oblique situation: here the author has elected to make this great eighteenth-century physician his life's work and understandably and naturally, has tried to turn over every document, re-edit it, reconsider it, reinterpret it, and then unite these reconsiderations into the form of books. But the others have been much less concerned with individual contributions, perhaps deeming this approach too narrow.

Books of the second variety are different. Recent examples include Lester King's two books on philosophical aspects of medicine – *The Road* to *Medical Enlightenment 1650–1695* (1970) and *The Philosophy of Medicine: the Early Eighteenth Century* (1978) – Joseph M. Levine's *Dr. Woodward's Shield: History, Science, and Satire in Augustan England* (1977), Margaret C. Jacob's *The Newtonians and the English Revolution, 1689–1720* (1976), and all Michel Foucault's books.[12] Each is important because it has 'discovered' something: this 'something' should be explained in some detail because it lies at the heart of the matter. Both King books apply a logical test of 'real and imagined cause' as well as 'sufficient and necessary cause' to medical theories and medical explanations generated between 1650 and 1780. No new texts have been discovered or even resuscitated. The texts studied – Sydenham, Willis, Baglivi, Pitcairn, Boerhaave, Cheyne, Stahl, Hoffman, *et al.* – are those well known to medical historians who work in the period. But no one has applied such logical rigour and method to these texts, thereby illuminating their strengths and exploring their weaknesses. Not even King would maintain that the contribution of any of these physicians should be elevated to that of a Galen or Hippocrates; on the other hand, he does not believe they should sink into oblivion and explains why. The degree of success of King's method will be gauged by his ability to persuade others that these logical and paralogical categories explain the flow or theoretical as opposed to practical medicine for a hundred years. The effect of the two books must be to cause future historians of medicine not to dismiss the eighteenth century so easily as they have done in the past.

Levine's book, as I have argued elsewhere,[13] is different, although it too is a 'thesis' book rather than a 'history' and despite its telling a story never before related. This is the narration of Dr John Woodward, an English virtuoso barely known today, who acquired in 1693 a bronze shield measuring fourteen inches on whose front was inscribed the invasion of Rome by the Gallic Brennus in the fourth century B.C. Woodward, believing he had discovered a rare antiquity of immense historical value, sent engravings of the shield to leading scholars in Europe, asking them if they thought it be a *parma equestris* or one of the *clypei votiva*. But eventually – after Woodward's death – the shield acquired a reputation as notorious as Woodward, proving in time to be a fake, a late sixteenth-century copy made in central Europe just one hundred years before Woodward acquired it. This intricate story had to be told. Yet Levine did not write to fill a gap; he did not write because he surveyed the entire landscape of eighteenth-century antiquarianism and concluded that this 'gap', rather than others, ought to have priority but because of 'an accidental encounter in the Manuscript Room of the British Museum'.[14] a few sentences later Levine writes: 'somewhere in the British Museum, almost forgotten and just a trifle rusty, there lies a small round iron shield, unpretentious enough and understandably neglected, yet notorious in its time'. Here then is the genuine origin of the book, not in any dream of completeness.

Margaret Jacob's book was also written as a result of discovery – the discovery that an all-important question had never even been asked: 'the crucial question became why did a particular natural philosophy, which I shall call the new mechanical philosophy, appear to certain influential Protestant thinkers as the only acceptable explanation of the natural order and hence the only valid support for what they understood Christianity to be?'[15] But the answer was not to be found in scientific literature, which she thought she would be exploring, but 'in the sermons and treatises of certain late seventeenth-century churchmen'. From this point forward her task seemed cut out: one sermon led her to another, and eventually Jacob discovered that 'Newton played a part in the formation of the first Boyle lectures' and in the almost instant adoption of Newtonianism by the Church of England. Jacob's work, following on the heels, as it were, of the most extensive editorial undertaking in Newton's three-hundred-year history,[16] of several biographies[17] and a sustained psychoanalytic study,[18] and after three decades of students tracking down the way 'Newton demanded the Muse',[19] and upon the completion of several comparative historical studies of his reception in different countries[20] – after all this no one could justify just another survey. But Jacob discovered crucial material that permitted her to disclose – for the first time – the secret of the seemingly instantaneous Newtonisation of England.

All these books have one thing in common: a *discovery* made by combing libraries or reconsidering relationships among documents. Each was written in an altogether different spirit from Garrison's gargantuan encyclopedia of the history of medicine and other similar works. It is true that publication of the books just discussed has advanced the authors' careers far more than it advanced Garrison's (who was a soldier and a librarian), but this advancement cannot have been the primary reason for undertaking the research. The degree of difference between these two classes of books approximates the status of literary criticism after the Great War of 1914; for the student of literary history searching his own field for parallels will discover it in a book written at approximately that time. This gap of more than a half a century (i.e. 1920–80) reveals something remarkably retrograde about the study of the history of medicine in our century.[21]

But the differences between the two classes are probably not as obscure as I may have intimated. The survey history – the first variety – is naturally more prolific, aims to fill in gaps, tries to synthesise and build on the works of other scholars, is chronologically rather than thematically arranged and enjoys unity solely by virtue of chronological organisation. Were it not a self-professed 'history' we might question whether it possesses coherence at all. Thesis-oriented books, on the other hand, tend to be short works, rarely extending beyond 200 pages, and attempt to be definitive only to the extent that a new idea has been generated or discovered.

Biography has plagued both varieties and has been, somewhat paradoxically, its curse and blessing. At the very least, it poses a special type of problem for each. For the Garrisons, biography is often used as an excuse for the lack of anygenuinely new material and, sometimes, in lieu of detailed or sophisticated analysis. For the Kings and Levines, biography is either irrelevant (as in the case of the former) or crucial, as in the case of the latter. Jacob's argument depends upon it for its very existence: without close study – almost day by day scrutiny – of the lives of her six or seven protagonists, the argument for the political dimensions of English Newtonianism cannot be made. King's argument about the philosophical (i.e., logical) aspects of certain Enlightenment *theories* (as opposed to *practices*) of medicine, survives very well whether or not we know anything about the 'philosopher's' life. The case is not dissimilar from the situation that obtained for certain types of formalists, especially the New Critics, until the 1950s. Levine's narrative, like Jacob's, cannot possibly be told without a complete reconstruction of Dr Woodward's life. That story had not been told anyway; so the author-historian in this case was not committing a redundancy. But surely the more essential conclusion to reach is that in none of these instances is biography a surrogation for something else, as it so often is in histories of the Garrison variety.

Some medical-historical books fall into neither category. The 'recognition book', for example, aims at correction because something – whatever it may be – has been overlooked. Examples of this variety in recent medical writing about the Enlightenment do not appear in book form but as articles. A perfect example is J. Rendle-Short's 'Infant Management in the 18th Century with Special Reference to the Work of William Cadogan', published in the tradition-minded *Bulletin of the History of Medicine* (xxxiv (1960), 97–122) presumably because the author believed that Dr Cadogan had been overlooked for far too long. Such works are usually born in an attitude of sore neglect, and their authors often become crusaders for this or that medical man. Another overlapping area, perhaps more interesting and consequential than the one just noted because it portends of what the future has in store, involves interdisciplinary books: books that cannot be strictly classified as belonging to only one discipline or subject. Only a handful of scholars (such as Jacob) dares to undertake this variety, perhaps because its success is necessarily a result of the capabilities of the author.[22] Under optimal conditions (i.e., an informed author who can write persuasively and clearly) it may be the most appealing form of contemporary scholarship in the history of medicine; under perverse ones it can verge on the brink of disasters, as in the case of much medical history written by practising physicians who have forgotten whatever little history they once knew.[23] No other variety hovers at such extremes; and no other variety is more controversial at the moment. Yet even in this disparity, what I am designating as optional and perverse conditions are influenced by the author's style. While his degree of historical, philosophical, and literary knowledge may distinguish him from the run-of-the-mill critic, his distinction, as I want to demonstrate, depends upon his writing, from sentence to sentence, from paragraph to paragraph.

Even so, the most significant difference between the two types of writing – the Garrisonian 'history' versus the 'thematic-oriented' study – is the author's degree of self-consciousness, that is, his consciousness of and subsequent reflectiveness about methodology. An interchange between two prominent contemporary literary critics, Wayne Booth and M. H. Abrams, is perhaps germane in this context. When Booth proffered commentary about Abrams's silence regarding his 'consciousness' and 'reflectiveness' in two books considered 'great' by Booth, Abrams made this reply:

> Wayne Booth is quite right: for all my interest in the methods of literary criticism, I say nothing about method in my two historical books, *The Mirror and the Lamp* and *Natural Supernaturalism*. The reason for my silence on this issue is simple: these books were not written with any method in mind. Instead they were conceived, researched, worked out, put together, pulled apart, and put back together, not according to a theory of valid procedures in such undertakings, but by intuition. I relied, that is, on my sense of rightness and wrongness, of doubt and assurance, of deficiencies and superfluities, of what is appropriate and what is inappropriate.[24]

Note that Abrams does not say there is no method in the books: merely that whatever method there may be is intuitive. The point of all his circumspection is probably the knowledge that self-reflexiveness and self-consciousness in themselves will never guarantee good books, good scholarship, good criticism. Abrams's two books are surely better than good: they are superb. But they do not enjoy – as Abrams is the first to guarantee – self-consciously imposed methods. Likewise, one could name any number of inferior books permeated with self-conscious methodology; indeed the trend at the moment appears to be in this direction. The strains developed earlier about ephebi and epigoni attempt to account, however superficially, for the reasons why American higher criticism seems to be moving in this direction. It may well be more crucial to notice in this connection, however, that the best (i.e., a value judgement made by an individual scholar about a type of medico-historical scholarship) books are still written by those who 'intuit their method.'[25] Authors who seem obsessed with method or who substitute reflections on method for solid content may be doing it as surrogation for another process: for the demanding process Abrams calls 'conceived, researched, worked out, put together, pulled apart, and put back together'. Where this remarkable trend will eventuate is anyone's guess. At this moment in the life of higher criticism it seems that the country – colleges, universities, institutes, conferences, symposia – is being overrun by scholars who feel a need to justify their activity and who consequently care more about method than about solid contents: clear, logical argument; factual accuracy; and, of course, an original hypothesis. I suppose I must also admit that I am very clearly indicting myself at this point.

A further point about style ought to be made in any discussion of the recent historiography of medicine in our period, one dealing with the value placed on different types of English prose styles. To begin with, these 'styles' have not as yet been clearly delineated, at least not taxonomically speaking.[26] No one to my knowledge who has studied the forms of English prose has authoritatively determined that there are, for instance, six or seven or eight styles (i.e., the clear, the opaque, the ambiguous, the ornamental, etc.). This fact notwithstanding, value is still placed on certain types, albeit intuitively, as Abrams has commented in a related context. And in the last decade or so, the decade following the one in which Marshall McLuhan contended that 'the medium is the message', more value has been placed, for whatever complex reasons, on the opaque than on the clear style. The consequence of this development, which has not been analysed to my knowledge in relation to the history of medicine, is the placement of a premium, a kind of bonus, on writing incapable of reduction to clear matter-of-fact statements: to wit, the recent fortune of the American brand of French structuralism. For medical history the significance of such value is paramount, demonstrating why the American epigoni today will tolerate medical history as a valid enterprise only when it is written in a particular style.

This style is not the one adopted by Susan Sontag in her recent book *Illness as Metaphor*, a journalistic approach to cancer and tuberculosis as 'metaphors' for vaster societal ills. It is one more clearly defined by viewing two extreme styles: the clear style of Fielding Garrison, already discussed, and the – for lack of a better phrase – opaque style of Michel Foucault.[27] A random passage selected from each author renders the point salient. First, let us consider Garrison on eighteenth-century psychiatry, the section dealing with conditions and cures:

> The cases treated were all of the dangerous, unmanageable, or suicidal type,

and no hope of recovery was held out. There was an extensive exhibition of drugs and unconditioned belief in their efficacy. A case that did not react to drugs were regarded as hopeless. Melancholia was treated by opium pills, excited states by camphor, pruritus by diaphoresis, and a mysterious power was ascribed to belladonna: if it failed, everything failed. Other remedies were a mixture of honey and vinegar, a decoction of *Quadenwurzel*, large doses of lukewarm water, or, if this failed, 'that panacea of tartarisatus.'[28]

The costly aqua benedicta Rolandi, with three stout ruffians to administer it, a mustard plaster on the head, venesection at the forehead and both thumbs, clysters, and plasters of Spanish fly, were other resources. Barbarities were kept in the background, but the harsh methods of medieval times were none the less prevalent. A melancholic woman was treated with a volley of oaths and a douche of cold water as she lay in bed. If purgatives and emetics failed with violent patients, they came in for many hard knocks, with a regime of bolts and chains to inspire fear. A sensitive, self-conscious patient was confined in a cold, damp, gloomy, mephitic cell, fed on perpetual hard bread, and otherwise treated as a criminal. The diet – soup, warm beer, a few vegetables and salads – was of the cheapest. There were some attempts at open-door treatment, such as putting the patients to mind geese, sending them to the mineral baths at Doberan, Toplitz, Pyrmont, Vichey, Bath or Tunbridge Wells, or sending them as harvest hands to Holland (*Holland-geherei*). Marriage was also recommended as a cure.[29]

These words conclude Garrison's single paragraph on madness in the eighteenth century. The 'conditions and cures' tally with some of the most detailed accounts of the period as does his repertory of panaceas: both range from common purgatives and emetics to such uncommon ones as the rarely mentioned and ineffectual 'Roland water'.

Garrison's last sentence is puzzling: it relates that marriage was often considered a cure for madness, but Garrison does not say on what authority and the primary literature is virtually silent on the matter, although recent social historians of the last two decades claim to have found examples. A few insane protagonists in major eighteenth-century English fiction are sent off to the marriage-stocks to be cured, despite Lawrence Stone's silence on the matter in the section on 'definition and types of marriage' (pp. 30–7) and in the passage on 'lunatic asylums' (p. 385) in *The Family, Sex and Marriage in England 1500–1800*, the fullest treatment to date. Even in the last sentence of the passage, which may or may not be factually accurate, Garrison relates psychological history in descriptive rather than analytic terms. Not in the slightest degree self-conscious or self-reflective, he has no inclination to scrutinise conditions or probe motives and reasons; nor does he relate why these abhorrent social conditions went unchallenged or take a stand against the progenitors of such intense human misery. It is history of medicine construed soley as fact, as the aggregate of facts: the more factual the pile-up, the better the historian. It is also history of medicine devoid of analysis, exploration, semantic inquisition, or the slightest sense that every verbal act, spoke or written, possesses a rhetorical strategy. The only valuative activity consists in the unwritten assumption that facts – or 'facts' – are good enough in themselves and require no justification. This is why Garrison can write six-page paragraphs, as in the passage quoted, which is part of a much longer passage stretching over many pages. At no point in the thousand pages or approximately 500,000 words is there a shred of reflectiveness about the grand enterprise.

If Garrison's mode is compared with Foucault's in *The Birth of the Clinic*,[30] a book dealing with the same subject, the difference is apparent. Foucault's language substitutes metaphor and metonymy for concrete detail; the narrator hides, as it were, behind a veil; and the direction of his paragraphs, written in the opaque style, is so unpredictable that any clear statement of their 'subjects' or 'themes' may suffer from reductio ad absurdum. Nevertheless, Foucault states at the start of his book that 'I should like to attempt here the analysis of a type of discourse – that of medical experience – at a period when, before the great discoveries of the nineteenth century, it had changed its materials more than systematic form. The clinic is both a new "carving up" of things and the principle of their verbalisation in a form which we have been accustomed to recognising as the language of a "positive science".'[31]

Foucault's inventory of themes and concepts is essentially rhetorical: the history of medicine construed as a branch of the history of ideas; excursuses on time and space that glance at the structuralist writings of Bachelard, Georges Poulet, and Roland Barthes; orations to persuade the reader that the crucial period in the development of the asylum in France and England occurs between 1750 and 1820; further orations contending that the period before 1750, after 1800, or even after 1850, is not nearly so essential; persuasion through the use of arcane materials (names, places, events) – so arcane that the reader grows distracted from the main subject; discourses in the self-reflexive mode; and the 'clinic', 'asylum', or 'hospital' as imagery used in discourse aimed at demonstrating that no subject in history or philosophy is so crucial (i.e, superlatively) as the announced 'principle of verbalisation'.

These 'inventories' do not answer the question whether or not Foucault is concerned with the 'straight' history of medicine – that is, history of medicine conceived in traditional terms and traditional moulds, and without preoccupation about the historian's rhetorical tropes, degree of self-reflectiveness, or other 'principle of verbalisation'. This is why the inclusion of a passage such as the following, appearing on the fourth page of Foucault's preface, is so curious and, in my view, intentionally ambiguous: 'In 1764, J. F. Meckel set out to study the alterations brought about by the brain by certain disorders (apoplexy, mania, phthisi); he used the rational method of weighing equal volumes and comparing them to determine which parts of the brain had been dehydrated, which parts had been swollen, and by which diseases. Modern medicine has made hardly any use of this research. Brain pathology achieved its 'positive' form when Bichat, and above all Recamier and Lallemand, used the celebrated "hammer, with a broad thin end." '[32]

It may have been true in the early 1960s, when Foucault was presumably writing these passages, that 'modern medicine [had] made hardly any use of this research'; and it may have been equally true that brain pathology achieved the 'positive forms' Foucault affirms it did under the conditions he describes after 1800 (i.e., Bichat, who died in 1802 and performed most of his important work at the end of his very short life of thirty-one years). But all this matter notwithstanding, it is Foucault's accuracy, or its lack, that constitutes the *least* interesting aspect of his method – the aspect that has least attracted students to his books.[33] The matter seems simple but should not be readily dismissed, for penetrating analysis of it reveals the clue to understand the recent historiography of scholarship in our field of enquiry. In the passage just cited Foucault derives his date of 1764 from a journalistic account of Meckel's experiment published in a popular Paris weekly.[34] Foucault probably knows the difference between the significant works of the elder Johann Friedrich Meckel and the insignificant, or at least less significant, work of his son by the same name. Foucault's book, *Naissance de la clinique*, offers no evidence that Foucault has actually read any of the elder Meckel's works, a charge that speaks for itself.

Moreover, the sentence about 'modern medicine', translated literally from the same two French words, is ambiguous. Does Foucault mean 'the practitioners of modern medicine', or recent scholarship in the history of medicine, or something else? Surely the basis of his claim here depends upon clarification of his ambiguity. After all, why *should* medicine rehabilitate Meckel's medicine?[35] The point about the 'celebrated hammer' is less controversial; every American, and probably European, medical student learns about it. Historians of medicine concur about its significance and usually there is little to discuss about it. But why are Recamier and Lallemand given more credit ('above all') than Bichat? Does Foucault know something medical historians do not? Is he withholding some information? the real issue in deciphering this prose is 'moving from sentence to sentence'; after sufficient decoding, it becomes clear that any ostensible subject of the discourse is ancillary to the pyrotechnics involved: the experience is almost that of reading Carlyle. But Foucault must be accredited when he discloses in the preface, already quoted, that he is most concerned 'with the analysis of a type of discourse'.[36]

Further study of *Naissance de la clinique* and similar Foucaultian books would demonstrate – especially the newest ones on the history of the concept of sexuality in modern civilisation – the reason for their success. The texts themselves have little to do with the history of medicine and are, as Jean Starobinski has noticed, 'an experiment in a new way of writing the history of science, a testing ground'.[37] I would add that the 'experiment' contains just the right amount of obnubilation, to use T. S. Eliot's term,[38] to mystify and confuse the unsuspecting reader. And, further accounting for the success of these historico-medical texts. Foucault also possess just what Garrison did not: a sufficiently brilliant imagination that connects disparate terrains. By so doing Foucault gives the illusion of having broken ground – i.e., developing a new epistemology – in the history of medicine. Furthermore, it is worthwhile to compare Foucault's remarks with what Garrison wrote about Meckel in his quasi-prosopographical study of the history of medicine: 'Johann Friedrich Meckel (1724–74), of Wetzler, graduated at Göttingen in 1748, with a noteworthy inaugural dissertation on the fifth nerve (Meckel's ganglion), became professor of anatomy, botany, and obstetrics at Berlin in 1751, and was the first teacher of midwifery at the Charité. He was the first to describe the submaxillary ganglion (1748), and made important investigations of the nerve-supply of the face (1751) and the terminal visceral filaments of the veins and lymphatics (1772).'[39] Foucault provides less information, but he inflates Meckel's position by rhetorical hyperbole ('Modern medicine has hardly made any use of this research,' etc.) and leads the reader to believe that a thoroughly new area of psychology in the late eighteenth century has been discovered.

The conclusions drawn from this brief comparison can be useful if interpreted in sufficiently wide contexts. Perhaps foremost is the observation that most *professional* medical historians (i.e., two dozen or so in America, about the same number in Europe – I use the term here in contrast to the hundreds of professional physicians who dabble in medical history as amateurs, as did the dilettanti of the eighteenth century) will not concede that Foucault's style is to be preferred at all. Upon occasions when they argue at all – medical historians rarely possess any consciousness about their methodology – they will maintain that Garrison's style is preferable, even if his so-called facts are now dated. The non-professional medical historians, literary critics like myself, and other ephebi and epigoni described at the outset, prefer Foucault. Depending on one's point of view, we are the villains, the culprits, the problem-children. Moreover, the main reason we literary types see little value in Garrison's variety of 'history writing in the clear style' is that those so-called medical facts in and of themselves have little significance. That is, the 'fact' must be contextualised and interpreted many times over in order to gather meaning and acquire value, to earn its board and keep among us, for we literary critics are a demanding lot; and nothing so purified or rarefied as 'straightforward medical history' will grasp our interest.

Precisely in this situation and constellation of factors lies the dilemma of contemporary medical history. On the one hand, the tribe of professional medical historians is so small – to call it an endangered species is to exaggerate the matter; it never was a species in the first place – that the tribe in itself is capable of determining virtually nothing within the wide contexts of culture or even within the vast network of interdisciplinary studies that flourishes today in American and European universities.[40] On the other hand, higher cultural critics, who are numerous and who enjoy a type of collective power either *ex cathedra* or by virtue of journalistic endeavours that shape the thinking of the élite members of our society, value medical-historical writing when it contextualises and interprets the subject almost by mystifying it and its students, not when it proffers another straightforward history of medicine authored by a contemporary Garrison with shabby interpretative abilities.

All this can be demonstrated in other ways too. Suppose we imagine a colloquium composed of professional historians of medicine and professional higher literary critics. The two groups have convened to debate recent writing of medical history in books, scholarly journals, professional periodicals.[41] The conferees have spent many hours classifying the writings. Two major categories of 'texts' have been established as representative of the extreme norms in contemporary writing in the field: the Garrisonian and – on the other fringe – the Foucaultian. The higher critics concede that it is worthwhile to have an updated history of medicine *à la* Garrison but cannot comprehend what difference such a history, no matter how accurate and exhaustive, would make to them. The medical historians argue that it makes all the difference to them, the professional historians of medicine; and they argue that such a solid work of reference would be consulted, as a further proof of its need and strength, by many types of students, not only those working in the subject. Yet the higher critics argue that the promise of ultimately such random consultation can have no appreciable effect on their own work, except to the degree that beginners and the general literate public may gain a better comprehension of the subject. Those medical historians, on the other hand, who profess to have scanned every page of Foucault's writings, cannot comprehend how anyone knowledgeable of the subject and trained in it can call this flimsy and scrappy form of higher journalism 'history of medicine'. Certainly the course of their own researches, they emphasise and continue to emphasise, will not be changed very radically by the reading of Foucault's books. They profess to have learned nothing about *the subject* medical history 1660–1800 from these books written in the opaque tropes already discussed above; and they maintain, furthermore, that Foucault's medical-historical content is either incorrect, derivative, outdated, or concealed behind such a deceptive veil that they cannot separate dark from light, obfuscation from clarity, abstraction from clear point. Nor can they imagine any professional historian of medicine wishing to call himself a 'disciple' of Foucault, professing to write as one of the 'School of Foucault', as a Foucaultian. These are harsh charges, either way.

We seem to have arrived at an impasse in our imaginary dialogue. Neither group can validate the other. Each speaks a different language, as it were; and each continues to maintain that the significance it has found in a single text (Garrison and Foucault) or a set of similar and related texts is precisely that and no other significance. Nor is the sense of value debatable: each knows what it considers solid; and the matter is hardly negotiable. Neither group is sanguine that it can convert the other. Each leaves the colloquium at the point where he started: convinced that he has learned very little, that, while others may disagree with him, he is utterly correct about the values placed on certain kinds of writing. Furthermore, the medical historians do not seek to become 'higher critics' and consequently do not search out conversion; the same is true of higher critics: the least aspiration the former harbour is to become medical historians with all that implies, especially assumption of the values of medical historians.

Do all these attitudes argue for a remarkable degree of profesaionalisation in our time?[42] Can this signify, once more, that the professions are indeed distinct and solipsistic, and that the possibility for communication among them is a pipe-dream, as limited as it was portrayed by students in the 1950s and 1960s of the 'two cultures'? Can it be that we – higher critics and medical historians – have been immured, like Foucault, behind a screen that binds us to the dire consequences of professionalisation?

It is hard to be sanguine when faced with disparities and dilemmas of the above variety. I suppose the contented but simple-minded folk among us will ask for a list of desiderata: as if the matter were a Simple-Simon case of what do we need, and how can we do it? One does not necessarily want to shatter the illusion of such hopeful types but, in Dryden's words, 'it wo'ld not be'. Nor is cause for optimism to be found in the accurate determination of 'lesser and greater value': no one living in such a relativistic, pluralistic, and politicised epoch as ours, especially within the universities, can determine what species (of anything) has greater 'value' than any other species. Therefore, it is not possible to decide whether the medical historians or higher critics are 'right' – impossible, moreover, to adjudicate whether history writing *à la* Garrison or in the opaque style of Foucault has *greater intrinsic worth*. It all depends on one's sense of value:

what *is* worth? and who determines it? Only the academic scholar retained by universities or institutes with unwritten but nevertheless understood built-in value systems will ask these kinds of professional questions: should one write in one mode or the other? Should one working in the field of medical history 1660–1800 attempt to write another but yet better (i.e., in one value system a more thorough and accurate) history of medicine or attempt to stumble upon manuscripts that will generate a theme? Should the student aspiring in the field diachronically trace medical response, for example, to the major social and economic upheavals in England (e.g., 1640, 1660, 1688, 1720, 1776, 1789) or limit these responses to strictly professional (i.e., medical) considerations (e.g., 1705, 1720, 1745, 1774)?[43]

None of these questions can be answered by anyone with a sceptical cast of mind. It all depends, the refrain goes; yet sceptical as I am, I am persuaded that the disappearance of subject – in this case the subject traditionally known as medical history – is taking and will continue to take a toll in the life of our American scholarship. This is why I have recently reconsidered the eighteenth-century analogy of 'the mirror and the lamp': because it is such an apt commentary on our present-day dilemma between the veil and what it conceals or, in other language, between subject and object. Nor is the development unique to medical history: the very same situation exists with regard to English literature, French literature, other literatures, philosophy, art history, general history, political theory, economic theory, and many other humanistic subjects. Students of the veil, to continue in this line of analogical reasoning, proliferate to a remarkable degree; and every indication is that they will continue to. For each 'serious' student of English literature 1660–1800 who tills the land because he or she wants to get close to the primary literature (Dryden, Pope, Gray, Fielding, Richardson, Johnson, *et al.*), there are probably a dozen today who till it in name only, not because the dozen have any genuine love of the literature but because they have lost their faith in it, because they both intuit and learn that the literature is bankrupt, for them as well as their students, and because they have turned to subjects other than the primary literature in order to survive.[44]

Therefore, when one discovers a passage such as this one amidst the meditations of a first-class historian of science, one necessarily laments the passing of an epoch; and one laments best, unlike the Puritan thinker George Fox, by lamenting with others:

> The workshop of the scholar in the history of science is the periods in which his authors lives. He should know those periods' ways of life and belief and education, both the common and the eccentric; their political histories; their variety in all aspects; their social and economic structures; their architectures, literatures, and arts. He should feel at home in houses of those times, sit

easily in their chairs, both figurative and wooden, and discern what was then most easily admired or rejected in painting and sculpture and decoration. He should have read not only the books that carried the intellectual products of his period but also those that were then the fare of young minds as they were taught, such books having been commonly of an earlier time. The student who does not command, as a minimum, the main episodes of Holy Scripture, classic mythology, and the corpus of golden Latin is glaucomatose in the modes of thought of Western men educated before 1900. In addition, the scholar in the history of science should know the lives of the scientists.[45]

But the dirge must cease; the time for lament, if lament there really should be, is past. Who among our students can measure up to these criteria 'as a minimum'? It makes no difference whether the workshop is that of the historian of science or the historian of medicine, the literary historian or the general historian: the important matter is that it is the workshop of the scholar. Can *scholarship* then be the genuine source of my discourse? For the American epigoni and their young ephebi have seen to it that these types of workshops are now empty, and no two people agree on how the evacuation actually occurred. Instead the workshops have been replaced with 'glaucomatose' polemicists who earn rank and fame to the degree they plunder the workshops of the scholar: his libraries, his texts, his subjects, and rob him of his identity.

Why write this at all? If silence is golden and reflection best when unarticulated, then why belabour the obvious and add to the already vast literature of dialogues from the dead? Why not surrender and hand over the keys to the traditional workshops? I suppose because some of us Americans who have been doing or who are now doing precisely that will not rest comfortably until we have persuaded ourselves that we have been purified by means of an Aristotelian catharsis.

NOTES

1   Each of the strains possesses something of a literature for the ephebi and their fortunes after the fifth century in Athens, see Alice Brenot, *Recherches sur l'éphébie attique* (Paris, 1920); Victor Ehrenberg, *The Greek State* (Oxford 1960); and M. Rostovtzeff. *The Social and Economic History of the Hellenistic World*, 3 vols. (Oxford, 1953). The locus classics for the Greek epigoni is Pausanias' Description of Greece (1.43.1); but much additional information is found in Aeschylus' *Seven against Thebes*, Apollodorus mythographus' *Apollodori Bibliotheca*, and the *Hygini Fabulae*. The formacalia, the 'moveable feast', is described by Ovid in the *Fasti*, 512–32, in the passage beginning 'Learn also why the same day is called the Feast of Fools', and is discussed in these works: Georg Wissowa, *Religion und Kultus der Römer*, 2d ed. (Munich, 1912); Richard Farnell Lewis, *Greek Hero-Cults and Ideas of Immortality* (Oxford, 1921); Sir James G. Frazer, *The Golden Bough: a Study in Magic and Religion*, 12 vols. (1911–15). In the notes below, page numbers have been included only when verification of a passage is in question. These strains are not discussed by Julian Jaynes in his provocative study of *The Origin of Consciousness in the Breakdown of the Bicameral Mind* (Boston,

1976), but one should see his discussion of consciousness in the Englightenment on pp. 436ff.

2   Nancy Streuver has studied some of these links in the Renaissance; see her excellent book, *The Language of History in the Renaissance* (Princeton, 1970) as well as her provocative essay entitled 'Humanities and Humanists', *Humanities in Society*, 1 (1978) 25–34. I know of no similar study of the 'language' of eighteenth century England or France, although valuable information about the theory of language is found in Murray Cohen's *Sensible Words: Linguistic Practice in England 1640–1785* (Baltimore, 1977).

3   Students of similar institutions – in Rome, medieval Gaul, modern Germany – have commented on the degree of fanaticism in the hero worship of the young disciples.

4   See V. Ehrenberg, *The Greek State*, 80–100 *passim*.

5   Ovid, *Fasti*, ii, 527 ff.

6   There is no need to name names for everyone can make his own list. Nevertheless, I am reasonably certain that if appointments to junior and senior posts in the period covered by this journal were rigorously studied, the point would make itself loud and plain.

7   Examples are provided in the discussion below.

8   Foucault himself is a satisfactory example of the point. For almost two decades he has continued to maintain that he is not a 'historian' of anything, certainly not a historian of science or medicine.

9   This aspect of the recent fate of historical studies is not dwelled upon by J. H. Plumb in *The Death of the Past* (Boston, 1970), a book about the death of history. Plumb is also silent about the French structuralists who play a role in the sad story he tells.

10 There are, of course, dozens of surveys that merely *include* this century amidst a couple of dozen other centuries but contain no extensive work isolating it and studying it alone. Roger French's *Robert Whytt, the Soul, and Medicine* (1969) is perhaps as near as one comes to a background of medicine in the period, although this perceptive study is by no means a 'history'. For an unexplained reason, the chronological arrangement of Charles Singer's *Short History of Medicine* (New York, 1962) – the single most often consulted medical history in the United States today – ceases after the Renaissance, besides, the treatment of the eighteenth century is shabby in quality and quantity. The real question is what is the history of medicine and what is 'a' history of medicine? Is a mere survey a history?

11 See G. S. Rousseau, 'Psychology', in G. S. Rousseau and R. S. Porter (eds.), *The Ferment of Knowledge: Studies in the Historiography of Eighteenth-Century Science* (Cambridge, 1979) (*Enlightenment Crossings*, chap. 4).

12 In my estimate they do not include: Max Byrd, *Visits to Bedlam: Madness and Literature in the Eighteenth Century* (Columbia, 1974); Michael V. Deporte, *Nightmares and Hobbyhorses: Swift, Sterne, and Augustan Ideas of Madness* (San Marino 1974); A. L. Donovan, *Philosophical Chemistry in the Scottish Enlightenment* (Edinburgh, 1975); or my own treatment of 'Science' in Pat Rogers (ed.), *The Context of English Literature: the Eighteenth Century* (1978), 153–207.

13 See G. S. Rousseau, *Journal of Modern History*, 1 (1978), 513–15.

14 See Joseph M. Levine, *Dr. Woodward's Shield: History, Science, and Satire in Augustan England* (Berkeley and Los Angeles, 1977), IX and 1.

15 See M. Jacob, *The Newtonians and the English Revolution* (Ithaca, 1976), 16, 153–4.

16 See Derek T. Whiteside, who is editing for the Cambridge University Press eight volumes of the mathematical papers alone. W. H. Turnbull is editing several volumes of correspondence (1959–).

17 Emminently worthy of notice in the last few decades is I. B. Cohen's *Franklin and*

18 *Newton* (Philadelphia, 1956) and Frank E. Manuel's *Isaac Newton, Historian* (Cambridge, 1963). See also the latter's *The Religion of Issac Newton* (Oxford, 1974).

18 See Frank E. Manuel, *Isaac Newton: a Portrait* (Cambridge, 1968).

19 See Marjorie Hope Nicolson, *Newton Demands the Muse* (Princeton 1949); W. P. Jones, 'Newton Further Demands the Muse', *Studies in English Literature 1500–1900*, III (1963), 287–306 and *The Rhetoric of Science: a Study of Scientific Ideas and Imagery in Eighteenth Century English Poetry* (Berkeley and Los Angles, 1966).

20 For example, see Valentin Boss, *Newton in Russia* (Cambridge, 1972).

21 This 'something' has not been revealed in the historiography of medicine. Whatever literature exists on the subject has been generally inadequate, written by scholars who do not consider medicine within large contexts. The literature itself is scant: see R. H. Shyrock, 'The Historian Looks at Medicine', *Bulletin of the History of Medicine* (1937), 887–94; G. Rosen, 'A Theory of Medical Historiography', *ibid.*, VIII (1940), 655–65; O. Temkin, 'Henry E. Sigerist and Aspects of Medical Historiography,' *ibid.* (1958), XIII, 485–99; G. Rosen, 'Levels of Integration in Medical Historiography', *Journal of the History of Medicine*, IV (1949), 460–7; E. H. Ackerknecht, 'A Plea for a "Behavioristic" Approach in Writing the History of Medicine', *ibid.* (1967), XXXII, 211–14; George Mora (ed.), *Psychiatry and Its History: Methodological Problems in Research* (Springfield, 1970); Edwin Clarke (ed.), *Modern Methods in the History of Medicine* (1971), especially the essays by E. Clarke and G. Rosen. A further limitation of this type of treatment is that it excludes all consideration of style and rhetoric from the discussion; so long as the methodology of a given medical historian relates medicine to social political, and economic history, these authors consider the approach satisfactory. All would profit from a course in metonymy and from a close reading of Hayden White's *Metahistory* (Baltimore, 1973).

22 Arthur Scouten's attack on interdisciplinary approaches to eighteenth-century studies is ill-conceived, see *Eighteenth-Century Life*, V (1979). Scholars knowledgeable of more than one area (e.g. Marjorie Nicolson, *Literature and Science*; Martin Battestin, *Literature and the Arts*; Ronald Paulson, *Literature and Painting*; Norman Holland, *Literature and Psychology*) are not the culprits but those lethagric members of the profession who cry out for ever more interdisciplinary studies in the name of revitalising the eighteenth century yet do nothing themselves to bridge the gap; and others, like Scouten, who are the relics of New Criticism, in a brave new pluralistic world to which they cannot adjust. Attacks such as Scouten's can, unhappily, only gravedig the eighteenth century deeper. Margaret Jacob has told us why in the preface (p. 16) of *The Newtonians and the English Revolution*: 'I have employed an *interdisciplinary method purposefully* to merge subjects traditionally reserved for the history of science, or church history, or intellectual history. For if science in the seventeenth century possessed social relations, and if churchmen had political interests, and if scientists could be millenarians, then our approach to these problems *must be interdisciplinary*. If the Newtonian natural philosophy gained acceptance and popularity because it effectively supported a particular social ideology, then *this methodology is the only route* open to the historian who wishes to uncover that historical relationship' (italics mine).

23 The generalisation is particularly valid for our period, its medicine has been harmed by book after book composed by physicians with almost no understanding of the large cultural developments of the age. But there is hope in the reverse tendency at the moment: now the social historians have the foreground and are seeking to redress the former imbalance.

24 See M. H. Abrams, 'Rationality and Imagination in Cultural History: a Reply to Wayne Booth', *Critical Inquiry* (1976) II, 447. This essay replied to Wayne Booth's 'M. H. Abrams Historian as Critic, Critic as Pluralist', *ibid.*, 411–45.

25 Foucault, once again, is a prime example: although he is self-conscious about his writings, his self-consciousness does not extend to open discussions of method in

his books: he pronounces without being impeded by a methodology that deflects him from pronouncing. Nor do I believe that he consults with his own ideas about method, and *then* puts his books together. Others may do that; but Foucault is generally happy about the Foucaltian way of writing and executes it without having 'any method in mind'. In this sense, he writes essays in a manner opposed to Emerson.

26 Everyone cites the conventional secondary literature, but scrunity of this literature shows that no one has worked out an 'anatomy', Morris Kroll has written perceptively about English prose style as has George Williamson about English 'Senecan style'; others have written about prose rhythm and tone and texture. But the point remains valid.

27 Here, once again, there is a lack of a clear statement about the development of the 'opaque style' in English prose and the value placed on it in different historical epochs. I am grateful to my colleague, Richard Lanham, for discussing with me aspects of the classification of English style.

28 Garrison does not give source for the quotation, but in the form it appears here it is similar to that given in John Quincy's *Pharmacopoeia Officinalis* (1726), 349–50, s.v. 'tartar tartarisatus'.

29 F. Garrison, *History of Medicine* (Philadelphia, 1929), 440–2.

30 Originally published as *naissance de la clinique* (Paris, 1963) and in 1973 under the above title and translated by A. M. Sheridan Smith.

31 *The Birth of the Clinic*, xvii-xviii.

32 *Ibid.*, xii.

33 The point has also been made by Hayden White in 'Foucault Decoded: Notes from Under-Ground', *History and Theory*, XII (1971), 52–7. See also Donald F. Bouchard (ed.), *Language, Counter-Memory, Practice: Selected Essays and Interviews by Michel Foucault* (Ithaca, 1977).

34 See *The Birth of the Clinic*, 20 n. 18. for an account of Foucault's source in the *Gazette salutaire*, 2 August 1764.

35 The standard German histories of medicine written since 1900 make no mention of Meckel nor do the histories of psychiatry. One would not expect American and British histories to comment on Meckel in view of the insularisation. The only treatment I have found is in obscure German periodicals.

36 *The Birth of the Clinic*, xii. Also see my attempt to categorise eight essential aspects of Foucault's presentation: (1) 'The Historically Meaningless'; (2) 'A Dialectic of Interrogation and Myth-Making Founded upon the Arcane'; (3) 'Analysis as Intentional Fiction in the Preterit'; (4) 'The Perceptive Insight Embedded in Ambiguous Syntax'; (5) 'Analogy, Personification, Cause'; (8) 'The Conventional Narrator'; (7) 'The Poet Speaks'; (8) 'The Medical Propagandist' in Robert Allen (ed.), *The Eighteenth-Century Bibliography for 1974* (Iowa City, 1975), 790–4.

37 See Jean Starobinski, 'Gazing at Death', *New York Review of Books*, 22 Jan. 1976, 18–22 and his various studies of the language of action and reaction in eighteenth-century psychiatric literature. See also his essay entitled 'The Word Reaction: From Physics to Psychiatry', *Psychological Medicine*, VII (1977), 373–86, which sheds further light about the historiography of medicine in the eighteenth century.

38 T. S. Eliot, *The Function of Criticism at the Present Time* (1923), in *Selected Essays of T. S. Eliot* (New York, 1960), 21–2. Eliot's analysis of the style of Rémy de Gourmont, the early twentieth-century author who wrote on scientific topics, has parallels with my analysis of Foucault. Both authors are in Eliot's words, 'master illusionists of fact' (21).

39 Garrison, *History of Medicine*, 334.

40 The number of departments or programmes in the history of medicine in America

41 is fewer than a dozen, of these there cannot be more than a handful who have offerings in the eighteenth century.

41 In 1978 I attended such a meeting in Pittsburgh devoted to discussion of 'Paradigms and Eighteenth-Century Science', especially in the writings of T. S. Kuhn, author of *The Structure of Scientific Revolutions* (1962; rev. ed. Chicago, 1970). The conclusion of the participants was not dissimilar from the imaginary dialogue described below.

42 Some of the origins of this development have been treated by Susan F. Cannon, 'Professionalisation', in *Science in Culture* (New York, 1978), 137–66.

43 One practically needs to argue *à la* A. O. Lovejoy for a 'discrimination of self-reflexivities'. It is one thing to be self-reflective about specific medical questions in relation to a given historical period and quite another to be reflective about the whole enterprise of what one undertakes. For example, consider the disparity between the idealistic dream of completeness (i.e., those who write histories in the manner described above) and of total originality (those who write 'theme-oriented' books). Suppose a student of the history of eighteenth-century medicine discovered a manuscript of Sir Hans Sloane, the Anglican naturalist and Tory president of the Royal Society, linking him with certain English Jacobites and Tory philosophers. This manuscript may tell us a great deal about the politics of a single naturalist – Sloane – or of several physicians (i.e., other naturalist-physician-scientists in the Sloane circle) or about the attitudes of the medical establishment in relation to conservative and radical politics of the day. It may even permit the generalisation with sufficient support, that there was a conservative, moderate, and radical fringe among English naturalists and physicians; but it would not provide a general history of the medicine of the time nor a picture of the whole class of physicians then or their social status, and it would fall short of the belief of the novelists (e.g. Fielding, Smollett, *et al.*) that a history must be a large canvas filling in the whole picture. Therefore, the question is not so simple as it seems. The questions begged are (1) why is there the desire for completeness of any sort and (2) what specific advantages are there in being intentionally incomplete?

44 It is obviously impossible to support, let alone prove, this contention without naming names and counting heads. I merely appeal to one's experience with one's own colleagues in the last decade.

45 C. Truesdell, 'The Scholar: a Species Threatened by Professions', *Critical Inquiry*, II (1976), 631–2.

4

Medicine and millenarianism

My interest in the historical anthropology of the human imagination (Enlightenment Crossings, chap. 1) and its neurophysiological ramifications (same volume, chap. 5) led me to an interest in Dr George Cheyne, the author of a best-selling book called The English Malady *(1733) who became one of the great 'nerve doctors' of the eighteenth century. At the zenith of his career he was so famous that Alexander Pope labelled him the 'immortal Dr Cheyne'. When I began to write about Cheyne around 1979, there was no biographical work that explained his curious life and best-selling book.*

Cheyne was indeed a unique phenomenon and would have been notorious in any era. Swelling to 32 stone (about 450 pounds) and then shrinking to just 9 stone (about 125 pounds), he claimed he had accomplished this incredible feat by using a special diet composed of vegetables, nuts, milk, and wine. He advocated eating little meat and drinking only wine, no late nights, high living, eating orgies. His physical misery at the peak of his weight had been so excruciating that he had had to be followed by a servant with a chair and a stool for support. After his successful weight loss he attracted dozens of rich and famous patients who could afford his exorbitant fees. He removed himself to Bath and established himself as the English physician of the aristocracy.

However, Cheyne still remained a living contradiction. He had been an ardent Newtonian follower, espousing mechanical philosophy, yet was at the same time a pious devotee of mystical sects which professed salvation in the apocalypse. I have tried to reconcile his science and theology (they were not perceived as divided in his time) and explain how relevant he remains for us over two centuries later.

The considerable disciplinary aspect of the material raises vexing questions about research. In which 'discipline' ought a figure such as Cheyne be studied? I don't supply a satisfactory answer to the question, but the essay does suggest that only by breaking out of traditional borders and boundaries can a figure of this sort be understood.

I delivered the essay as a formal Clark Lecture at the University of California in 1981 and expanded and published it in 1987. Apart from the labyrinth of its detective work, its main point – that Cheyne is of great interest to us today – still seems valid to me.

> . . . if there might not, I say, be higher, more noble, and more enlightening *Principles* revealed to Mankind *somewhere* . . . (George Cheyne, 'The Case of the Author' (1733), 331)

... this material *Metaphysicks* of a *Regimen*. (George Cheyne, *ibid*., 367)

'IMMORTAL DR CHEYNE': THE LITERARY RESPONSE

If contemporary anthropologists and sociologists are correct in believing that cult figures in every age are trusted by their devotees while viewed suspiciously by the rest of the world, then George Cheyne deserves to be categorised as a cult figure. For almost everyone who knew him well trusted him and and liked him, whereas those who did not either disliked or despised him. And most educated people knew who he was and were aware of his reputation as a celebrated physician and author. But here the agreement ended. The only other consensus was his weight. Cheyne swelled to 448 pounds – '32 stone'[1] – and whereas no handy eighteenth-century *Guinness Book of Records* is extant to report whether he was the fattest man of the Enlightenment, he certainly must rank among the heaviest.[2]

Literary evidence demonstrates the attitude of his contemporaries better than any other source. For example, Pope knew Cheyne well and relished his eccentric blend of reason and madness: '. . . so very a child in true Simplicity of Heart, that I love him; as He loves Don Quixote, for the Most Moral and Reasoning Madman in the world.'[3] This reference caresses Cheyne and is one of many vignettes of the doctor that abounds in Pope's correspondence. 'He is', Pope later affirmed, echoing the book of St John, 'a kind of living Parson Adams, in the Scripture language, an *Israelite in whom there is no Guile*, or in Shakespeare's, *as foolish a good kind of Christian Creature as one shall meet with*'. All Pope's estimates of the childlike Cheyne portray the doctor as a fundamentally good man; deluded, eccentric, confusing windmills and giants, but nevertheless a man who was as good as the salt of the earth. Fielding's patron Lyttelton agreed with Pope: '. . . Immortal Doctor Cheyne. . . . The Doctor is the greatest Singularity, and the most Delightful I ever met with.'[4] Every Scriblerian, even Swift, adored Cheyne, the only famous physician in England who escaped their venomous collective pen.[5] John Gay continued to be overwhelmed by 'Cheyne huge of size' whom he greets in *Mr. Pope's Welcome from Greece*;[6] and Edward Young nostalgically immortalised Cheyne in a central passage in the *Epistle II. to Mr. Pope* when inprecating for 'three ells round huge Cheyne'.[7] Fielding was far too robust for Cheyne's lettuce and milk diet, but even he nodded at 'the learned Dr. Cheyne' in *Tom Jones*.[8] Richardson was much more devoted. He and Cheyne were in constant correspondence for many years – certainly throughout the composition of *Pamela* – indeed for such a long time that Richardson may not have enjoyed the necessary perspective to determine precisely who Cheyne was. And Richardson relied monolithically on Cheyne for professional medical advice, as well as for literary guidance in the composition of *Pamela*. He told Stephen Duck, the 'thresher poet', that Cheyne 'was so good as to give me a Plan [*in Pamela*] to break Legs and Arms and to fire Mansion Houses to create Distresses'.[9] Few patients who were also novelists ever trusted their doctor to this degree of compliance.

The catalogue of comments is endless. In January 1742 Thomas Gray concocted an 'Imaginary Conversation' between ancient and modern geniuses, including Aristotle, Virgil, Locke, Swift, and Cheyne, wherein Cheyne is made to recite his own aphorism: 'Every Man after forty is either a fool or a Physician'.[10] The young Hume turned to Cheyne for medical advice about his mysterious illness in 1734, entreating Cheyne as if he were Galen or Hippocrates.[11] Lord Chesterfield intensely disliked metaphysics but valiantly wrote to Cheyne to say that, if ever he were compelled to choose 'a system', he would select Cheyne's as the most probable.[12] Lord Hervey, Pope's rival in politics and love, considered Cheyne the most eminent physician in England and was not abashed to say so.[13] The wealthy and lovely Countess of Huntingdon became the patron of Cheyne, the only man in the realm – except for the Methodist preacher George Whitefield – fortunate enough to have captured her attention.[14] Even Wesley, the great reformer, converted to Cheyne's diet and advocated it in his popular handbook of medicine, *Primitive Physick*.[15] William Somerville the poet versified Cheyne's best-selling book *The English Malady* in his poem *The Hip*. John Hill, a notorious if eccentric *enfant terrible*, extolled Cheyne in *The Construction of the Nerves* (1768) and praised his theories of nervous physiology. The medical writings of the too-little known William Porterfield are permeated with approving glances at Cheyne.[16] Porterfield was the first secretary of the Edinburgh Literary and Philosophical Society; from him many who were later to become lights in the Scottish Enlightenment first heard about 'immortal Doctor Cheyne'. The 'Great Cham' told the lecherous Boswell that he ought to read Dr Cheyne's books on temperance and health, advice Boswell probably never heeded but which abundantly adumbrates Johnson's very high esteem.[17] Thomas Tyers, who was apparently one of Johnson's favourite people, wrote the first biographical sketch of Johnson a day after he died on 13 December 1784 and while Tyers mourned he composed a 'Set of Resolutions' founded on Cheyne's principles: 'especially to make exercise a part of one's Religion', and 'to be religiously observed'.[18] This material represents just the tip of the iceberg that constitutes the literary evidence:

> Not all the Gemmy Treasures of the East,
> Nor yet the Spicy Odours of the West;
> Not all the Glorious Trophies of the Great,
> Would please so much, or form one joy compleat,

> Like that I feel, great wond'rous Genius, when
> I scan th' amazing Beauties of thy Pen.

Thus an anonymous poet rhapsodised Cheyne in 1733 on reading his works.[19] Lyrical though the mode is, the praise is specific:

> Long did the Sacred Art in Bondage mourn,
> Become the Jest of Fools, or else their Scorn;
> Till Heav'n, to set the fetter' Science free,
> And pit'ing abject Man, created Thee.
> Made Thee to act of Gods the healing Part
> And live a Pillar to the Noble Art,
> To be the only shining acting Sage,
> Not giv'n, but lent from them to heal this Age.
> Great Wonder from above, thou Boast of Men,
> Accept these Offerings from a Namesake's Pen.

The fact that this catalogue of response can be extended considerably is only the first of my points today, as is the repeated strain about Cheyne's corpulence and medical eminence, although it is intriguing to notice in history how often physical size is equated with heroic greatness. Cheyne's contemporaries were no doubt amazed by his weight reduction from 448 pounds to 130.[20] The further fact that he practised what he preached about the relation of weight and diet also lent him credibility lacked by many eighteenth-century physicians. But the public's image of Cheyne in the eighteenth century was nevertheless jaded and distorted – was not at all the sense of himself he had. Obesity vanquished; a long life of 72 years in an epoch when so many died in youth or early adulthood; national fame as a writer; a close friend of so many famous and influential Britons: these unassailable facts were important to Cheyne, but they were distant from the centre of his intellectual and private emotional life. Cheyne's idea of selfhood and the niche he had carved depended to a certain extent on these incontrovertible facts, but rested equally, if not more so, on other private beliefs which can only be understood in the light of his chronological biography. Without this crucial background, the flow of his ideas over five decades remains a muddle and a mystery.

CHRONOLOGICAL BIOGRAPHY AND INTELLECTUAL DEVELOPMENT

Cheyne was born in Aberdeen in 1671 in a Scottish Episcopalian family the intended him, like his father and both grandfathers, for the Church. His early education was classical, as were his university studies in Edinburgh where he read mathematics. One teacher alone, Archibald Pitcairne, the illustrious mathematician and physician, enjoyed remarkable sway over him. Both had an Episcopalian religion in common, and Pitcairne urged Cheyne to follow in his iatromathematical footsteps: to apply mathematics in the service of medicine.[21] Cheyne ardently followed the advice, obtained a medical degree (at Aberdeen), and even became Pitcairne's staunchest defender in fierce paper wars about iatromathematics. Yet Pitcairne's impact on Cheyne extended beyond this sphere. As Boerhaave's most important teacher at Leyden, Pitcairne had attracted the best minds there and was known throughout Europe as a towering intellect, as the most distinguished iatromathematician of the seventeenth-century *fin de siècle*.[22] His protégés rapidly became fervid Newtonians – especially Freind, Mead, and James Keill – and built a kind of 'school of iatromathematics' around him in which Newtonian calculus, or fluxions, was applied to medical theory. But Pitcairne was also an enthusiast in religion, and generated his medical theory with the zeal of an apostle, a characteristic of personality Cheyne discovered to be temperamentally compatible with to his own personality.[23] Calculus, geometry and medicine filled only part of Cheyne's imagination during his twenties[24] – intellectually his most formative period – the rest of his energy occupied by a deep-seated religious mysticism. He joined a group of Scottish mystics in the 1690s centred on George Garden, the Quietist. Through them he obtained a post as tutor to the young Earl of Roxburgh; but more importantly, he made friends who remained loyal to him for the rest of his life and who joined him later on in his endeavour to become one of Britain's main distributors of Quietist literature.

Why did the young Cheyne have such conflicting aspirations: iatromedicine on the one hand, and mystical religion on the other? The 1690s was understandably the great decade of chiliasm and millenarianism in Western Europe,[25] and although Cheyne was too young personally to have witnessed the events of the 1650s, he was attentive to those of the 1690s. He heard much millenarian talk at Roxburgh House, the great country estate where he lived in comfortable circumstances during those years. He had also heard millenarian talk at home and in Aberdeen.[26] He knew about Jane Lead and the English Philadelphians, and about her prediction of an imminent millennium commencing in 1700. At Roxburgh he had read about the chiliastic interpretations of the year 1697: that the Treaty of Ryswick which finally brought peace to Europe was evidence – 'public testimony' – of the Deity's intention to commence the millennium. And he certainly read John Craig's *Theologiae Christianae Principia Mathematica* (1699), which applied Newton's inverse square law to derive the precise year of the Second Coming, for Cheyne continued to refer to it and quote from it.[27] Cheyne was also related to Thomas Burnet, the author of the *Sacred Theory of the Earth*, and knew Burnet's descriptions of millennial life. Although Burnet had not dated his predictions, he implied they were soon to occur. Furthermore, through Pitcairne Cheyne had been introduced to the most devoted young Newtonians – especially to David Gregory and the Keill brothers – and discovered, if he had not already realised it, that iatromathematics could be compatible with mystical religion. This discovery cannot be precisely dated, but it must have occurred sometime around 1699 or 1700, a crucial millenarian moment. Finally, there was the example of Pitcairne himself. Pitcairne was not a hardened mystic, but he showed mystical tendencies. He was anything but a solid member of the Scottish Church ('the Kirk'), a fact that hindered his academic career after he returned to Scotland from Leyden,[28] and which caused him to be attacked by the medical profession.

Cheyne's career drastically changed after he migrated to London in the winter of 1701/2.[29] Now, daily, he saw before his own eyes the millenarian fervour about which he had heard and read so much in Scotland. If there were relatively few Philadelphians of Quietists in Roxburgh, or even in Edinburgh, there were many in London. Here was millenarianism of another magnitude. Medical degree in arm, and letters of introduction too, Cheyne arrived in a city riven by diverse opinion about the politico-religious development at the turn of the century. This was especially true among the Fellows of the Royal Society, with whom Cheyne was closely associated when he arrived. He had been resident in London only a few weeks when the War of the Spanish Succession broke out. To some mystics this gruesome event signified that the deity had interrupted his millenarian intentions, and demonstrated his dissatisfaction with English national behavior. Cheyne had not yet arrived in London when the Philadelphians read in public their famous Proclamation on Easter Sunday 1699,[30] but he certainly heard accounts of their radical prophecies. During his first spring and autumn in London – 1702 – he focused his energy, as he recounts in his autobiography, on establishing a successful medical practice. He pursued this goal by appearing in the 'right' coffee-houses, and by cultivating the wealthy and the great in their private saloons and drawing rooms.[31] Years later, Cheyne described this period of his life as one of immense 'luxury, gluttony, and upper-class vice without exercise', and typified it in his memoirs by the act of forever 'taking snuff out of a ponderous gold box'.[32] (An essay is not the place to discuss the sociology of medicine in the eighteenth century; but it should be noted that while Cheyne's method of gaining patients was common, his degree of application was not.) During these years he also sat in Batson's and Child's, and in the town houses of dukes and duchesses; and he wrecked his health through drink and gluttony. The talk Cheyne heard in these places must concern us as preponderantly as his dissipation. Here he was apprised of the aspiring physician's need for written credentials and word-of-mouth recommendation. But he also heard about the mounting war on the Continent, and the radical prophecies interpreting it. On street-corners he saw freethinkers and mystics chanting about the millenium come or interrupted, and heard tales about the hysterical uprising of the Camisards in the Cévennes.[33] Cheyne's urban hedonism was extravagant, but apparently not so extreme as to prevent him from writing a mathematical treatise in 1703,[34] which so annoyed Newton that he dropped Cheyne from the circle of young disciples to whom his bounty was given. Cheyne suffered the Newtonian fall miserably. He repressed it, and never again commented on it in any of his autobiographical memoirs.

Newton's dispraise however, did not prevent Cheyne from setting to work shortly thereafter – probably in 1703 – on another book that proved far more theological than the previous two. Two years earlier, late in 1701, Cheyne proclaimed the need for a *Principia Medicinae Theologia Mathematica* based on Newton's *Principia* – one that would integrate medicine and mathematics. Cheyne's idea derived, in part, from John Craig's 1699 *Theologia . . . Mathematica* which Cheyne read and acknowledged in the preface of the new book. But Cheyne's desideratum is not what he wrote. Hoping to reingratiate himself with the Newtonians – by 1703-5 the most powerful scientific coterie in the Royal Society – Cheyne abandoned mathematical medicine, and composed a type of 'Boyle-lecture' which he called *Philosophical Principles of Natural Religion* (1705), modelling the title, as well as the book, on Newton's *Mathematical Principles*. Not surprisingly, *Philosophical Principles* was greeted by the Newtonians with more hostility than Cheyne's previous book.[35] Dislike was based on two grotesque and unpardonable errors: first, that Cheyne had misunderstood the essence of Newtonian gravity, and, then, that his analogical method of reasoning was altogether unscientific. Cheyne replied to the first charge that Newton had 'stolen' certain points in the *Queries* appended to the Latin *Opticks* from him 'in private conversation',[36] an argument no one then seems to have construed seriously. At least it did not persuade the Newtonians, old or young, that Cheyne understood anything about gravity, or that he ought to be readmitted to the clique. The second charge – unscientific analogy – appeared less critical in 1705, but this is the aspect of Cheyne's writing that renders him such a unique figure in the physico-theological world of the early eighteenth century. It is also the strand of his thinking that leads directly to his curious doctrines of early eighteenth-century millenarianism.

Cheyne's analogies derive from a 'Universal Law of Attraction, whereby all the parts of Matter endeavour to embrace one another'.[37] From this given 'Law' he reasons that a 'Divine Providence permeates' both the natural and supra-terrestrial universe. Yet almost every conclusion he draws from this point forward is at odds with the basic assumptions of Newtonian thinking. Moreover, Cheyne's inability to grasp the inconsistencies and obliquities of his own principles in relation to those of Newton constitutes the best comment on his scientific abilities. Such defect of talent certainly did not go unnoticed by the English Newtonians, who now began to wonder if Cheyne was a scientist at all; and the mere fact of an earned medical degree counted for nothing – especially inasmuch as it was granted in Aberdeen – in an epoch when medicine was commonly anything *but* scientific. Geoffrey Bowles has argued that Cheyne's discussion of short-range attraction is the most interesting feature of *Philosophical Principles*, observing as well that Newton had not pronounced publicly on this matter until 1706.[38] Bowles's contention is that Cheyne's method of analogical reasoning permitted him to make an intuitive leap: reasoning from long-rang to short-range attraction. This may be true, but the English Newtonians hardly saw the matter in this light. They grasped on to Cheyne's mathematical errors, and were troubled by the Stoic undertones of his concept of Providence. Only Jean LeClerc, head of the Remonstrant – Arminian Seminary in Amsterdam and himself an ardent theologian, reviewed Cheynes book.[39] Otherwise, the book went unnoticed and bitterly disappointed Cheyne. In this capacity, it did not matter whether or not his medical practice was a success, or whether his urban hedonism had wrecked his physical health.[40] He had lost the support of the Newtonians and other fellows of the Royal Society whose approval he direly sought. A second attempt to re-enlist himself proved futile. Now he elicited their fury twice.

COLLAPSE AND CRISIS

The result was breakdown and collapse in 1706. It may never be known whether this condition was primarily physical or mental. But Cheyne's account of the collapse is so detailed, that he must be believed when commenting that at this time (1706) he 'went about like a *Malefactor* condemn'd, or one who expected every Moment to be crushed by a *ponderous* Instrument of Death'.[41] However, Cheynes attribution of the collapse to his London hedonism, and to the defects of his physiological constitution, is probably incomplete, although not inaccurate. 440 pounds of human flesh will afford even the soberest human being with a perfect rationalisation for anything that ever happens, or happened, to him! What counts for more is the curious way that Cheyne permitted his breakdown to determine the course of his whole future career.

He swiftly departed from London – from the hub of luxury and glut – and fled to the country, hoping to die in pastoral simplicity. He was uncertain about many things; but he was sure, at thirty-five, that death could not be far away. He also 'fix'd on one, a worthy and learned

Clergyman of the Church of England, sufficiently known and distinguished in the *Philosophical* and *Theological* World (whom I dare not name, because he is still living, tho' now extreamly old'.[42] This may have been Whiston whose Arianism and disavowal of the coeternity of the Father and Son were notorious by late 1706. Ill and despondent, Cheyne, following Whiston's example, 'resolved to purchase, study, and examine carefully such *Spiritual and Dogmatic Authors*, as I knew this *venerable Man did* most approve and delight in'.[43] These were works of primitive Christianity, 'a *Set of religious Books* and *Writers*, of most of the *first Ages* since *Christianity*'. They confirmed Cheyne's developing sense that the material world was proximate to dissolution and the New Jerusalem imminent. They encompassed the writers of the first four centuries who had not been contaminated by the Council of Nicea and the Apostolic Succession. Cheyne and Whiston probably did not meet: Cheyne was in Bath, Whiston in Cambridge. But there is a good deal of circumstantial evidence to suggest that Whiston is the 'now extreamly old' – now as Cheyne writes in 1733, not 1706 – scientist and philosopher whose 'primitive Christianity' subdued his misery in illness.

But Cheyne did not remain in 'the country' (wherever that may have been) throughout 1706, the year of collapse. He very often returned to London, and may have been there when the first French prophets arrived that autumn.[44] Fatio, Newton's disciple, quickly enlisted himself in the service of Elie Marion, their leader, and introduced David Gregory, and possibly Cheyne, to the prophets. But whereas Gregory was resistant, Cheyne was sympathetic. By Christmas 1706, the prophets predicted that the millennium had arrived and that the 'hidden keys of Divine Wisdom' were daily being revealed to women, children and common folk. During these months at the end of 1706, Cheyne – ill, despairing, believing he was near to death – began to connect the apocalypse with medicine, and started to realise that his life could serve a higher purpose than he ever dreamed.

Strangely, he soon began to improve, this after six or seven months. His near-fatal illness, he concluded, was clear revelation: not only that he should instantly mend his ways – his whole style of life then – but also that he should serve his Maker by delivering 'a message' to mankind. When Cheyne reflected (in 1733, thinking about the curve of his whole life) on the validity of the cosmological picture he had painted in *Philosophical Principles*, he was altogether dissatisfied. 'I found', he writes, 'that *these* [Philosophical Principles] alone were not sufficient to quiet my Mind at that Juncture.'[45] Then he describes the new vision acquired since his illness:

> ... especially when I began to reflect and consider seriously, whether I might not (through Carelessness and Self-Sufficiency, Voluptuousness and

Love of Sensuality, which might have impaired my Spiritual Nature) have neglected to examine with sufficient Care: If there might not be more required of those, who had had proper *Opportunities and Leisure*: if there might not, I say, be higher, more noble, and more enlightening *Principles*[46] revealed to Mankind *somewhere* . . . and lastly, if there were not likewise some clearer Accounts discoverable of that State I was then (I thought) apparently going into, than could be obtained from the mere Light of *Nature* and *Philosophy*.

This 'mere Light of *Nature* and *Philosophy*' – especially Cheyne's new perception of the limits of science – constitutes the source of his mental frame during the new decade. Now, more than before, he believed that religion was revelation; and that the body of man had been the most sorely neglected source within the Book of Nature.

HEALING AND REBIRTH

Still believing himself close to death in 1707, Cheyne heard the voices of another type of natural revelation than those he had heard on London street-corners or read about in books. Medicine, like mathematics, had been part of the deity's grand plan from the start;[47] but now Cheyne understood how the deity would reveal his wisdom and might through suffering and healing. The body of man, like the book of nature, was a major seat of revelation; and anatomy and physiology its correlatives within 'natural philosophy.' Cheyne's illness – his 'crisis of 1706' – was then a part of a larger providential plan. Had there not been evidence of revelation through the body of man in recent social events as well? Sudden healing of the sick poor; the unprecedented establishment of almshouses; other medical services that rescued men and women who would have been given over for dead only a few years ago?[48] It seemed to Dr Cheyne that the millennium had commenced or soon would, and that he had been chosen to be instrumental in the establishment of the New Jerusalem in England. This was a far more important calling, he reasoned, than the previous Newtonian one.

Now persuaded that the body could not be overlooked, Cheyne turned elsewhere in the apocalypse than to mathematics or iatromathematics. His primary task, as he recounts in his 'case history', was to recover. He persuaded himself that by practising the most vigilant temperance he could avert further collapse. He increasingly renounced London, returned there less frequently, and abjured its indolence and luxury. Precisely why he did not join Fatio and the other prophets remains a mystery,[49] unless his decision owed something to the personal or public intervention of his mentor Whiston. By the autumn of 1706 Whiston's *Essay on the Revelation of Saint John* had appeared, announcing that certain of his earlier prophecies had been fulfilled and that others were yet to come. Cheyne may not have read this work, but he probably heard or read Whiston's Boyle lectures delivered in the next winter (1707–8) and printed the following summer. Here Whiston argued that scriptural prophecy is capable of one and only one interpretation; he also fixed the precise date of the millennium as 1736, which he later updated to 1766. More urgently for Cheyne, Whiston warned against the placing of trust in the French prophets,[50] a position that may have weighed somewhat in Cheyne's decision not to join them. One further consequence of possible Whistonian influence was Cheyne's apparent realisation that he (Cheyne) had been wrong about the source of the 'Universal Attraction' discussed in *Philosophical Principles*. At least, this line of argument – that Cheyne revised his theory as a consequence of Whistonian influence – is more likely than the arcane explanation that for a third time he tried to regain the bounty of the Newtonians.

Alone then in the country, ailing but improved, and no longer near death, Cheyne set about to revise his 'philosophical principles' in a manner that would take account of his 'great crisis' of 1706. 'If my life was to be sav'd,' he comments in his memoirs looking back at these years between 1706 and 1709, 'it was only be this [temperate] Regimen.' Cheyne's solace in his illness was that he had learned to understand Grace in a new light: 'if my Time of *Dissolution* was come, I knew I should die under Misery . . . [rather] than by an other Means.'[51] By 1709 he settled in Bath, close to the mineral waters in case further crises of health should arise. Here he could practise medicine if he recovered, but the main attraction was the pure quality of the air and the proximity of the spa. He arose early and retired early; his diet consisted exclusively of vegetables, milk, and seeds; he sought out no patients and lived frugally. In the terms of modern psychoanalysis, his ego underwent radical redefinition. If a patient visited him, he would treat him or her, but he no longer craved to be a fashionable London physician. James Cuninghame 1665–?), one of the four main French prophets, sought him out while recuperating from his own illness in Bath early in the spring of 1709, and filled Cheyne's ears with talk about the prophets' activities and the imminence of apocalypse.[52] Cuninghame read Augustine Baker's *Sancta Sophia* during recuperation, while Cheyne while Cheyne still scanned the works of the early Christian fathers, and possibly of Boehme.[53] It must have occurred to them how remarkably parallel their lives were. Both men were Scots who craved worldly recognition in England; both had been introduced to mystical religion by the Aberdeen – Garden group; both had recently been afflicted with a near-fatal illness; the recovery of each coincided with new insight into the nature of Providence and resulted in a major conversion of life style. Immediately thereafter, in 1709, Cuninghame joined the French prophets, whereas Cheyne renounced his previous life. Cuninghame resolved to work for the prophets in Scotland,[54] while Cheyne aimed to convert Englishmen to the 'New Jerusalem' by the same doctrines of abstemiousness in diet he himself was rigidly following. Around 1709 or 1710 Cheyne may not have considered himself a 'Quietist', but an observer of his daily life in Bath would have concluded he was one. Well-read in the works of Mme Guyon, Bourignon and other Continental pietists, and possibly by now of Boehme, Cheyne believed that the millennium had begun, that recent political and social events were sufficient proof, and that he bore a special mission in its commencement. We do not know if he agreed with Whiston that the millennium would not begin until 1736 or 1766,[55] but Cheyne was confident that the important day could not be very far away. Besides, Cheyne's conversion had clearly occurred in 1706, in the very same months when the French prophets landed in England and when many were prophesying that doomsday was close at hand. What evidence, all seeming to convene, could be more explicit from Cheyne's point of view?

During the next five years – 1709–14 – Cheyne revised *Philosophical Principles* and practised his body in rigid diet, regular exercise, plenty of sleep, pure air – non-naturals, as the eighteenth century called them, the abuse of which had been a primary cause of his collapse. In 1711 the French prophets began to roam the West Country and to proselytise more actively than they had in London. Cheyne probably heard them in Bath or Bristol, even if he resisted them. During these years he also associated with Richard Roach (1662–1730), their foremost apostle who, like Cheyne and Cuninghame, had experienced a pattern of illness, healing, conversion, and redemption through new works.[56] Roach also published a diary which would have aroused Cheyne's sympathy if Cheyne could have read it. 'Divines and Physicians, Literal and Mystical', Roach cryptically wrote, 'There is a world of Science, Soul of the Science unknown to the former.'[57] Roach also scrutinised the Kabbala, and may have introduced Cheyne to the interpretations he, Roach, would publicise before his death in 1730 in *The Imperial Standard of Messiah Triumphant* and *The Great Crisis*. For all three men, millenarianism and medicine were related, and even if no one of the three was searching for a universal panacea[58] – as Fatio was – each had learned that extreme illness followed by healing was itself the highest form of revelation: the basis for a philosophical natural religion based on the body of man. Then, in 1713, the German Baron Metternich published an explication of Boehme under the guise of an attack on John Locke, entitled *Fides et Ratio – Faith and Reason*. Cheyne acquired it and sent it to William Law, the author of the popular *Serious Call to a Devout and Holy Life* (1728) which implored mankind to renounce the hustle and bustle of material life in preference for a quiet world of constant religious devotion.[59] By 1715 the second edition of Cheyne's *Philosophical Principles* appeared, reasoning again 'by way of Analogy' but now espousing a more mystical, if indeed somewhat neo-Platonist, theory of attraction than before. Creatures of the world were now direct reflections, or embodiments, of the Creator. Because of this similitude, Cheyne argued, one could reason exclusively by analogy and without hesitation from the material to the spiritual realm.[60]

Yet Cheyne's argument is not Shaftesburian. Attractions between living creatures are merely another form of attraction than that between the deity and his material creation, but are no less valid or real. Therefore, spiritual love between man and fellow man is attraction of as noble a type as that between man and God. The 1715 revised edition also contained reflections on God and the 'Divine Essence'. Here Cheyne argued that forms of 'divine things' exist as well as of material things, the material ones having been 'Copied out' in the process of original genesis.[61] Gods creatures, man included, become 'Images, *Emanations, Effluxes*, and *Streams* out of his own *Abyss* of Being',[62] a position that appears closer to the *Book of Urizen* than to the *Principia* or *Optics*. By 1720 Cheyne prepared another book championing Stoic abstemiousness entitled *Observations concerning the Gout*. This was a book less about gout than about the healthful effects of a lettuce, milk and seed diet, one promoting plenty of sleep, good air, and complete avoidance of luxury in diet.[63] It was well received by the medical community, which was so obsessed with gout in the 1720s that almost any book by an M.D. would have been viewed seriously. But since the book proclaimed nothing about cosmology or physico-theology, the Anglican Newtonians overlooked it or shunned it altogether.

THE SECOND REVELATION

But Cheyne relapsed. Again in 1723, now over fifty, Cheyne once more swelled to enormous size – 'I exceeded 32 Stone', about 450 pounds – and grew so ill that 'if I had but an Hundred Paces to walk, [I] was oblig'd to have a Servant following me with a Stool to rest on'.[64] This time Cheyne was better prepared for dire calamity than in 1706, and could rely more on the resources of his acquired mystical millenarianism. Certain that misery is the mother of salvation, he bore up to his 'perpetual *Sickness, Reaching, Lowness, Watchfulness, Eructation* and *Melancholy*'[65] for almost two years, and diagnosed gout as the source of these melancholic conditions. Returning to a diet of 'lettuce, little wine, and water best', this second protracted illness caused him to grow increasingly hermetic in his theory of diet. Contra Mandeville and Nicholas Barbon before him, Cheyne argued against the virtues of luxury. Yet like them, he considered luxury to be a psychological state as well as a physical reality (i.e. the presence of sugar in the poorest household). And he linked himself with others who related psychic health to daily diet. He would surely have encountered a kindred spirit if he had then read Thomas Tryon's works, the author of the book on the mystical divination of dreams.[66] But if Tryon advocated a similar rigid vegetarianism, he possessed little of Cheyne's mystical faith – 'naked faith', as John Byrom later referred to it – nor was Tryon medically trained. Yet both men, to be sure, were of the hermetic tradition of Boehme: a lineage descended from Paracelsus to Boehme, and from Boehme to the baptists, pentecostalists, and other versions of English pietism that flourished in Cheyne's most formative years. Actually, Cheyne would probably have found himself in greater spiritual agreement with Thomas Byfield, the Anglican physician who turned prophet in 1707 after the arrival of the Camisards in England.[67] Byfield too came to Bath in search of health, where he may have sought out Cheyne to diagnose his case. While there, Byfield joined forces with the English prophets based in Bath – Bristol, and converted the vicinity into a stronghold of radical millenarianism.

In the midst of all this religious tumult, Cheyne was following his old pattern of publication accompanied by illness. In 1723 or 1724 James Leake, Richardson's brother-in-law whom the Earl of Orrery described as 'the Prince of all the Fraternity of Booksellers' in Bath, opened a business on one of the parades, a few yards from Cheyne's house, and printed as his first book Cheyne's *Essay on Health and Long Life*: yet another Cheyne treatise work advocating abstemious diet, this time referring historically to the writings of Cornaro, Lessius, and other early vegetarians. It is impossible to know – at least Cheyne's memoirs offer no clue – whether this publication bolstered Cheyne's spirits sufficiently to cure him. But by December 1725 he was well enough to travel to London to consult with the most 'distinguished physicians' alive about his ailments.[68] Cheyne's habit, then, of growing seriously ill just before and shortly after the publication of his books seems by now to have hardened into a confirmed pattern.

Geoffrey Bowles has called attention in the essay already cited to a correlation between Cheyne's theory of 'attraction' and his attitude to the medical profession.[69] It is equally plausible that a correlation exists between the reception of Cheyne's books and his health. If this approach is valid, it would have to include Cheyne's anticipation of the reception he would receive. Every time the Newtonians decimated him, his health declined. When no one took notice of *Philosophical Principles* in 1705, he grew dangerously ill. Now, in 1723, he again blistered into mania and fever while writing the *Essay on Health and Long Life*, only to be cured by the eventually favourable reception of the book. The *Essay* received more attention than any previous work by Cheyne, and within eighteen months was translated into several languages.[70] Gilbert Nelson, then an authority on gout, wrote approvingly of Cheyne, claiming that he was the only physician in England to be ranked with Sydenham on the subject.[71] Arbuthnot himself was willing to be deflected from other pressing professional work, and studied Cheyne's theories about vegetarian diet in *An Essay Concerning the Nature of Ailments* (1731). As these medical estimates and literary appraisals appeared, Cheyne improved; by 1729 he claimed 'complete Recovery'.[72]

Yet it falsifies the known facts to sketch a picture of praise without blame for Cheyne's reception. As the negative criticism mounted and continued to surpass the positive,[73] he began to see medicine and millenarianism – together and apart – in a different light from the view he held during his first collapse. In 1729 – his second medical *annus mirabilis* – Cheyne was fifty-eight, no longer young. According to Whiston, the millennium was now only seven years distant (1729–36), or if Whiston altered his view by then, thirty-seven years (1729–66). Unfortunately no evidence exists to learn whether Cheyne still extolled Whiston as he had in 1706 – as a beacon of primitive Christianity – and the lack of any reference to Cheyne in Whiston's *Memoirs* complicates the matter further. But whether or not Cheyne was still reading authors of primitive Christianity, by 1730 he was certainly grieving the death of Richard Roach, perhaps the most inspired of the English prophets who has converted to the cause of the French.

As Cheyne's books continued to be reprinted and as the negative record began to accumulate,[74] his evangelical mission to connect medicine and millenarianism increasingly obsessed him. Doubt too crept in: perhaps he had not accomplished enough in the conventional medical sphere. Furthermore, as social and economic conditions suddenly changed in England after the disastrous South Sea Bubble, and as the tide of luxury dramatically increased, Cheyne believed that his energy was urgently needed to combat this appallingly widespread condition. Accordingly, sometime around 1730 he set about to write a treatise on scurvy, the only major disease he had not written about before. He noticed that it had peaked during the last two centuries, concomitant with melancholia. He also observed that there had been a drastic upsurge of dyspepsia in England – the main symptom of 'flatulent melancholy' – during the decade of the 1720s, as well as an increase in suicide. By the mid-1720s the medical profession was calling suicide, even more so than gin, 'the English vice'. Cheyne imaginatively combined all these current ideas into a single work, and in the same year (1733) in which Pope published *An Essay on Man* he brought out *The English Malady*, certainly his best known book today.

MEDICINE AND MILLENARIANISM

Cheyne made clear in his preface, first of all, that he was writing for a particular audience. 'Such a *Diet*', he insists at once, 'is only proper for the *thinking, speculative* and *sedentary* Part of Mankind, and not for the *active, laborious* and *mechanical.*'[75] Yet this 'thinking Part' – clergymen, scholars, writers, artists, the whole intellectual establishment – constituted practically the whole group that was committing suicide with such astonishing rapidity. What had these types in common? A weak constitution, Cheyne concluded; one whose blood and, more significantly, whose nervous system, was either congenitally defective or imperiled through abuse. His own condition had been a case in point: from youth onwards he had observed a weak physiological constitution composed of 'feeble Solids' and 'excessive Juices'.[76] Yet the nervous system – the constellation of nerves, spirits, and fibres – was the ultimate culprit; and Cheyne concluded that the only remedy, given that 'none can choose his own Degree of *Sensibility*', was the spartan lifestyle he had been recommending since his visionary experience in 1706. The assiduous reader who was willing to peruse all 300 pages of *The English Malady* would discover how the nerves and fibres actually produced the melancholy about which so many Britons (including Hume) complained to their physicians and which drove many of them to suicide. Theoretically and methodologically viewed, *The English Malady* by no means endorsed 'the mechanical philosophy', but it infused Newtonian and mathematical learning to a degree Cheyne had not used since 1715. Only a novice reader who had not followed Cheyne's bizarre scientific and personal career could reasonably have concluded in 1733 that Cheyne was still the dyed-in-the-wood iatromechanist he had once been. Yet he was not. If anything, his preface anticipates precisely the opposite charge: that he has now 'turn'd mere *Enthusiast*, and resolv'd all Things into *Allegory* and *Analogy*, advis'd people to turn *Monks*, to run into *Desarts*, and to live on *Roots, Herbs,* and *wild Fruits*'.[77]

The English Malady was an instant success. Within fifteen months it went into six editions. Cecil Moore, the literary historian, was so awed by its reception that solely on the basis of it he wrote of the mid-eighteenth century: it 'deserves to be called' not the Age of Reason, Enlightenment of Exuberance but 'the Age of Melancholy'.[78] An element of Moore's attitude was influenced by Cheyne's own explanation of the title: 'the title I have chosen for this treatise is a reproach universally thrown down on this island by foreigners, and all our neighbours on the Continent, by whom spleen, vapours, and lowness of spirits are in derision the English Malady'.[79] But *The English Malady* is not an apology for eccentricity or a justification of a dangerously high rate of English suicide. It is a cultural treatise embodying many of the unwritten assumptions of the age, and it synthesises a whole range of current medical *topoi* (melancholy, spleen, vapours – the whole repertoire of then current psychosomatic illnesses) and controversial physiological assumptions (the nervous system in relation to the rest of the body and its behaviour), as well as fundamental laws about the nature of man.[80] In view of Cheyne's ingrained iatromechanical and Newtonian beliefs, it is not surprising that many of the explanations in the book are mechanical and mathematical. But it would be a serious error to interpret *The English Malady* merely within the development of English mechanistic theory, especially because the deepest explanations – answers to the question, what is life? – are remarkably non-mechanistic. *The English Malady* also has historical value because it assembles so many prevalent mid-eighteenth-century biases and discusses them within a social and topographical context, an instance of which is the effect of English climate on human health.

Viewed, however, from the perspective of millenarianism, *The English Malady* is less significant than the revelatory treatise appended to it: *The Case of the Author*, a fifty-page memoir delineating Cheyne's life, his various crises and his conversion to 'more enlightening *Principles*'.[81] Ten years later, in 1742, Cheyne told Richardson that he wrote this work to prevent his patients from believing he was 'really mad'. But this was recollection in hindsight, and it may be that Cheyne himself did not fully realise how well his *Case* formed a companion piece to *The English Malady*. Cheyne's memoir of his spiritual life provides a context for *The English Malady* and demonstrates how 'nervous physiology' – one of his favourite scientific subjects – lies directly in the service of these 'more enlightened principles'. Here Cheyne argues, vulgarly, that we are our physiology, and that our constitution will predetermine most aspects of our behaviour; yet he does so as justification of his own 'naked faith'. For more than anything else he wants to remind his readers that they possess nerves. But he also hopes to impress upon them the larger claim that health depends upon 'Simplicity', and that every physiological type – whether a robust Fielding or a willowy Richardson – can improve his condition by adopting a spare diet.[82] '*Simplicity* is the greatest Contradiction to *Laziness, Foreign Studies, Negligence, Incuriosity* and *Ignorance* in the Profession; but such a *Simplicity* . . . is worth a *Million* of these false and *foreign Art* sometimes us'd to rise in it; for it [Simplicity] is, in Truth and Reality, an *Eminence of Light* and *Tranquillity*.'[83] Cheyne's final trope is characteristically mystical; it is the clue to his whole physico-theology. If others looked at the stars and heavens to discover 'the Book of Revelation', he gazed inwardly at the body.

Yet a need to confess – to lay one's heart bare – is as crucial to the intentions of the author narrating this 'case history' as are any observations about temperance and abstinence. This is why Cheyne locates abstinence within the Stoic or Quietist life as subservient to that 'universal attraction' he thought he now (1733) understood better than ever. 'For the Means us'd by *infinite Wisdom and Goodness* towards reclaiming his *wandering Creatures*, seem only to be either *Love or Punishment*: that those whom Love will not draw and allure, *Punishment* may drive and force.'[84] Both types of revealed 'attraction' – love and punishment – have strayed from any scientific model: Newtonian or otherwise. Now Cheyne belittles the progress of science and the revolutionary value of the Newtonianism he had formerly championed, postulating that the 'physic' (i.e. medicine) of the early Christian fathers had achieved equally good results, 'tho' not quite so soon perhaps as well by all our *Mathematicks, Natural Philosophy, Chymistry,* and *Animal Oeconomy*'.[85] His tropes are consequently those of the anti-scientist who discovers who he 'really is' by deconstructing his former scientific life; and there is very nearly something Sartrean and Barthean about Cheyne's concept of autobiography. Yet Cheyne proclaims himself to be no open enemy of 'those *Divine Sciences*', and explains that luxury is the culprit. Luxury has outpaced science, as it were, and rendered its wisdom ineffectual in England. He concedes that Cornaro and Lessius, previous diet theorists discussed in *The English Malady*, lived in earlier and simpler times, when science and theology were not so intertwined. But saliently, they wrote long before the onset of the apocalypse. Their ideas of health, Cheyne reasons, were not coloured by the onset of the millennium. As a consequence, they could not perceive man's ultimate needs so clearly as he could. Nor did physiological necessity – there, again, was Cheyne's law of physiological determinism – cause them, as it forced him, to remain apart from society while contributing to it.[86]

Cheyne was now (1733) in his mid-sixties. He professes not to worry if death be close; all his goals have been accomplished and his mood is ironic. He has become the most dedicated spokesmen of the age for the medico-millenarian analogy. Of this he seems practically certain.

FAME AND MYSTICISM

Cheyne remaining years (1733–43) displayed no evidence of mental decline but a rather marked intensity of belief in mystical religion. Amazingly, he had lost two-thirds of his weight, and remained thin and relatively well (relative to what he had been) until his dying day. This last decade, the 1730s, was the period, ironically, when his medical practice soared. Fashionable ladies – dowagers, duchesses, princesses – from everywhere sought him out in Bath where he was now a legendary figure. By 1734 he was treating the wealthy Countess of Huntingdon and continued in constant correspondence – 'pious conversation' – with her.[87] Her own letters to Cheyne portray her attitude as that of a worshipper in a temple.

In Cheyne's brand of mystical millenarianism she discovered the reflection of her own ideas; simultaneously, as she obeyed her cult hero, her health mended.

By 1738 Cheyne was also deep in constant correspondence with Richardson – his 'literary patient' – and trying to persuade him that extreme abstinence was the only salvation for someone physiologically as 'nervous' and 'delicate' as he was.[88] The source, Cheyne reasoned, lay in Richardson's 'defective nerves'. But Richardson's 'nervous paroxysms' and 'paralytic tremors', Cheyne recognised, were those of a creative artist; therefore they could not be treated as if Richardson were another aristocratic lady in Bath. Cheyne expended much energy and more ink to persuade Richardson in dozens of extant, and long, letters that his malady could be constructive – no less constructive than Cheyne's illnesses had been. Cheyne had something specific in mind, though it did not surface for three or four years.

While corresponding with Richardson, Cheyne published another book, *An Essay on Regimen* (1740), attempting to delineate 'the principles and theory of philosophic medicine [sic] and [to] point out some of its moral consequences'. This was the medico-moral analogy Cheyne had established long ago; only now it was extended more explicitly into a medical arena. A year before his death in 1743, he produced another long essay arguing that 'disorders of the mind' depend 'upon the body', and that care should taken to keep the body healthy.[89] This position was the reverse of the psychosomatic one gaining ground at mid-century: namely, the notion that disease of the body were owing to mental distress. Yet, however receptive to psychosomatic theories of illness Cheyne may have been, by the end of his life he was more ardent than ever about the body as an instrument of divine revelation and as the source of the truest revealed religion. If there were such a state as 'Enlightenment' in the England of the 1740s, this was Cheyne's most enlightened credo. Cheyne grew so fanatic about the matter that he could not imagine any 'revelation' that circumvented the body, an intellectual stance that ought to cause Romantic scholars to be far more interested in him than they have been.

Richardson knew his correspondent well and was aware of his bent. As late as August 1742 – eight months before Cheyne's death – he continued to 'bribe' Cheyne with gifts of 'Boehme bound'.[90] A few volumes would elicit the free medical advice Richardson direly needed. Sometimes, Cheyne, rather than Richardson, drove the bargain. By September 1742 Cheyne was imploring Richardson to print a 'Catalogue of Books for the Devout . . . and Nervous'.[91] Actually believing that it 'would be of greater Use in England than any Books', Cheyne advised Richardson that he hoped to model it on 'the Catalogue of the mystic Writers published by Mr. Poiret'.[92] Richardson was not altogether unreceptive to the idea, but

Cheyne could not have known he would be dead within a few months. When Richardson niggled and procrastinated, Cheyne conceived yet another 'project in mystical religion'. 'Pray be so good', he begged Richardson, 'to inform me if you know any Person having a Taste of Spiritual Religion that could translate a little French Book into clean English, entitled "L'Essence de la Extract de Religion Chretiene" '.[93] Richardson apparently found a translator of whom Cheyne did not approve. So this ultimate dream, like some of its awe-inspired predecessors, went the way of all flesh.

During this period at the end of his life Cheyne also corresponded with William Law, although many of their letters have disappeared, as Stephen Hobhouse, Law's recent knowledgeable student, has discovered. The subject they discussed most was religion and science: especially the diffusion of 'mystical religion' among the growing numbers of naturalists. Cheyne was amazed at the number of young Newtonians who continued to carry on the work of their real and symbolic 'father'; Law was intent, for obvious reasons, to prove that Newton himself had been a mystic of profound dimensions. Yet Cheyne wondered why Law claimed that Boehme *in particular* had been the source of much of Newton's science. On 31 March 1742 Cheyne put the question to Law in a letter, asking him to substantiate what Law had just claimed in his recently published *Appeal to all that Doubt* (1742): '. . . that he [Newton] had been a diligent Reader of that wonderful Author [Boehme], that he made large extracts out of him . . .' Law replied by repeating his claim in the Appeal, and assured Cheyne that these 'large extracts' had been among Newton's papers at the time of his death. Law's reply to Cheyne was not published in the lifetime of either man – Law died in 1761 – but appeared in the September issue of the *Gentleman's Magazine* (p. 329), and was later republished by Christopher Walton, Law's Victorian biographer, in *Notes and Materials for an Adequate Biography of . . . William Law* (1854). Cheyne may have been satisfied by Law's letter: at least he had no reason to deny what had been included among Newton's manuscripts at the time of his death in 1727. But it is also possible that Cheyne was too preoccupied with Richardson and his invalids at Bath, who he continued assiduously to treat, to pursue the intriguing question about Boehme and Newton.[94]

Throughout that summer of 1742 – Cheyne's last – and during the next autumn he continued to search for a translator. Precisely why he relied so preponderantly on Richardson, and why he could not locate a translator by himself, most be something to preoccupy Cheyne's future biographers. Of greater concern here is the situation of the septuagenarian millenarian – dreamer knocking at death's door in this precise stance: still diffusing mystical but 'more *enlightening* Principles' by scattering books throughout the British Isles, still compulsively imposing on Law, still enticing Richardson. According to so many commentators in his own day, Cheyne had been a brilliant medical mind, a caring doctor, a personality totally worthy of the notice he would no doubt receive in futures ages. Yet time has somehow managed to obscure the very aspect which his contemporaries deemed to be so original to the Bath physician: his unique blend of medicine and mysticism, as even his bodily corpulence and exiguity demonstrated. It is, then, one matter to depict Cheyne in his own milieu and against the backdrop of his own times and quite another to rescue him now. If greater emphasis is placed on the second concern, then a different question ought perhaps to be put to the modern student. In this case the absolute historical portrait loses some of its thunder as we wonder – today – if Cheyne's career does not pose some major paradoxes for the intellectual historian of our times who happens to be interested in the eighteenth-century Englightenment. Cheyne, Richardson, Newton, Law, Boehme, Pope, Poiret: what a strange lot of bedfellows! Surely, we wonder, this is a jumble worthy of commemoration in a polished neoclassical English couplet. But the constellation may reflect our own sense of 'jumble' according to principles of 'Enlightenment' we have inherited more than historical truth warrants. Perhaps our sense of *the* Enlightenment and its attributes requires some radical adjustment. This is the issue I want to discuss in conclusion, isolating it in relation to Cheyne's demise.

DEATH IN THE APOCALYPSE

Cheyne died in April 1743, having failed to convince the English public about a balanced diet, and valetudinarians like himself, about the terrific value of abstinence. His programme for scurvy – *the* disease of the seas at the time – had to await the late eighteenth century before gaining public recognition; and his plan for a balanced diet containing plenty of vegetables and nuts rather than meat and potatoes, has lingered into the twentieth century before partial adoption. His attack on luxury, it is true, impressed many of his patients, but 'Estimate' Brown, Smollett, Goldsmith and others in the 1750s and 1760s were needed before the war against luxury could be formulated, let alone combated.[95] When Goldsmith reported that Beau Nash 'would swear, that his [Cheyne] design was to send half the world grazing like *Nebuchadnezzar*',[96] we view the comic strain of Cheyne's programme. Indeed there is a sense in which he must have appeared to many of his contemporaries as if he had been – or *ought* to have been – a caricature in a Smollett novel. Viewed solely as a type, he was a celebrated but decidedly eccentric physician who had become a best-selling author. But his contemporaries could not view him from within. If they had been able, they would have found a complex man who firmly believed that he had resolved intellectual dilemmas through a doctrine of 'Universal

Attraction' based on analogy. Furthermore from within, that he had faithfully served his Maker by carrying forward the supreme message about man's nervous body. And, despite the weight of a quarter of a ton, he had managed to live to seventy-two! As he told Richardson near the end of his life: 'I [who] have gone the whole Road had one of the most cadaverous and putrified Constitutions ever was known, and I thank God am returned safe and sound at 70 every way well . . . and surely he knows the Road better who has gone to and come from the Cape of Good Hope, all the Surroundings, Rocks, Shelves, and Winds, than they who have only seen them in a Map.'[97] In other words, he had been a good physician but a better Christian, and by prioritising the two in this way he had practised what he preached: 'medicine begins where philosophy ends'. Despite his unorthodox Christianity it would be wrong to see Cheyne – whether from without or within – as a hermit, even though he had retreated from city life to encourage the apocalypse. In my view it is equally incorrect to portray him merely as a zealot who repressed his earthly needs by rationalising them in the name of millenarian enthusiasm.[98] Temperamentally, Cheyne was as social and clubbable as he was irenic and retiring. Socially, he was not so recalcitrant as his memoirs suggest: we know this from his medical activities in Bath. His personality was outgoing, permeated with a constant sunny cheerfulness he never abandoned. But he had been born with chronic obesity which played havoc with his physiological constitution to such a degree that he never expected to live to more than twenty or thirty. From youth onward his religious tendencies had been mystical, but when the crisis of 1706 broke, followed by the subsequent conversion-experience, something new in his apocalyptic and millenarian imagination jelled. Mentally and emotionally he was never again the same. The Cheyne who wrote to the ailing Richardson, 'it is true you are not a Physician, but I hope you are a Christian',[99] was the mature, ultimate Cheyne – anything but a Hogarthian or Smollettian caricature – who was persuaded he had found the way to Grace and eternal redemption:

> Our Saviour bids us fast and pray and deny ourselves without Exception, but for this there is no need for Revelation Advice. If you read but what I have written in this last in the Essay on Regimen in long Life and Health or Cornaro's or Lessius' little Treatise your own good sense would readily assure you; but you puzzle yourself with Friends, Relations, Doctors, and Apothecaries, who either know Nothing of the Matter, or whose Interest it is, or at least that of the Craft [,] to keep you always ailing . . .

Richardson, though reliant, as Cheyne affirms, on social relations, heeded the advice personally and artistically. Having been exposed to Cheyne's medicine and millenarianism for two decades, as his familiar correspondence with the pious doctor shows, he became thoroughly knowledgeable of the quietist world in which Cheyne was so deeply enmeshed. This familiarity permeates Richardson's writings, especially his great, tragic novel *Clarissa*. The hermetic symbols on Clarissa's coffin are only a few of the many clues suggesting Richardson's awareness of the mystical circles in both the north and south of England that practised a religion opposed to the rational sects of the era. Cheyne – as I have tried to show – stood at the centre of this circle, especially for his sentimental religious doctrine of 'naked faith and pure love'. He and Richardson corresponded enthusiastically about this version of pious sensibility, which intrigued Richardson the sentimentalist enough to infuse this mystical spirit into the fabric of his great narratives and the character of his greatest heroine. Clarissa, above all, is the purveyor of these Cheynean doctrines, relying as she does on spiritual sense rather than fallen reason, and never diverting her attention too far from the glow of mysticism that hovers closer to her personal sensibility that any sparks of Enlightenment rationality. She conveyed them so well that it would be fascinating to know what her readers on the Continent made of her brand of mysticism, towering symbol of the English sentimental heroine that she remained for the world of Rousseau and the German Romantics.

CHEYNE AND THE CULTURAL MAP OF HIS TIMES

These conclusions about Cheyne's temperament and career may be valid, but they are inadequate in themselves unless Cheyne is properly related to the temper of his times. In this sense, it is irrelevant whether he was a major or minor figure. His activities as a representative man of the Enlightnment are far more crucial, and not merely his intellectual thought but his frenetic energy in dispersing Quietist literature from the Continent. It is necessary, then, to ask two or three large questions to understand precisely how he is a representative man and how he relates to the map of his time. Prominent among these topics is his relation to the overall science – especially to the 'natural philosophy' – of the Enlightenment. Precisely how does Cheyne relate to eighteenth-century science?

Clearly, he was one of the more interesting early Newtonians: not only because he was, Arbuthnot notwithstanding, the physician closest to 'the Wits' – he himself was something of a 'scientific wit' whose numerous bestsellers made his name a household word among the 'Hackney Scribblers' he described in the preface of *The English Malady* – but also as a result of his hermetical way of reasoning. Cheyne may not be a 'scientist' when viewed from our perspective today, or when judged by our criteria of science; but he was certainly considered a 'scientist' in his own time, although – as we have seen – a poor one by the Newtonians. The foundations of his system thrive on a hermetical concept of analogy that is neither logical nor accessible.[100] Rather than pitched at mathematical logic or secular accessibility, Cheyne's analogies were exercises to derive the love of God in a hostile, yet hardly void, universe. Yet Cheyne's life and works show positively no contradiction between science and theology, although it is perfectly clear that most historians of science would be more comfortable with his career if it had evolved in the early seventeenth, rather than the early eighteenth century. If it is true that in the seventeenth century a good scientist also had to be a good theologian, this law applies integrally for Cheyne, although he lived a century later. In this sense, though, the relation of his science and theology may compel contemporary students of the Enlightenment to ask some hard questions about the so-called 'rational century' or 'Age of Reason'. It is true that Cheyne stands apart from many of his medical brethren – the Arbuthnots, Cheseldens, and Olivers – who were less interested in theology than he was, and that he appears instead to be a harbinger of Hartley and Priestley.[101] But the point is not at all that Cheyne was born too early or too late; but rather that he, like Fatio, was 'stricken' by the French and English prophets, and that this seizure impelled him to integrate 'mathematics' and 'naked faith' in a way that scholars have yet to describe. Perhaps there is an even larger point to be gathered. Cheyne was not alone in his mystical millenarianism. He had his brand, just as Pitcairne, Fatio, Byfield, and so many others had theirs. Yet Enlightenment scholars have remained largely oblivious to this huge underbelly of their so-called Age of Reason.

Moreover in the relation of science and theology, Newton was not the messiah for Cheyne who would have agreed with those in the 1730s who interpreted Pope's famous couplet in the *Essay on Man*, about 'showing Newton as we show an Ape', as a satiric barb. And he probably would have argued that it was directed specifically at those of his scientific brethren (FRS?) who gadded about portending that the messiah had arrived in the same of Sir Isaac Newton. As Cheyne gradually retreated throughout his life from a system that may crudely be called 'Newtonian metaphysics', he substituted a set of poetic analogies derivative from – pure love. At least nothing in the empirical universe could even begin to corroborate these similitudes. They formed the basis of a metaphysics that increasingly denied the basis of physics. But this is precisely why Cheyne is so interesting, and why the *literati* were so attracted to him. In this sense Cheyne's career violates the paradigm referred to earlier about 'a good scientist also having to be a good theologian', yet it shows a man continually striving to wed science and theology.

In this evolving drama Newton is represented as an anti-hero who continues to lose ground to the more potent Boehme. Even Pitcairne recedes, although Cheyne could never reject the Symbolic Father. Newton had been a pillar of Cheyne's early intellectual life – but not because he was any type of 'Saviour'. Furthermore, there is no evidence whatever that Cheyne read Newton's posthumously published prophecies, neither the *Observations upon the Prophecies of Daniel and the Apocalypse of St. John* nor the manuscript about the conversion of the Jews,[102] nor is there any evidence that Cheyne was curious about this aspect of Newton's thought. Besides, even if Cheyne had read these works, it is doubtful that his quasi-Behmenistic attitude to the laws of universal attraction would have changed. Newton, for Cheyne, was too whimsical in his wedding of science and religion, perhaps as a consequence of the way he was lionised by the whole of England. Cheyne had enjoyed no such instant success; he argued that the millennium was here, that there could be no doubt it had started. He had been old enough in the 1690s to witness, and then to remember, its first appearances. He had personally watched the events of 1706 and reasoned that they coincided with unprecedented brilliance in the mathematical sciences. Newton's appearance as the most perfect mathematician the world had ever known seemed to be evidence of the Deity's providential intentions, even though Cheyne was personally (and obviously) less awed by Newton's achievements than were most Englishmen; and he viewed the rapid succession of several mathematical geniuses – Pitcairne, Newton, the Bernouillis – as an important millenarian clue.

What remained, Cheyne believed, was to integrate medicine – 'queen of the sciences' – into this state of mathematical perfection.[103] Finally in the scientific sphere, Cheyne's role in the development of medicine is clearer than is his precise millenarian niche. I would even go so far as to contend that it is perilous to omit him from any 'Whig history' of eighteenth-century medicine: for this is one category in which he shines constantly. In what we today approvingly call holistic medicine, he may be the most important spokesman of the century. He not only developed a theory but advocated a therapy as well.[104]

We must also ask what Cheyne's career reveals about the theology of the period. This issue is far less equivocal than the scientific one because the emphasis of recent eighteenth-century studies has been on the century's so-called ever-increasing secularism. Historians have been willing to concede to the occasional appearance of Quietism, chiliasm, and millenarianism as a backdrop on the stage of ordinary life in the period; but few historians other than historians of religion have acknowledged these appearances as the period's underbelly. Yet Cheyne's career demonstrates that more radical enthusiasm existed – even among 'the wealthy and the great', as Pope might have said – than has been thought. The important question, then, about Cheyne's theology is not precisely of what version it was, but rather how it related to that of his contemporaries, and how it grew hand-in-hand with his scientific and medical hypotheses. Clues must be drawn from his life-long attachment to the Aberdeen Quietist group centred on George Garden and James Keith, as well as from Cheyne's tropes which thrive on a principle of analogy and which suggest a symbolic rather than scientific imagination.[105] Yet Cheyne does not fit the labels currently used by historians: Platonist, Quietist, Chiliast, Philadelphian, Behmenist, French Prophet, English Prophet. In a sense he was all, yet paradoxically none of these; and his personal theology, to the degree that it can be isolated, was a blend of these. But he cannot be cavalierly labelled by any of these tags without explanation of his life and his particular constellation of beliefs. For example, if one must label, then it is equally accurate to consider him a neo-Pythagorean or neo-stoic, for Cheyne certainly practised aspects of Pythagorean and Stoic religion in his daily life. Perhaps the point to be gathered without belabouring it is that we have been coerced into dividing the religious sensibility of the early eighteenth century into opposed camps of traditional versus dissenting religions, while neglecting what I am calling 'the underbelly of religion': the great *diversity* of types of radical enthusiasm. Moreover, we are willing to acknowledge the influence of Shaftesbury as a Platonist and Stoic, but not of a Cheyne, altogether different though his influence was. Our new sense of the widespread activities of the French prophets in England will eventually change this bipolar thinking, but it may be a decade or two before this recent research is assimilated into eighteenth-century studies.[106]

The final matter pertains to Cheyne's peculiar brand of millenarianism: to his sense of life in the apocalypse as well as to Christ's role in man's eventual redemption. But only his peculiar biography, its startle of ups and downs, can begin to account for the contours of his piety. At the moment of greatest crisis in his life, Cheyne turned to the early Christian fathers and to Madame Guyon and her followers rather than to the traditional Church. Like the Quietists with whom Cheyne associated, he extolled natural and revealed religion in place of the teachings of Jesus, attitudes that earned him the reputation of enthusiasm, and even of Arian heresy. This is why he was 'indicted for heresy' shortly before his death in The Arraignment of George Cheyne . . . for . . . logical heresies.[107] But Cheyne was not an Arian, despite his one-time worship of Whiston. He was a millenarian fanatic, or more accurately, a medico-mystical millenarian. His portrayal as such has not been made, pre-eminently because his early students (Marx, Greenhill, G. D. Henderson) knew almost nothing about his medical career, and conversely, because his close ties to all types of enthusiasts have been overlooked by those (Viets, C. A. Moore, R. Schofield, G. Bowles) who have studied only his scientific career. Cheyne, of course, has been connected to William Law, then the leading British exponent of Boehme, but not to the Quietists, chiliasts and millenarians from the Cévennes, or to their English converts whose influence on British soil, as well as on Cheyne's career, was far more extensive than has been thought. Cheyne's career demonstrates the trend I am attempting to delineate: the effect of radical millenarianism on early eighteenth-century England has been neglected to such a degree that most scholars of its literature and science write and think as if it never occurred.[108] Contemporary Newtonians, for instance, explicate much about Fatio, but say little about his role in the radical millenarianism I have been describing.[109] This Quietist – mystical context is the one in which Cheyne belongs, with one exception: the neo-Stoical cults of the period.

Elsewhere I have written that neo-Stoicism is the least understood intellectual development of the Restoration and early eighteenth century.[110] Professor Funkenstein has brilliantly attempted to describe the Cambridge Stoics (Stoa) rather than the inaccurately labelled Cambridge Platonists. Likewise for Cheyne, aspects of his radical theory of abstinence derive from Stoic and sometimes neo-Pythagorean attitudes, rather than from neo-Platonic beliefs. But Cheyne usually does not name his sources, as a consequence of which his commentators have overlooked his relation to the Stoic and Pythagorean cults of the time.[111] I do not want to engage in unnecessary hermetical classification; I see no reason to classify Cheyne as a 'Stoic medico-millenarian'. But I think it is important to stress the affinities he has with neo-Stoic thinking, and I would want to add that he came by his Stoicism through reading of the early Christian fathers and the Quietists, and when prompted by near-fatal illness and chronic suffering, rather than by reading of the so-called Cambridge Platonists.

In conclusion, Cheyne's radical millenarianism is ultimately paradoxical. On the one hand he advocates extreme abstinence in diet; on the other, he tries to convert a Richardson with all the ardour of a Christian missionary in China: his approach is anything but stoic or passionless. Every restraint Cheyne espouses in diet is contradicted by apparent excess in mystical proclivity. His career is replete with other paradoxes as well. He begins to write, as we have seen, as an avowed mechanist (iatromechanist) and ends as an animist, although this shift, too, has not been studied in the light of his religious beliefs. All the Stoic fervor about relinquishing the needs of the self are contradicted, it would seem, by his intense search for a professional identity in an age (the early eighteenth century) when the physician could appear in almost any typology; his blend of personal interests and misfortunes would not readily fit any profession. A perfect example of these paradoxes is found Cheyne's unrelenting need to take stock and confess: whenever he studied himself he uncovered layer upon layer of Providence that had tended to favour him.

I agree with Robert Schofield that Cheyne's 'progress from kinematic mechanism toward vitalist materialism, by way of Newtonian dynamic corpuscularity, was ... occasioned by religious considerations'.[112] But this explanation does not extend far enough. It omits consideration of these all-important religious contexts in relation to his personal and professional life. Moreover, Cheyne's death in the early 1740s had been said to make of him a transitional figure: his career – the argument goes – lies on 'the boundary' of a vast continental shift between apparently opposed sets of values.[113] But these are not merely the differences of neoclassicism and Romanticism, Mechanism and Vitalism (Animism), or Mechanism and Organicism. Cheyne's mysticism deserves to be studied precisely because of the way in which it accommodates iatromechanism, Newtonianism, and animism. A close look at his theology demonstrates its affinities not only with animism, but with a pantheism of the type the Romantics, especially Coleridge, were to invoke. The Cheyne who at the end of his life espouses a pantheism in which every living creature embodies the specific attributes of the godhead, is hardly the same thinker who wrote mechanistically about fevers or analogically about 'philosophical principles'. Inconsistency is a venial sin for a confirmed mystic; intellectual growth is not. But Cheyne's was intellectual growth incapable of adequate explanation unless his millenarianism is also described.

The Cheyne who wrote just before his death that 'Man is a diminutive *Angel*, shut up in a Flesh Prison or Vehicle',[114] has more in common with Blake's visionary physics and Coleridge's pantheism than with his own early thought. Does this alteration make of him a 'Romantic thinker'? The Cheyne who argues as he approaches his Maker that man's creative powers are somewhat analogous to God's – are 'Something *analogous* to Creative Fecundity'[115] – sounds more like Wordsworth on the creative imagination than the mechanistic Pitcairne or the mathematical Newton he served at the start of his career. Does this hermetic reasoning of Cheyne's render him a transitional figure in the shift from neoclassicism to Romanticism? The Cheyne who laboriously anatomises the mystical revelations he has experienced surely deserves to be considered as more than merely 'one of the early Newtonians'. Yet he has consistently been described as no more than another early Newtonian. The Cheyne who was anonymously extolled by a rhyming scribbler in the 1730s as having tamed 'the Sacred Art ... the Fetter'd Science',[116] was someone far more prone to romantic agony and temperamental pantheism than any labels such as neo-Platonist or early Newtonian suggest. Again, paradoxically, the same Cheyne who finally claimed to have understood himself so well, seems never to have comprehended to what an extent he had been one of England's staunchest anti-luxury campaigners – perhaps *the* fiercest opponent of luxury anywhere.[117] Yet our most recent scholarly survey of luxury never – not even once – mentions Cheyne's name.[118] Finally, the Cheyne who wrote so prolifically about nervous diseases seems never to have realised to what an extent the very diet of abstemiousness he was proposing would directly lead to nervous tension.[119] An age that practically ate itself into the grave will certainly not grow calm and steady if ninetenths of its daily diet is suddenly removed by the likes of a Cheyne.

Perhaps my subtitle indicates the ultimate paradox: 'immortal Doctor Cheyne'. Immortal he has hardly been, for most scholars today do not know who he was. Yet in his own day one could flirt with the idea that history would keep his name alive because he managed, despite his colossal weight, to live on for so long. A type of myopic immortality, then, was granted to him by his contemporaries, perhaps for the wrong reasons.

NOTES

1. George Cheyne, 'The Case of the Author', *The English Malady* (1733), 342, henceforth cited merely as *CA*. It is best to state – here at the start – that I consider this a reliable source for the shape of Cheyne's life. Given the romantic excesses and mystical turns of Cheyne's temperament, it is possible that he exaggerated the pain he suffered during the various crises he endured. It is also possible that during the composition of these memoirs in the early 1730s, his memory distorted or confused events that had occurred three decades ago. But no reason exists to believe that Cheyne intentionally distorted the facts of his past life in order to endow posterity with a better image of himself, or that the agony and anguish of repeated breakdowns and collapses had put him out of touch with a minimal reality. The failure of scholars who have not relied on these memoirs must be attributed to their lack of interest in Cheyne's life, or to their sheer neglect and ignorance of the work. I am grateful to Professors Richard Popkin, M. E. Novak, James Force, and Donald Greene who commented on several versions of this essay, although my acknowledgement should not imply that they agree with my conclusions about Cheyne.

2. There is no twentieth-century biography in any language, although several short articles exist. The fullest of these is H. R. Viets, 'George Cheyne', *Bulletin of the History of Medicine*, XXIII (1949), 435–54. The only lengthy account is W. A. Greenhill's *Life of George Cheyne, M.D.* (Oxford, 1846), but it is anecdotal, based on secondary sources, and often unreliable. An estimate of Cheyne as an 'eccentric genius' is found in J. H. Burton, *History of the Reign of Queen Anne*, 3 vols. (Edinburgh, 1880), III, 429. Though it is limited in its explanations about Cheyne, W. G. Hiscock's (ed.) *David Gregory, Isaac Newton and their Circle: Extracts from David Gregory's Memoranda 1677–1708* (Oxford, 1937) is far more useful. Valuable information is also found in G. D. Henderson, *Mystics of the North-East* (Aberdeen, 1934), still the best source on Cheyne's relation to the circle of George Garden; M.D. Altschule, *Origins of Concepts in Human Behaviour* (Washington, 1977), 53–74, and in L. S. King, 'George Cheyne: Mirror of Eighteenth-Century Medicine', *Bulletin of the History of Medicine*, XLVIII (1974), 517–39. Need exists for a detailed modern scholarly biography.

3. George Sherburn (ed.), *The Correspondence of Alexander Pope*, 5 vols. (Oxford, 1956), IV, 208.

4. *Ibid.*, IV, 46.

5. Cheyne is nowhere mentioned in *The Memoirs of Martinus Scriblerus* (1714), a natural locus in view of the satires of science and the lampoons of doctors and their hypotheses found in this collaborative work, but he is discussed as a 'mathematical authority' in William Wotton's *Defence of the Reflections upon Ancient and Modern Learning* (1705), 10. The reference in Wotton makes it clear that by 1705 Cheyne was well-known to Bentley and others whom Swift attacked in *A Tale of a Tub* (1704).

6 John Gay, *Mr Pope's Welcome from Greece* (1725), p. 133.
7 Edward Young, *Epistle to Pope* (1757), I, 199. Young read at least one of Cheyne's books. In February 1746 he visited Richardson, and 'inadvertently stole one of [his] books'. A few days later he wrote to Richardson: 'On turning over my cargo, I find Dr. Cheyne among my other books.' See H. Pettit (ed.), *The Correspondence of Edward Young 1683–1765* (Oxford, 1971), 231. It is impossible to know which book of Cheyne's this is, but Young apparently knew of Cheyne's *Philosophical Principles* (1705 edition) in *The Force of Religion* (1714).
8 M. Battestin (ed.), *Henry Fielding: The History of Tom Jones, a Foundling*, 2 vols. (Oxford, 1975), II, 605, and n. 73 below.
9 See J. Carroll (ed.), *Selected Letters of Samuel Richardson* (Oxford, 1964), 52 and C. F. Mullett (ed.), *The Letters of Doctor George Cheyne to Samuel Richardson (1733–1743)* (Columbia, 1943), cited henceforth as *Letters to Richardson*.
10 See P. Toynbee (ed.), *The Correspondence of Gray, Walpole, et al.*, 2 vols. (Oxford, 1915), II, 19.
11 J. Y. T. Greig (ed.), *The Correspondence of David Hume*, 2 vols. (Oxford, 1932), I, 12. Ernest Mossner attributes this letter to Dr Arbuthnot rather than to Cheyne but his reasons as given in *The Life of David Hume* (Oxford, 1980), 84, are unconvincing. Hume's self-diagnosis was 'melancholic mania', the eighteenth century term for manic depression, and he wrote to Cheyne as the leading authority on melancholy.
12 Cheyne dedicated his last book – *The Natural Method of Cureing* [sic] *the Diseases of the Body* (1742) – to Chesterfield.
13 Hervey, MSS., 31 Jan. 1738.
14 See C. F. Mullet (ed.), *The Letters of Dr. George Cheyne to the Countess of Huntingdon* (San Marino, 1940). On p. 17 of *Letters to Richardson*, Mullet notes: 'There is no reason to suppose that the letters printed by me constituted the entire correspondence of Cheyne with her ladyship.' A limited search has failed, however, to produce further letters.
15 *The Works of John Wesley*, 2 vols. (1865), XI, 493, entry for 12 March 1742.
16 See the many references, for example, in W. Porterfield's *A Treatise on the Eye, the Manner . . . of Vision*, 2 vols. (Edinburgh, 1759).
17 G. B. Hill (ed.), *Boswell's Life of Johnson*, rev. L. F. Powell, 6 vols. (Oxford, 1934–50), III, 26–7; see also H. R. Viets, 'Johnson and Cheyne', *TLS*, 5 Feb. 1954, 89. Johnson admired Cheyne's *The English Malady, or a Treatise of Nervous Diseases of All Kinds* (1733) as the best account of modern melancholia, frequently cited it, and owned a copy. See D. J. Green, *Samuel Johnson's Library: an Annotated Guide* (Victoria, 1975), 48.
18 John Nichols, *Literary Anecdotes of the Eighteenth Century*, 8 vols. (1812), VIII, 82.
19 'To Dr. Cheyne of Bath. On Reading his Works', *Gentleman's Magazine*, III (1733), 205. For other similar poems, see *Letters to Richardson*, 126–37.
20 Cheyne himself is probably the best source for this figure; see *CA*, 342.
21 Cheyne describes this influence in *CA*, and in *The English Malady* (1733), and there is mention of Cheyne in Pitcairne's correspondence; see W. T. Johnston (ed.), *The Best of Our Owne: Letters of Archibald Pitcairne* (Edinburgh, 1979). See also W. G. Hiscock (n. 2), 23–6.
22 For example, when someone calling himself 'Sir Edward Ezat', attacked Pitcairne as 'Apollo Mathematicus' in *The Art of Curing Diseases by the Mathematicks, According to the Principles of Archibald Pitcairne* (1695), Cheyne counter-attacked brilliantly in several pamphlets and showed how effectively he could demolish the enemy. Cheyne's first book, *A New Theory of Fevers* (1701), which originally appeared as a Latin dissertation, was an attempt to combine Newtonian mathematics with Pitcairnean

mechanics, but the Newtonian disapproved of it and their dispraise in the years 1702–6 eliminated Cheyne from consideration as a future Boyle lecturer.

23 In precisely which year in the period 1695–1706 Cheyne absorbed, rather than paid lip service to, Pitcairne's influence, it is difficult to say.

24 According to Cheyne (*Essay on Health and Long Life* (1724), 47) during the 1690s he had read the works of: Sir William Temple, Willis, Glisson, and Borelli. In *The English Malady* (1733), 78–80, he claims also to have read Glisson, Bernoulli, Molière, Sydenham, and, of course, Newton and Pitcairne.

25 See D. P. Walker, *The Decline of Hell* (1964); E. Tuveson, *Millenium and Utopia* (Berkeley, 1949); K. Thomas, *Religion and the Decline of Magic* (1971); R. T. Vann, *The Social Development of English Quakerism 1655–1755* (Cambridge, 1969); M. C. Jacob, *The Newtonians and the English Revolution 1689–1720* (Ithaca, 1976), chap. iii; P. Toon (ed.), *Puritans, the Millennium and the Future of Israel* (Cambridge and London, 1970); Desirée Hirst, *Hidden Riches: Traditional Symbolism from the Renaissance to Blake* (1964); and still a standard work for the radical millenarian sects of the 1690s, Nils Thune, *The Behmenists and the Philadelphians* (Uppsala, 1948). Craig believed that the millennium would commence when faith left the earth; and by applying the law of the inverse square he calculated the rate at which faith was disappearing.

26 Religious talk abounded in his house, especially on his paternal side; yet there were clergymen on both sides of the family, and Cheyne's mother was Thomas Burnet's second cousin.

27 See the prefaces of both versions of Cheyne's *Philosophical Principles* (1705; rev. 1715). Hisock (n. 2) briefly discusses the reception of work by the Newtonians, and shows how they laughed it staight out of court.

28 See W. T. Johnston (ed.), *The Best of Our Owne: Letters of Archibald Pitcairne* (Edinburgh, 1979), 42.

29 Edinburgh University MSS. 38,305 (anonymous) describes reasons for his departure from Scotland, as well as the basis for his medical degee: 'He's not only our owne countryman, and at present not rich, but is recommended by the ablest and most learned Physicians in Edinburgh as one of the best mathematicians in Europe; and for his skill in medicine he hath given a sufficient indication of that by his learned Tractat *De Febribus* [see n. 22 above], which hath made him famous abroad as well as at home; and he being just now going [*sic*] to England upon invitation of some of the members of the Royal Society.' The invitation may have come from Arbuthnot, whom Cheyne met upon arrival and with whom he quickly became closely associated. The two Scots remained associates for three decades.

30 *A Declaration of the Philadelphian Society of England* (1699), which is discussed by Cheyne's friend Richard Roach in *The Great Crisis* (1725–7), a mystical work containing a description of millenial life. Roach's description on p. 181ff. also bears similarities with Thomas Burnet's position in *De Statu Mortuorum & Resurgentium (Of the State of the Dead and of Those that are to Rise)*. But few copies of Burnet's book were printed before 1727, as Frances Wilkinson. Burnet's literary executor, explains. It would be interesting to know whether Cheyne had seen a copy before 1727. For Roach, see his six volumes of diaries in the Bodleian Library, Oxford.

31 Dr Richard Mead's medical career had rocketed to fame in just this way, and may have been held up as a paragon to Cheyne by his friends in the Royal Society. In any case Cheyne and Mead were friends by 1703.

32 *Letters to Richardson*, 73.

33 A picture of religious street-life at the time is found in D. P. Walker, *The Decline of Hell*, 245ff. and in Hillel Schwartz, *The French Prophets: the History of a Millenarian Group in Eighteenth-Century England* (Berkeley, 1980), 37ff.

34 *Fluxionum Methodus Inversa* (1703). For the response of the Newtonians, see Hiscock (n. 2), 15 and R. Schofield, *Mechanism and Materialism: British Natural Philosophy in*

Medicine and millenarianism

an *Age of Reason* (Princeton, 1970), 59, who comments: 'Scorned by Gregory and attacked by deMoivre, it provoked Newton into publishing his *Tractatus de Quadratura Curvarum* as an appendix to the 1704 *Opticks*. This was its only virtue and it was quickly forgotten . . .'

35 For the reception, see Hiscock (n. 2), 24–5.

36 *Ibid.*, 35. In fairness to Cheyne, it must be noted that the extant evidence comes primarily from Gregory, who was no friend of Cheyne's. Gregory also persuaded Arbuthnot that Cheyne was a poor scientist, and attempted to intercept their association.

37 *Philosophical Principles* (1705), 104.

38 See G. Bowles, "Physical, Human and Divine Attraction in the Life and Thought of George Cheyne", *Annals of Science*, XXXI (1974), 473–88, especially the discussion on 481. Other commentators have been less sympathetic to Cheyne. H. Metzger discovered little of scientific value in *Philosophical Principles* and categorized the book, disparagingly, as 'Neo-Platonist' in *Attraction universelle et religion naturelle chez quelques commentateurs anglais de Newton* (Paris, 1938), 139–53. D. Kubrin is more sympathetic, but has discovered little that is original in Cheyne's 1705 *Principles*; see 'Newton and the Cyclical Cosmos: Providence and the Mechanical Philosophy', *Journal of the History of Ideas*, XXVIII (1967), 325–46. Cheyne's scepticism about 'the mechanical philosophy' is as pervasive in these 1705 *Principles* as is his hermetic notion of analogy. See, for example, *Philosophical Principles* (1705), 12: 'But if any one can tell by what Laws of Mechanism, any one Animal or Vegetable ws produc'd, or from what Mechanick Principles the Planets describe Elliptick Orbits, I shall for the sake of these allow their [i.e. the mechanists'] whole Scheme to be true.'

39 J. LeClerc, *Bibliothèque ancienne et moderne: Pour servir de suite aux bibliothèques universelles et choisies* (Amsterdam, 1715), III, 41–157. Although LeClerc waited ten years to publish the review, he reviewed the 1705 first edition of *Philosophical Principles*.

40 Years later, Cheyne confided to Richardson a progression throughout his life of composition and publication, followed by illness. See *Letters to Richardson*, 69: 'I never wrote a Book in my Life but I had a Fit of Illness after.' There are many other versions of this self-confession in the Cheyne–Richardson correspondence.

41 *CA*, 327.

42 *Ibid.*, 332. George Garden (1649–1733), the religious centrepiece of the Aberdeen mystical group with which Cheyne had been involved in the 1690s, was also 'now extream'ly old', but had not been scientifically distinguished. Besides, Cheyne would not have described the mystical Garden in this way. Other possibilities – Samuel Clarke, John Craig, Newton himself – were either dead or incapable of this description in 1733. Andrew Michael Ramsay (1686–1743), the direct link between Madame Guyon and George Garden's mystical group in Scotland, was hardly old in 1733; see D. P. Walker, *The Ancient Theology* (1972), 232. Five or six others may be possible candidates, but no one fits the whole description and context of the allusion as well as Whiston.

43 *Ibid.*, 332. By this time – 1706–7 – Cheyne probably read much that Whiston had written, and may have heard about or read Whiston's theory of attraction. But Whiston's Boyle Lectures, *The Accomplishment of Scripture Prophecies*, were not delivered until the following year – 1707 – and were not published until 1708. Useful information on Whiston's Boyle Lectures is found in M. Farrell, *The Life and Work of William Whiston* (New York, 1981), 262–6.

44 The most authoritative study to date is the work by Hillel Schwartz (n. 33) from which I have learned much. See also M. C. Jacob, "Newton and the French Prophets", *History of Science*, VI (1978), 134–42, who concludes on the basis of manuscript evidence that Newton was not altogether hostile to the prophets during the first few years of their residence in Britain.

45 *CA*, 331, i.e., narrating the events of the summer and autumn of 1706.

46 I.e., than those "principles" Cheyne had studied in his last book, *Philosophical Principles* (1705).

47 Suffering and healing had, of course, entered everywhere into millenarian discussion, especially in sermons, but medico-theologies had not been delivered in the Boyle Lecture series. There are many reasons for this absence, not least the fact that iatromathematics was unpopular with the Newtonians. It may be, then, that Cheyne now (1706–9) returned to an old project, one whose idea had germinated *c.* 1699–1700, at the turn of the century, when talk of the millennium peaked and shortly after Cheyne read and discussed John Craig's *Theologiae Christianae Principia Mathematica* with the author. The profuse acknowledgements to Craig in the preface of *Philosophical Principles* suggest this chronology and development. It is also possible that John Freind of Christ Church, Oxford, with whom Cheyne had been in correspondence before 1704, played some role in the genesis of *Philosophical Principles*. Whatever the case actually was, Cheyne's statement in the preface that he composed the work 'to record his dialogues in the 1690s, with his former pupil, the Earl of Roxburgh', is inadequate as explanation. The likelier reason is that Cheyne wrote the book to win back the support of the Newtonians.

48 H. Schwartz (n. 33) is right to remind historians of science on p. 250 that many scientists in the apocalypse 'sought a tincture that would cure every disease because it was in essence a microcosm of the soul's union to the body and of God's relationship to Christians. The panacea was the apex of medicine just as the perpetual motion machine was the apex of physics. Universal perfect health, like universal perfect motion, was as close as the apocalypse.' See also such works as anon., *Universal Health . . . made possible for the Poor* (1697), a rare treatise in the Wellcome Institute Library for the History of Medicine.

49 While Fatio was swiftly converted, Gregory was hostile and sceptical of the new prophets; Cheyne's role in the early days (1706–9) is unclear, as is that of Dr James Keith and several other members of the Aberdeen–Garden mystical group who had by now migrated to London, especially Cuninghame (a friend of the Garden group) and Roach, and was engrossed in reading mystical literature. See n. 52 below.

50 In his *Boyle Lectures*, 2 vols. (1739), II, 329, Whiston warned his countrymen to beware of the 'dangerous and false' prophets: 'If any person in this age, who pretend to a prophetic spirit do foretell events, whether of mercy or of judgment, which do not come to pass according, we have the warrant of God himself for their rejection.'

51 *CA*, 349.

52 On 21 May 1709 Cuninghame wrote from Bath that he 'had recovered to a miracle', and did not recognise himself 'to be the same man I was some weeks ago', see National Library of Scotland MSS. 493, 73, and G. D. Henderson, *Mystics of the Northeast* (Aberdeen, 1934), 192.

53 It is impossible, on the basis of extant material, to determine with accuracy when Cheyne first read Boehme, but he certainly knew his works by 1714–15; and I suspect, on the basis Cheyne's friendships and associations in Bath in 1709, that he knew Boehme's works by 1709. Cuninghame may have introduced Cheyne to Boehme's works when they met during that spring. For the dissemination of Boemhe in England during this time, see S. Hutin, *Les Disciples anglais de Jacob Boehme aux XVIIe et XVIIIe siècles* (Paris, 1960), chap. 2.

54 See H. Schwartz (n. 33), 157–8.

55 The date had been announced in Whiston's 1707 Boyle Lectures, publicized in 1708 in his *Account of Scripture Prophecies*, and repeated as an accurate calculation in his *Literal Accomplishment of Scripture Prophecies* (1724).

56 Roach, like Cheyne and Cuninghame, suffered a major illness in his early adulthood which he interpreted as Providential, as 'an Internal Call to a more silent Attendance

Medicine and millenarianism   111

on the Powers of the Work of the Kingdom to come', see the unpublished Roach Diaries, II–VI, *passim* and Schwartz (n. 33), 195–8, where they are discussed. Cheyne and Roach may have been brought together by Cuninghame sometime in 1709; some of their correspondence, still unpublished, was at Calladen, Scotland, in the Garden archives, until the house burned down in 1985.

57   Roach Diary, II, 304ᵛ.
58   A salt compound based on *sal ammoniac*.
59   Cheyne probably obtained the book from his old friend Dr James Keith in London, the main link between Pierre Poiret in Leyden and Cheyne in Bath. The typical route for Cheyne's dissemination of Quietist literature was this: Pierre Poiret (Mme Guyon's secretary and disciple who now wrote prolifically in semi-seclusion in Leyden) → the firm of J. H. Wetstein (the Swiss Protestant printer and bookseller in Amsterdam–Cheyne could not pronounce or write his name and continued to refer to him as 'Western') → Paul Vaillant (the French Huguenot printer and bookseller in London) → Dr James Keith (the Scot and London friend of Cheyne) → Cheyne (Bath). Cheyne then circulated these books throughout England, as is known from the correspondence of James Keith and Lord Deskford. In his diary for 28 May 1743 John Byrom, Law's loyal disciple, commented on this fascinating Anglo-Dutch network in his journal: 'Dr George Cheyne . . . was always talking about naked faith, pure love', and Byrom explained that Cheyne had been 'the providential occasion of his [Law's] meeting or knowing Jacob Behmen, by a book' which Cheyne had sent to Law; see H. Talon (ed.), *Selections from Byrom's Journals and Papers* (1950), 221. The 'book', Metternich's *Fides et Ratio collatae, ac suo utraque loco redditae, adversus J. Lockii* (Amsterdam, 1708), was translated into English in 1713 as *Faith and Reason Compared; Shewing that Divine Faith and Natural Reason Proceed from Two Different and Distinct Principles in Man*. See also S. Hobhouse, '*Fides et Ratio*: the book which introduced Jacob Boehme to William Law', *Journal of Theological Studies*, XXXVII (1936), 350–68, where Cheyne's role is acknowledged. Cheyne's enormous activity as a transmitter of Quietist mystical literature from Holland to England, and from Jews to Christians, has been overlooked, even by the most erudite recent scholar of the Dutch book trade; see I. H. van Eeghen, *De Amsterdamse Boekhandel 1572–1795*, 6 vols. (Amsterdam, 1960–78). In fact, he, together with Dr Keith, was actually the main disseminator in the early eighteenth century, as Samuel Richardson, the printer–novelist, well knew.
60   But it is important to ask whether this assumption in itself, and without further consideration of the contexts and facts of Cheyne's life, renders Cheyne a Platonist or neo-Platonist. Cheyne did, of course, write a short poem in rhymed pentameters 'On Platonism', which deals with conventional Platonic love, but it refers to neither analogies nor causes nor a Platonic cosmology. See the manuscript collection of poems collected by Charles Parr Burney in the British Library, Burney MSS. 390 fol. 8b.
61   *Philosophical Principles* (1715), part II, 46.
62   Ibid., part I, 47.
63   The book was in part autobiographical, as Cheyne was now 'in a regular Fit of the Gout' (*CA*, 346). Perhaps the stinging criticism of the Newtonians in 1705–6 had not yet been forgotten, for Cheyne avoided all metaphysical claims here.
64   *CA*, 343.
65   Ibid., 346.
66   See *Pythagoras his Mystick Philosophy reviv'd, or the Mystery of Dreams Unfolded* (1691). Tryon, a constant reader of Boehme, sustained a 'crisis of the spirit' in 1657 partly as a result of his reading of Boehme, after which time he recommended vegetarian diets similar to those later advocated by Cheyne. In the 1690s Tryon joined up with some of the London Philadelphians; a splinter group formed calling itself 'Tryonists',

reading the works of Madame Guyon, and practising abstinence in diet; see *Memoirs of the Life of Mr Thomas Tryon, later of London, Merchant* (1705), which appeared a few weeks after the publication of Cheyne's *Philosophical Principles* (1705) and on the advent of Cheyne's 'great crisis'. Cheyne refers to several Pythagorean cults (*CA*, 368), but I have found no evidence to suggest that he had heard of Tryon or the Tryonists or read their works. Benjamin Franklin recounts in his *Autobiography* how he became 'a Tryonist' during his youth. Presumably, Franklin absorbed the vegetarian aspect and neglected the Behmenistic–Quietist strain of Tryon's thought.

67 In old age Byfield published *Directions tending to Health and Long Life* (1717), a book advocating a modicum of the abstemiousness Cheyne insisted upon. Byfield was also the author of a number of medico-millenarian works such as *The Christian Examiner* (1720).

68 See *CA*, 349. These included the luminaries one would expect: Arbuthnot, Noel Broxholme (Pope's physician in London), James Douglas, Richard Mead, and John Freind: Cheyne's old friend at Christ Church in Oxford who had been one of the first Pitcairnean proteges and whom Cheyne mentions with gratitude in the preface of *Philosophical Principles*. But Cheyne apparently did not consult John Freke, by 1723 Richardson's physician and a great friend of Dr James Keith, Cheyne's main supplier in London of Quietist literature arriving from Wetstein's firm in Amsterdam. I have searched in vain for manuscript notes these physicians may have scribbled while treating Cheyne, on the grounds that such materials could illuminate the specific nature of Cheyne's ailment. My own diagnosis is that Cheyne suffered primarily from what we would call manic depression, but that as he aged this psychiatric condition was aggravated by chronic cardiac arrest. I date the onset of cardiac-pulmonary arrest c. 1722–4, around the time of his second 'crisis', when Cheyne was in his early mid-fifties.

69 See n. 38 above.

70 Clifton Wintringham translated in into Latin with extensive commentary. It also appeared in French and German. Seven years after Cheyne's death in April 1743, Edmond Litton, a self-styled disciple of Cheyne's synthesised its argument in *Philosophical Conjectures on Aereal Influences, the Probable Origin of Diseases* (1750).

71 See G. Nelson, *The Nature, Cause and Symptoms of the Gout: as stated by Dr. Sydenham, Cheyne* . . . (1728).

72 *CA*, 352: 'Upon the Whole, as in my *Nervous* and *Scorbutical* Disorder. I had continued my Milk, Seed, and Vegetable Diet, with proper Evacuations, for above two Years [1727–9], before I obtain'd a compleat Recovery, so in this last illness, I had observ'd the same Regimen near twice as long, before my Health was perfectly established.'

73 It had been building up for four years. In 1724 two anonymous books hostile to Cheyne appeared: *Remarks on Dr. Cheyne's Essay on Health and Long Life. By a Fellow of the Royal Society* and *A Letter to G. C., M.D., Occasion'd by his Essay on Health*. In 1725, Edward Strother, M.D., vigorously attacked Cheyne in *An Essay on Sickness and Health in which Dr. Cheyne's Mistaken Opinions in his late Essay are . . . taken notice of*, and someone merely calling himself 'Pillo-Tisanus' published *An Epistle to George Ch – ne, M.D., F.R.S. Upon his Essay on Health*, which ridicules every aspect of Cheyne's writings, especially his 'stilted style'. Also in 1724, John Wynter, a somewhat jealous Bath rival, published a tepid appraisal of Cheyne's milk and seed diet in *Cyclus Metasyncriticus*, which Richardson's brother-in-law, James Leake, printed. Other works discussing Cheyne during the late 1720s are listed by F. Shum in *A Catalogue of Bath Books* (1913) 5–9. Negative criticism of this type continued to be published to the end of Cheyne's life. In three separate numbers of *The Champion* (15 Nov. 1739, 17 May and 12 June 1740), Fielding ridiculed Cheyne's ungrammatical style, although he dropped the charge when referring to Cheyne fifteen years later in *The Journal of a Voyage to Lisbon* (1755), 'Sunday, July [14]'. Such harsh criticism extended into the 1760s and 1770s, especially in unpublished correspondences and

after Cheyne had long since been dead. For example, John Rutty, the Quaker physician, condemned Cheyne's diet to William Clark, the Wiltshire physician and author of an interesting treatise on psychosomatic illness; see Rutty's letter to Clark dated 8 Aug. 1773 in Rutty MSS., Society of Friends, Friends House, London, case 32.

74 Four years later, in 1733, he wrote in the preface of *The English Malady* (iii): 'I have been slain again and again, both in verse and prose.'
75 *The English Malady*, iii.
76 CA, 325.
77 *The English Malady*, ii.
78 C. A. Moore, *Backgrounds of English Literature 1700–1760* (Minneapolis, 1953), 179.
79 *The English Malady*, i.
80 Although a best-seller, *The English Malady* was virtually unknown in certain quarters for over a decade. For example, John Morris wrote in his *Observations on the Past and Present State of the City of London* (1751) that 'hypochondriasis' and the 'hysterical passion' were clearly the two main 'forms of lethargy', then known, but he wrote unaware of Cheyne's analysis in *The English Malady*.
81 This work is not the posthumously published compilation entitled *Dr. Cheyne's Account of Himself . . . (1743) His Remarks upon Pythagoras, Cornaro, Sir Isaac Newton, the famous Mr [William] Law . . .* (1743), which was edited by John Campbell (1708–75). Cheyne had read Campbell's *Voyage to the Levant . . . and the Abyssinian Empire* (1739) as a consequence of his belief that 'our [best] diet is *Eastern*' (*Letters to Richardson*, 121). Early in 1743 Campbell had finished a translation of Johann Heinrich Cohausen's (1665–1750) Latin *Hermippus redivivus: or, the Sage's Triumph over Old Age and the Grave. Wherein, a Method is Laid Down for prolonging the Life and Vigour of Man*, and wished to publish extracts before his complete translation appeared in 1744. Cohausen may have used Cheyne's death in the spring of 1743 as the rationale for a brief compilation which he called *Dr. Cheyne's Account*, and which concludes with a sample of his forthcoming translation.
82 Although Cheyne and Fielding never met, so far as is known, Cheyne took all sorts of personal liberty with Richardson, even reassuring him that he was physiologically *beyond* madness (*Letters to Richardson*, 94), and contending that Richardson's physiology had predetermined his literary destiny and medical condition: '. . . your constitution is not like Dr. [Stephen] Hales's: you are short, round, and plump; he is taller, and very thin and uses a good deal of Exercise' (*ibid.*, 70). In *The English Malady* (366–7), Cheyne generalises a monolithic law of physiological determinism: '. . . none have it in their *Option* to choose for themselves their own particular *Frame* of Mind, nor *Constitution* of Body; so none can choose his own Degree of *Sensibility*. That is given him by the *Author* of his *Nature*, and is already determined . . .' The consequences of this 'law' have yet to be absorbed by students of the cults of eighteenth-century sensibility.
83 CA 370
84 *Ibid.*, 367
85 *Ibid.*, 367
86 H. Schwartz (n. 33) perceptively inquires if 'one might, applying Erik Erikson's developmental schema, associate kinds of millenarian ethos with stages of psychological development. For example, the ethos of judgment might be attractive to those who wish to resolve the issue of trust vs. mistrust, the ethos of cataclysm might appeal to those who must resolve the issue of initiative vs. guilt, the ethos of pentecost might be advocated by those resolving the issues of identity vs. identity confusion, and the ethos of the New Jerusalem might be taken up by those perplexed by the issue of intimacy vs. isolation' (261–2). Such a suggestion, no doubt, is fraught

with peril, but it is interesting to notice how well Cheyne's career fits the schema of the last category. *Intimacy* v. *isolation* continued to be the major dynamic issue of his adult life, even at the geographical level: whether to live in Scotland or England, whether to live in Bath far away from the booksellers and printers, or to expose himself to the excesses, luxury, and illness-producing conditions of London, etc. The only caveat is economic dislocation. In the Middle Ages and Renaissance, millenarian fervour, especially in its hysteric versions, was often the result of severe and sudden economic deprivation, even within one generation. Cheyne had certainly been 'dislocated' in this sense: moving from his parents of middle income to the estate of the fabulously wealthy Earls of Roxburgh, and then on to the Newtonians in London and to poverty in Bath where his finances fluctuated as much as his health. Only in the last two decades of his life, from about 1730 onward, did economic stability manifest itself.

87 See n. 14 above. In the winter of 1741–2, the Countess of Huntingdon wrote to her husband from Bath that she had been engaged 'in most pious and religious conversation' with Cheyne, who had been 'talking like an old apostle. He really has the most refined notions of the true spiritual religion I ever met with.' See *Hastings MSS*, III, fol. 32. Cheyne sent her many of Poiret's Quietist books.

88 *Letters to Richardson*, 104.

89 *The Natural Method of Cureing* [sic] *the Diseases of the Body and the Disorders of the Mind depending on the Body* (1742).

90 *Letters to Richardson*, 107.

91 *Ibid.*, 111.

92 I.e., Poiret's *Bibliotheca Mysticorum* (Amsterdam, 1708), which had been translated into French and which printed excerpts from the Baron de Metternich, the German adherent of Mme Guyon who has already been mentioned. Later in the letter Cheyne states that he owns Poiret's 'Catalogue of Mystic Writers', which he describes as 'finely and elegantly painted in a small Octavo in Latin' (*Letters to Richardson*, 111). Cheyne then explains to Richardson where Poiret's book can be obtained, pointing to the specific network described in n. 59. The significant matter here is not the specific work by Poiret, but rather Cheyne's intentions in the project. For many years now, Cheyne had dreamed of continuing in the footsteps of Garden and Ramsey by disseminating Quietist literature throughout England. Now he hoped to obtain Richardson's assistance – it must not be forgotten that Richardson was first and foremost a prolific printer – in the somewhat underground Anglo-Dutch network already delineated. Cheyne's intention was not merely reaching the William Law to whom he had already sent many Quietist books. Now, in 1742, he also hoped to convert to Quietism naturalists such as the young Richard Symes, eventually the author of *Fire Analysed* (Bristol, 1771), and a large group of scientific disciples in the Bath-Bristol area. Thus, when Law published an *Appeal to All that Doubt* in 1742, the stage was set for Cheyne. Law's book introduced Behmenism unequivocally into natural philosophy. What remained was Cheyne's persuasion of Richardson to print an English translation of Poiret in a cheap single volume that could conveniently be sent through the post. As Cheyne was dying in March – April 1743, he continued to dream of the fulfilment of his plan. It was his last project, his private version of 'rational Enlightenment', or as proximate to Enlightenment as he would come while on earth.

93 *Letters to Richardson*, 124. This work is another of Poiret's Bourignonist compilations; it describes the life and works of Mme Guyon and was printed by Wetstein in Amsterdam.

94 The state of Newton's papers at the time of his death, and the precise number of manuscripts left, remains mysterious although his relation to Boehme has now been admirably studied by Betty J. T. Dobbs in *The Foundations of Newton's Alchemy* (Cambridge, 1975) pp. 9–12. Also important is Stephen Hobhouse's discussion of

Medicine and millenarianism 115

Cheyne as the link between Newton and Law in *Selected Mystical Writings of William Law . . . and an Enquiry into the Influence of Jacob Boehme on Isaac Newton*, with a foreword by Aldous Huxley (2nd ed. rev., New York, 1948), 397–422. Walton's manuscript copy of his *Notes and Materials* is found in Dr Williams's Library, London; for Cheyne see Walton MSS. Book 1118 (I, i.38).

95 As J. Sekora has demonstrated in *Luxury: the Concept in Western Thought, Eden to Smollett* (Baltimore, 1978), although he omits medical literature, the all-important messianic and millenarian tradition of luxury, and, perhaps more consequentially, neo-Stoic and neo-Pythagorean attacks on luxury by Cornaro, Lessius, Tryon, Byfield, and – of course – Cheyne. Sekora does not acknowledge to what an extreme degree luxury is a psychological rather than a physical state, and therefore why its religious strains and components are of crucial concern to the historian of luxury.

96 A. Friedman (ed.), *The Life of Richard Nash* in *Collected Works of Oliver Goldsmith*, 5 vols. (Oxford, 1966), III, 364. For Cheyne in *The Bee*, see *ibid.*, I, 400.

97 *Letters to Richardson*, 81.

98 Karl Marx – not the Marxist but the illustrious nineteenth-century professor of medicine at the University of Göttingen and the prolific commentator on Blumenbach – wrote a Lucianic 'letter-to-the-dead' Dr Cheyne, which James Mackness translated into English and published in *The Moral Aspects of Medical Life, consisting of the 'Akesios' of Professor Karl Marx* (1846), 34–46. Here Marx incorrectly addresses Cheyne as a fanatic 'Quaker, who belonged to that respectable body of Quakers', Marx claims to have been profoundly moved by Cheyne's two most outstanding qualities: his 'Quietist aversion to all personal strife' (36) and his 'peace-loving disposition' (38). Neither quality perfectly tallies with the facts of Cheyne's diversified career, but the notion of an irenic personality at the root of his temperament is worthy of consideration.

99 *Letters to Richardson*, 81.

100 I use the vexed term hermetical as it has recently been developed by Dobbs in *The Foundations of Newton's Alchemy*, already cited and F. Yates in *The Rosicrucian Enlightenment* (1972). Unfortunately, there is no book such as K. Thomas's *Religion and the Decline of Magic* (1971) which deals primarily with England in the period of Cheyne's adult life (1695–1743), nor is R. S. Neale's *Bath: a Social History 1680–1850* (1981) of any help in these matters.

101 Cheyne was in fact more self-reflective about his use of analogy than any other millenarian thinker I have encountered in the early eighteenth century. His writings abound with comments about his self-consciousness in the use of analogy, and even his rhetorical tropes ae worthy of scrutiny when he is in this reflective mood. See, for example, *The English Malady* (1733), ii.

102 See F. Manuel, *The Religion of Isaac Newton* (Oxford, 1974), 99–104. Manuel comments on p. 35 that Cheyne's 'new-found principle of Reunion with God, analogous in the system of intelligent beings to the principle of attraction in the material universe, was too saturated with religious Neoplatonism for his [Newton's] taste'. True, but Manuel seems unaware that Cheyne's pietism led him to far more byzantine beliefs than mere neo-platonism.

103 The idea that mathematics and medicine, in the Greek sense both Apollonian activities, represented the pinnacle of the sciences was a commonplace of seventeenth- and eighteenth-century thought. See Gideon Harvey, *The Vanities of Philosophy and Physick* (1699).

104 Cheyne continued to argue that *both* body and mind – in this priority – had to be sound for health to obtain. In the *Essay on Health and Long Life* he comments: 'When I see a gloomy, melancholy, heavy, stupid, thoughtless, joyless creature, much more a whimsical, anomalous or libertine . . . I conclude him in a bad state of health, under a dangerous bodily disease, or under a perpetual mal-regimen, which will

soon terminate in one, whatever appearances be to the contrary, and, sooner or later, I have been always confirmed in the justness of this opinion . . . For I am convinced that calmness, serenity, cheerfulness, and common-sense . . . are the constant attendants and only infallible symptoms of perfect bodily and intellectual (or of *sana mens in corpore sano*) health.'

105 The analogical frame of mind and the type of imagination stimulating it has been ignored for scientific thinkers in the period 1680–1780. Here Cheyne is a natural candidate who ought to be included in the continuum of thinkers pronouncing about the natural world from Newton and Whiston to Priestly and Erasmus Darwin.

106 A clear example is found in literary historiography of the period. Even the most serious scholars of English literature who toil today in this period write as if there had been only the extremes of traditional and dissenting religion. This situation may perhaps change when the work of H. Schwartz (n. 33) is assimilated. Schwartz's monograph, *Knaves, Fools, Madmen, and that Subtile Effluvium* (Gainesville, 1978) is also important in this context, but it does not exhaust this response, pro and con, to the French prophets. Much more work remains to be done.

107 The work is pseudonymously signed by 'T. Johnson', and is ultimately disappointing, despite its title, because the 'heresy' focuses on Cheyne's grammar rather than his mysticism: '. . . the English language has had more Violence done it by a very great and eminent Physician, George Cheyne . . . [who] hath so mangled and mauled it, that when I came to examine the Body, as it lay in Sheets in a Bookseller's Shop, I found it an expiring heavy Lump, without the least Appearance of Sense' (34).

108 Yet another example is found in the commentators on *A Tale of a Tub* (1704) who discuss this complex satire as if it had been written in a religious milieu that consisted only of Anglicans and Puritan dissenters, without appreciating the cults of mysticism and millenarianism that flourished while Swift was writing (1696–1704). Another instance is found in secondary writing about Swift's published predictions (e.g., *Predictions for the Year 1708: a Famous Prediction of Merlin* (1709); etc.), which fail to understand the millenarian context of these prophecies. Swift's adult life from 1690 to 1710 needs to be reconsidered against this background.

109 See F. Manuel, *A Portrait of Isaac Newton* (Cambridge, 1968), 274, who merely notes this about Fatio 'By the time Newton became President of the Royal Society, Fatio had fallen into disfavor, though he lingered on the sidelines for a few years.'

110 See G. S. Rousseau, 'Science', in Pat Rogers (ed.), *The Context of English Literature: the Eighteenth Century* (1978), 192.

111 Scholars such as H. Metzger (n. 38) and F. Manuel (n. 102) considered Cheyne a neo-Platonist for four main reasons: (1) they were unaware of his central role in the Anglo-Dutch dissemination of Quietist literature; (2) they were apparently unaware of his involvement with the leaders of the French prophets; (3) they overlooked or were unaware of his ties to the mystical Garden – Ramsey group and to the type of millenarianism is fostered; (4) they wrote with little appreciation of Cheyne's health and bizarre personal life. In bewilderment, then, they grasped at the label neo-Platonist in the hope that this catch-all would sum up the many conflicting tendencies they despised. All they knew for certain was that Cheyne had fallen out of favour with the Anglican Newtonians.

112 R. Schofield, *Mechanism and Materialism, British Natural Philosophy in an Age of Reason* (Princeton, 1970), 62.

113 Ibid., 61–3.

114 *The Natural Method of Cureing* [sic] *the Diseases of the Body* (1742), 79.

115 *Essay on Regimen* (1740), 270.

116 *Gentleman's Magazine*, VIII (1738), 136, anonymous poem.

117 Vicesimus Knox, the physician and commentator who wrote at the end of the

eighteenth century, seems to have comprehended Cheyne's extraordinary contribution to countering the immense gluttony of the epoch, as well as the degree of hostility shown him by his contemporaries. See *Personal Nobility* (1793), 90.

118 See Sekora (n. 95).

119 Why did so many nervous diseases proliferate after mid-century? The situation vis-à-vis nervous ailments grew so serious by the 1780s that James Makttrick Adair, the physician with whom Burns the poet took walking tours, replaced Cheyne at Bath as a 'nervous doctor' and wrote book after book about the reduction of food intake as a direct cause of depression and anxiety.

The answer to the question about proliferation involves the growth of personal stress and the newly intensified lifestyles of the Industrial Revolution. The matter is that the phenomenon semiotically captured by the word 'nerves' and its cognates (nervous, neurasthenic, neurotic, etc.) touched on most aspects of life then and was not confined to a few zones, nor did it affect merely the wealthy and professional classes. As the pulse of daily life quickly gathered momentum in the eighteenth century, as every form of transportation and communication rapidly improved, as cities began to experience their first waves of urban sprawl and the growth of crime, as personal health made strides and longevity increased, most people – even poor people – found themselves living better and more dangerously and paying a high price for such over-extension and over-commitment. The phenomenon represents the beginning of what we call, in the vernacular, stress and stress-related conditions, but even now, in the 1990s, the topic stress has not enjoyed an academic discourse outside medical schools and health clinics. The humanists have been silent on the matter, not even recognizing that the history of stress awaits writing, just as the discourse of the body, gender, sex, sexuality, race, health, and diet did in the 1980s. For discussion of the semiotic and linguistic aspects of stress, see G. S. Rousseau, 'Cultural History in A New Key: Towards a Semiotics of the Nerve', in Joan Pittock and Andrew Weir (eds.), *Cultural History* (1991), pp. 25–81.

5

Rationalism and empiricism in Enlightenment medicine

This was one of the first essays I wrote about the history of medicine. I composed it at Harvard at the end of the 1960s just before reading theory and before my encounters with Foucault and his works. I had been engaged in dialogues with various traditional historians of medicine in which we discussed the great potential of their developing discipline to liberate modern medicine from many of its myths and to interrogate its so-called objectivity. I published it several years after writing it in Paul Korshin's anthology, Studies in Change and Revolution: Aspects of English Intellectual History 1640–1800 *in 1972.*

I cannot imagine myself dealing with the material in this way today, now that theory has caused me to doubt the 'truth' of the old history of ideas. I do not abjure the binary categories established here or their conceptual authority in the Enlightenment, but I think my contexts are too narrow. They neglect social history and class ideology to a perilous degree. This writing represents the old history of ideas in which I was trained but which I have since come to distrust.

INTRODUCTION

Any discussion of change in eighteenth-century medicine, theoretical and practical, must take into account two dominant trends: rational and empirical. Both trends are riddled with problems, not the least jagged among them definitional ones, and while such labels as rational and empirical are ultimately inconsequential, the actual changes they describe are not. The historian surveying these changes, however clever and elusive he may be, is perpetually tyrannised by the labels, knowing full well that he cannot really escape them if he is to fulfil his task – charting out the medicine of an entire century. This is all the truer in the case of eighteenth-century medicine because medical historians for over a century have been calling it an age of empirical growth.

A modern philosopher, J. W. N. Watkins of the London School of Economics, writing in another context altogether, has admirably stated the objection: 'Like all questions of classification, the question [when are statements empirical?] has no intrinsic interest. To someone with no axe to grind, what matters about a statement is not whether to label it "empirical" or not, but what it says.'[1] A medical historian, Dr Lester King, has raised similar objections in a recent study of Sydenham:[2] 'The important feature is not whether [Sydenham] was an "empiricist" or not, but rather whether he examined his description and inferences in a critical fashion . . . Sydenham was not nearly so *critical* as we might have expected in one who had such a strong reputation as an empiricist.' While some will object to the term 'critical', it is none the less more meaningful than 'empirical'. The question then is whether one permits modern philosophical criteria to influence one's estimate of the empiricism of eighteenth-century medicine, or whether an historical approach is adopted, thereby permitting the age to speak for itself. Since many modern criteria are irrelevant to medicine of that epoch, the latter approach – after all aspects are considered – is preferable.

Other significant objections exist. Suppose a meaningful definition of empirical criteria as opposed to rational criteria could be established,[3] how then would one go about answering the question 'how empirical was eighteenth-century medicine?' The 'important' doctors presumably would be examined and their medicine classified as either empirical or rational, or 'more empirical' or 'less empirical'; and if a majority of these doctors was deemed empirical (or 'more empirical' rather than 'less empirical') then the medicine of the age would presumably be labelled empirical. The difficulties of such a methodology are immense. How are the 'important' doctors selected? Should 'doctors' be interpreted literally, i.e., as licensed physicians in contrast to surgeons and apothecaries, or should they also include empirics and quacks (Joshua Ward, John Hill, Joanna Stephens)? Should scientists who are not strictly medical men (e.g., Boyle, Newton, Priestley, Wesley) be eliminated although they contributed, sometimes considerably, to the development of medicine? What about borderline cases like Locke, a practising physician and a philosopher, and Smollett, first a surgeon, later a doctor, and throughout his life a publishing writer? Even if agreement existed about the 'important' doctors, how representative of their age were they?[4] Pioneering doctors, like all scientific pioneers, are usually decades ahead of their times. Yet even a single medical man is not all of a piece. Suppose he were empirical in his experiments but rational in his theoretical writings? Or conversely, rational in the preliminary structure of his experiments and an empirical propagandist in his writings? Or suppose – as was often the case – that he was extremely rational in his writing and empirical in his daily medical practice, prescribing medicines by trial and error without the slightest trace of influence from his rational discourses?

Such possibilities are not fanciful: Harvey, for example, now considered by historians of science as one of the two or three leading men of seventeenth-century science, was almost thoroughly empirical in his research but – in the words of his most recent biographer, Sir Geoffrey Keynes – 'still Aristotelian in his practice of medicine'.[5] But Harvey was a thorough-going traditionalist (i.e., Aristotelian) in his methodology despite his insistence on the necessity of observation for solving anatomical problems. Still more incongruous was his cavilling at Bacon, another supposed 'empiricist', of whom he is known to have said 'He writes Philosophy like a Lord Chancellor',[6] in other words, like someone who legislates without knowing his terrain. Furthermore, Harvey, while observational in the cardiovascular problems he explored, was contemptuous of the 'New Science' whose empirical devotees he called 'neoteriques' and outright 'shitt-breeches'.[7] The point illustrates difficulties involved in assessing a single doctor's work (i.e., his experimental research, theoretical writings, daily practice, public communications and personal views), let alone assessing the accumulation of 140 years (1660–1800).

If time and space would allow, studies of every medical man, à la Namier, could be made and the objections stated in the last paragraph partly corrected. It might be possible to determine the interpenetration of rational and empirical methods in each man's work and the disparity, if any, between his professed intention (i.e., statements of purpose, prefaces, introductions, etc.) and actual execution (in research, writing, and actual medical practice). But the difficulty of establishing an adequate definition of the term 'empirical' would remain,[8] as would the problem of disentangling its meaning from older disputes concerning nominalism and realism, universals and particulars, and certain problems in logic that had not really been settled. The overlapping of empirical and rational methodology in eighteenth-century medicine is so extensive that one searches in vain for purists of either camp. There is no mysterious reason for such intimate intermingling: deductive reasoning requires experience from which abstractions (i.e., first principles) must be drawn, and certain abstractions must be known before one can begin even the crudest experimentation and observation. Sydenham, for example, the so-called 'English Hippocrates', derides rationalists in search of first principles (e.g. Riverius's four humours and four temperaments), but in daily practice uses reason and in medical experimentation relies on logic. Conversely, hardly a rationalist lived from 1660 onwards who, although professedly rational, did not indulge in some form, however rudimentary, of empirical observation and who did not reason inductively (i.e., from facts to abstract theories). The matter is one of degree, although more than degree is at stake.

A crude schematic representation of some medical men of the period reveals the folly of any enterprise undertaking to answer the question originally posed about medicine and empiricism:

Doctor	Rational	Empirical
Boyle		X
Sydenham		X
Leeuwenhoek		X
Baglivi	X	
Hoffmann	X	
Stahl	X	
Hunter		X
Cheyne	X	
Boerhaave	X	
Robinson	X	
iatromechanists[9]	X	
iatromathematicians	X	
iatrochemists	X	
Cullen		X
Whytt		X
Haller		X
Brown (J.)		X

This tabulation is severely misleading on many counts: (1) in each case interpenetration of rational activity (i.e., reasoning deductively from first principles) and observation (i.e., some form of experiment or induction from facts) exists; (2) an even number of so-called 'rationalists' and 'empiricists' has been purposely selected, permitting no conclusion about the prevalence of either, while in point of fact chronological trends exist and generalisations about decades or clusters of years can be made; (3) no allowance is made for the disparity of intention and execution in a single man's work nor for different approaches in his different medical activities such as daily practice, laboratory experiment, and writing; (4) most important, the table does not tell us anything about the man's overall medical work, his critical grappling with problems, and his significance (then and now) within the medical community. I would even question the validity of an elaborate scheme that managed to take account of these factors because it would not tell us the two things we most need to know – what the man said and how he evaluated, step-by-step, the observations, descriptions, and inferences he drew.

Dr Lester King, a professional medical historian concentrating his attention on the seventeenth and eighteenth centuries, has written three articles[10] examining the 'intimate intermingling', as he calls it, of rationalism and empiricism in the writings of three important medical authors: Sydenham (1624–89), Hoffmann (1660–1742), Boerhaave (1668–1738). He has demonstrated the interplay of reason and observation in the medical practice and theoretical writings of each, showing us the type of interpenetration and synthesis reached. There is no reason to present his arguments here, for they are available to those who wish to consult them.[11] One need comment only that if King extended his researches to the remaining figures in our diagram, he would very probably find 'intimate intermingling' in each and all schools of medicine, although the precise nature of interpenetration must vary from man to man. We learn from his studies that the reputedly 'empirical' Sydenham 'had many of the very faults that he condemned in the "rationalists" ' and that despite his 'concerning himself less with high abstract concepts and more with specific observations . . . he failed to apply the same degree of critical acumen to his own low-level generalisations and to his own acceptance of fact'.[12] More specifically, Sydenham the theorist, who considered diseases in terms of species and genera, 'was impelled to the concept of "cause" as the hidden essence or quiddity which made a thing what it was, directed its course and development, produced its manifestations and properties' – which is an entirely rationalistic approach. The point to be gathered is that Sydenham's rationalism was intrinsically bound up with his empirical proclivities; he did not know the 'hidden essence' of diseases but urged doctors to observe the symptoms, manifestations, response to medication and the 'proximal cause' of each illness. Hoffmann is summed up as a synthesiser of both tendencies: 'his explanations rested on abstractions that did have some basis in experience and that he thought derived logically from concrete observation.'[13] Boerhaave, the physician most influential on doctors during the period 1700–50, also demonstrates the extent of intermingling especially in his *Academical Lectures on the Theory of Physic* (1751–7). As King shows, '[Boerhaave] accepted a basic distinction between something more or less "given" and something else added thereto by "reason" '.[14] Boerhaave believed that facts learned in experiments do not lie about the truth and that error arises when the researcher incorrectly reasons about these facts; but he never clearly sets forth the ways in which facts and inferences are related in a given experiment. All three scientists stress the need for observation but appear unaware of the spectrum of results that arises when different researchers undertake the same experiment. All three emphasised the importance of 'right reasoning' about these facts and all engaged in experimentation. But for every Sydenham, Hoffmann, and Boerhaave dozens of other medical men, some well-known and almost as influential, reasoned otherwise and never performed laboratory experiments although they observed their patients' response to different medicines.

What King has not done, nor to my knowledge anyone else, is account for the precise types of blended rationalism and empiricism that appeared during the period 1660–1800: the varieties of 'intimate intermingling' in the total medical activity of practitioners, and the reasons explaining how and why one blend yielded to another.[15] This is essentially an historian's task. It chronologically explores the rise and decline of ideas and movements in medicine and shows that other areas of culture (philosophy, religion, economics, politics, aesthetics) influence medical theory and practice. During this period the interaction of seemingly disparate parts of one culture was notable and even an inchoate account of that epoch's medicine must examine it. Naturally the two great antitheses, 'rationalism' and 'empiricism', must be isolated for reasons of clarity, but even more important is an appraisal of those not purely medical factors that enhanced or limited the blend and caused it to take particular forms. In this regard some of what follows is a continuation of R. F. Jones's study of 'Ancients and Moderns', which terminates about 1700.[16]

RATIONALISM

Bacon's influence on the late seventeenth century is still a mystery. Whether one speaks of the actual man, of 'Bacon', or of an elusive 'Baconian spirit', it is almost impossible to know what effect, if any, he and his works had on doctors in the Restoration.[17] Sydenham, once again, is a good example: although rational in his basic assumptions and contemptuous of quacks who had no rationale for their remedies, he called for more accumulation of facts. In so doing he paid lip service to Bacon's championing of observation, but there is little indication in his daily practice and medical writings that the actual teachings of Bacon reached him and his medical colleagues. Boyle, his contemporary, supports the point as well. Although he was not a practising or licensed physician, his influence on Restoration medicine was considerable. Boyle, a first-rate experimentalist, presumably had read some of Bacon's works, and more than likely his methodology was influenced by Bacon's scientific method. Yet despite this influence, Boyle in metaphysics thoroughly embraced an older philosophy that was more corpuscular and atomistic than Bacon's. If other medical men of Jones's 'Bacon-faced Generation' are studied – Thomas Willis, John Locke, John Mayow, Walter Harris, Edmund Halley, Richard Wiseman, etc. – it becomes clear that no hard generalisations can be drawn about the quantum of observation each performs. Some doctors, like Sydenham, distrust reason; others, like Richard Wiseman, a leading man in surgery, are 'empiricists' in the operating room and 'rationalists' in their medical writing: Wiseman surveyed the history of such diseases as gonorrhoea and tuberculosis without performing experiments or producing new 'facts' about these diseases. And yet he radically differs from Dr John Mayow, who demonstrated in a series of experiments that dark blood is changed to bright red when it incorporates igneo-aerial particles, and who confined his writings to publication of these experiments in his *Tractatus Quinque* (1674). Does one label as 'rational' or 'empirical' the writings of Drs John Graunt, William Petty, and the astronomer Edmund

Halley, all of whom compiled vital statistics without which eighteenth-century medical research would have floundered? It would be interesting to know what Dr Freind would have written about the 'rational' element of the 'Bacon-faced generation', especially writers of aphorisms, if he extended his *History of Physick* (1725–6) beyond the sixteenth century. He and other contemporary historians (Le Clerc, Robert James, Lamotte) did not use 'rational' at all, nor ought we except in the most rigorous sense to describe a logical, deductive process in explicitly theoretical writings (e.g., Boerhaave's *Aphorisms Concerning the Knowledge and Cure of Diseases*, 1715), but not to describe statements of intention in letters or in the works of self-professed rationalists who unequivocally derogated experimentalism and distrusted discoveries made through the senses.[18] 'Rationalist doctors' here refers to those men whose writings are in some profound way explicitly anti-experimental. It says nothing about their personal and temperamental allegiances (e.g., Harvey's traditionalist obeisance to Aristotle), nor claims to distinguish those who were or were not aware of their rationalistic leanings.

Such medical rationalists were not uncommon in the Restoration and early eighteenth century, although no adequate means (other than purely statistical ones) now exist to determine their prevalence in the medical world at large. They were often attacked by followers of the 'New Science' who expatiated on 'The Vanity of Dogmatizing',[19] but such attacks indicate how much of a threat they still were, for if their numbers had truly been diminishing no reason would have existed to challenge their methodology. Medical literature published in England 1680–1700 demonstrates how many works were 'rational' in intention, but nothing caused an increase in their numbers more than Newtonian mechanics. At a time when many publishing physicians were adopting Cartesian principles and making harmonious systems out of the principles of anatomy and the physics of mechanics, Newton's mathematics acted as a veritable spur to the rise of new schools of rational medicine – iatromathematics, iatromechanics, iatrochemistry. Beginning in the 1690s, shortly after the publication of Newton's *Principia*, iatromathematical works written by medical men began to appear annually.[20] These treatises quantifying anatomy and bringing Newtonian calculus to bear on physiological problems were written by doctors of all sorts, ranging from men influential in the profession (i.e., in the various Royal Colleges of Physicians and Surgeons) to untrained hack journalists who became the vanguard of a scientific movement then fashionable. The selection of a Briton, Dr Archibald Pitcairne, as a spokesman for the iatromathematicians by the medical faculty of the University of Leyden (then the most progressive medical school in Europe) indicates some of the appeal of these British rationalists.[21]

Basically their attempt was a worthy one, for they could not have known in 1700 that the exactness of mathematics could not ultimately apply to medicine – especially at a moment when relatively little was known about physiology. But their numbers increased despite professional hostility and they had the support of Newton. Like Newton, they tried to revert to a reconsideration of first principles by the introduction of advanced mathematics (especially geometry). Rather than discourse about the metaphysical substance or *vis viva* contained in the animal spirits, they applied geometric equations to determine the rate of flow of animal spirits within the blood. Their activity was referred to by enemies as 'mechanical reasoning', and by the mid-1720s the belief that incompetent doctors had misapplied the method was so prevalent that Dr John Quincy, known for the *London Pharmacopœia*, set out to reform all medicine:

> Mechanical Reasoning is what is much talk'd of now in Physic, and by some perhaps more than it is well understood; but the greatest Number of Professors of Medicine are declared Enemies to it, and make nothing of breaking their Jests upon Angles, Cylinders, Cones, Celerity, Percussion, Resistance, and such like Terms, which they say have no more to do with Physic, or a Human Body, than a Carpenter has to do in making Venice Treacle, or curing a Fever. It is therefore for the Information of both these, that I have been at the pains of showing what *Mechanical* Reasoning is, and proving, that all *Physical Certainty* depends upon the same Principles.[22]

Doctors like Quincy were firmly convinced that '*Mechanical* Reasoning' alone would reveal '*Physical Certainty*' – the incontrovertible, abstract truths about medicine. Their desire was to emulate Newton and endow medicine with the same degree of exactitude achieved in other sciences – mathematics, astronomy, physics. Because 'Newtonianism' was so fashionable in medical and other circles in the first three decades of the eighteenth century and because there were no 'Sydenhams' to oppose these doctors, the iatromechanical movement gained momentum. Doctors who did not make use of mathematics were considered reactionary and those who piled up facts in their laboratories were unnoticed. Such rationalists as the iatromathematicians lavished on 'theories' a prestige not previously enjoyed.[23] The formulation of hypotheses, criteria, and first principles became a professionally accepted activity. Doctors formulating systems of medicine based on mathematics were highly regarded by their colleagues, slightly suspected by non-medical scientists (i.e., biologists, embryologists), and contemptuously viewed by such satirists of 'systems' as Swift, the Scriblerians, and journalists who earned their daily bread by lampooning the 'professions.' For at least thirty years (1695–1725) rationalists maintaining that nonquantified laboratory research was a waste of time reigned supreme among English medical theorists. During the next quarter-century, 1725–50, they continued to claim important doctors as adherents – Peter Shaw, George Cheyne, John Arbuthnot, Nicholas Robinson, Richard Blackmore, Daniel Turner, Edward Strother, and David Hartley.

Rationalists other than iatromathematicians, especially aphoristic compilers, continued to comment on the 'Fallibility of Experience'. Their writings, often disguised as treatises on scepticism, were actually part of a revival of medical dogmatism based on a tradition extending to the Greeks but interrupted by certain seventeenth-century observers. The titles of these works indicate something of the approach of their authors: typical treatises are Theophilus Lobb's *Rational Methods of Curing Fevers* (1734); Hillary's *A Rational and Mechanical Essay on the Small Pox* (1735); Jeremiah Wainwright's *Mechanical Account of the Non-Naturals* (1707; 5 editions by 1737); S. J.'s ['Surgeon'] *Mechanical Dissertation upon the Lues Venera* (1731); William Wood's *Mechanical Essay Upon the Heart* (1735); James Parson's *A Mechanical Enquiry . . . into Hermaphrodites* (1741); and an anonymous *Mechanical Enquiry into Diabetes* (1745). Despite the fact that these and dozens of other similar works prefer clear reasoning to accurate observation, extreme caution must be exercised when generalising about them. Lobb believed that reasoning from first principles was the task of 'every medical author' but also stressed the need of extensive factual knowledge. Hillary, Wainwright, and Parsons performed experiments but insisted that medical systems must be founded on 'a just mechanical way of reasoning'. Even in Hoffmann's writings, which profess to distribute evenly the emphases laid on reason and observation, the result is lopsided: as soon as allowances are made for terminology altered from seventeenth-century Galenists one realises that the observation, dogmatism, and critical rigour brought to bear on his system are surprisingly similar to the theories of older rationalists.

We should not be surprised, however, to discover the apotheosis and beatification, as it were, of 'mechanical reasoning' in an age when Newton proved to the world that the heavens were as regularly ordered as the parts of a machine and when theologians discoursed on the 'mechanical creations of God in the universe'. Dr Richard Mead, a leading practitioner, firmly believed, as did several of his colleagues in the Royal College of Physicians of London, that the best 'remedies of the age were pointed out by mechanical reasoning',[24] an allusion to Boerhaave, the single most influential theorist 1700–40. It is significant that the English translated the words (*Theatro Critico Universal*, 5 vols. (Madrid, 1733–40)) of the greatest living Spanish priest, Benito Feyjoo, under the title *An Exposition on the Uncertainties in the Practice of Physic* – a title revealing as much about the state of English medical thinking as about Dr. Feyjoo's system. True, Feyjoo objected to experiments because 'many conclusions are founded upon only single Experiments, which [are] fallacious in many Respects',[25] and in section vi he presents the pros and cons of bleeding, demonstrating why experiments cannot prove that the practice is legitimate because doctors 'pick the Experience they wish'. His variety of scepticism was prevalent among dogmatic rationalists but so was an opposite tendency to proclaim the efficacy of untested remedies. Dr John Allen wrote to Dr John Woodward, the 'salts and mercury' man satirised in *The Memoirs of Martinus Scriblerus*, about a new cure for visceral disorders: 'By the little experience I have had of this Method on myself and some few others, I am persuaded, it would be of extensive use in the cure of many Distempers, was it practicable.'[26] Historians and compilers surveying the progress of medicine noted that seventeenth-century observationists had been the real pioneers: 'The Experimental Philosophers, who by frequent and well-made Trials and Experiments, as by Chymistry, &c. sought into the Natures and Causes of Things: And to these almost all our Discoveries and Improvements are due; and much more would they have done if they had not fallen into *Theories* and *Hypotheses*, which they forced oftentimes their Experiments to maintain, whether they could or not.'[27] Such decay into the complicated '*Theories* and *Hypotheses* of dogmatic rationalists, Stone believed, obstructed medical advancement.[28] During the 1730s and 1740s opposition mounted in certain camps and what had been in Bacon's lifetime the seeds of 'rational suspicion' (as Boyle later called it) now gathered momentum.

If the mid-century witnessed a decline in iatromathematical medical theory, this was in small part due to 'Newton's' waning popularity and in larger part to a competing trend reacting to speculative medicine, i.e., systems based on deductive reasoning, not on inductive generalities derived from fact. And yet, the very same objections had been voiced all along but not so loudly nor by so many. As early as 1710 Mead (not a pioneering Boerhaave but still an influential physician, especially in supporting inoculation) wrote in the preface of *The Influence of the Sun and Moon upon Human Bodies*: 'I praise reasoning [in medicine], when it is grounded on such principles as fall under our senses, or are proved by experiments, and draws conclusions from manifest premises.'[29] The difficulty in assessing such prefatorial statements is apparent.[30] 'Reason' here includes faithful recording of observation, experimentation, and inductive reasoning from 'manifest premises'; but practice did not always bear out the theoretical assumptions of doctors. There is, in other words, a disparity between the intention of medical men, as stated in the introductory portions of their treatises, and the derivative medicine described in the rest of their books. In the same work Mead noted that 'the two great pillars of medicine are experience and reason' and that 'a *rational theory* will teach a man to apply his experimental knowledge to the various cases that occur'.[31] But for Mead the converse was untrue: experiments in themselves could not teach a man to apply his reason rightly and make it submit to 'the various cases that occur' (i.e., experimental data). Mead, like Riverius and other seventeenth-century doctors, could not imagine that new empirical data would require a new conceptual framework. The limitation of many of these 'rationalists' lay in their sometimes futile attempt to connect observed data and outworn hypothetical concepts. Though they considered themselves 'empiricists' and emphasised the importance of observation, close scrutiny of their writings shows that many were as 'rational' in practice as the iatromechanists. Often those protesting loudest against 'hypotheses' and 'first principles' themselves spent the least time performing experiments.

As late as 1760, Dr. Hillary unabashedly wrote in defence of the rationalist approach to medicine on grounds that experiments were misleading and because of the variety of species:

> Since it is as possible that the *Divine Being* may have given different laws of motion to different *species* of matter, as to have created and given existence to different kinds of matter: And that *Light*, as well as *Fire*, are composed of different elements from those of all other kinds of matter, will appear from a more full and proper examination of their peculiar properties, tho' they are two different beings; and we cannot say that there is not almost as great a variety of species of matter existing in different parts of the universe, as there is a variety of other beings.[32]

Such speculative endorsements of rationalism as this, and others even more garbled, abound in the heterogeneous medical writings of the eighteenth century, but as time passed the observationists gained strength and eventually overtook the allegedly pure rationalists. These doctors pointed out that Nature was irrational and that grave errors arose by assuming she was logical or even reasonable. Religion, especially sermons, supported, however indirectly, the empiricists who stressed the need for gathering facts about 'Nature's way'. And when doctors like Dale Ingram, a typical practising physician of no distinction, commented that every 'Physician is impatient to know wherein the genuine and latent *Principle of Life* consists',[33] he was rebuffed by observationists who told him that the question could not be answered. The assumptions behind the experimental advances of the Hunters, Trembley, Cullen, and on the Continent, of Haller, Wolff, Morgagni, and Sauvages, began to permeate the writings of dogmatic rationalists. Rationalism remained – for example, in the systems of Bordeu and other vitalists – but it could no longer help taking cognisance of the new empiricism.

EMPIRICISM

Dr John Brown, commenting in 1787 on the two main trends (observational and rational) of eighteenth-century medicine, noted that the former was responsible for whatever little progress had occurred and the latter for its retrogression. Echoing the Bible (Hosea, 8; 7), he likened rational systematists to the makers of the Sumerian calf who 'sow the wind, and reap the whirlwind' in one of the most eloquent indictments written by a physician of that century:

> Widely different [from the empiricists] . . . is the mode of inquiry which the philosophers of another description pursue. Superficially surveying, or totally neglecting, the investigation of particular facts, they begin with an inquiry into their ultimate cause [first principles], and, after tedious and fruitless attempts to define, describe, and explain to others, a proposition of which they themselves have no adequate idea, their whole after aim is to reconcile it to the detail of facts. But in this preposterous occupation, they sow the wind, and reap the whirlwind: For they not only find a perpetual repugnance between the phenomena of nature and their imaginary cause, but even, when by much art and labour, they at any time seem to succeed in forcing a connection between their fundamental proposition and a few of the phenomena, still the far greatest part of these reject reconciliation.[34]

Such an affirmative stand would not have been impossible but unlikely during the period 1700–1740 – and the reader will soon note why 1740 is regarded as the very rough and approximate end of a period. By 1740, not much before, a belief grew that observationists were the genuine pioneers in the advancement of medical knowledge. Before that the idea was inchoate, appearing only occasionally, without significant consequences in the medical world, and was underdeveloped in contrast to the elaborate propaganda and printed dogma of rationalists. Mandeville, in his *Treatise of Hysterick and Hypochondriac Passions* (1711), a dialogue between an 'empirical' physician and his sceptical patients, had voiced some of these beliefs years before they became fashionable. During the subsequent fifty years (1711–60) such assertions were more widespread. At the other end of the period, indeed as late as 1770, John Millar, a physician with staunchly anti-rationalist inclinations, still deemed it imperative to attack medical rationalism in his *Observations on the Prevailing Diseases in Great Britain*: 'Every system, however admired, hath been overturned to make way for others, which have only been preferred because of their novelty. Their [the rationalists'] fallacy is now universally acknowledged; physic is cultivated upon a more rational plan, and faithful observation is allowed to be the sole foundation of medical knowledge.'[35] Not surprisingly, Millar's 'more rational plan' was a deathblow to the old 'rationalism'; it was less rational and rather more empirical; it aspired to few *a priori* premises. By 1770 Galenic presuppositions were demolished (they had been in gradual decline for more than a century but would not be quickly obliterated) except in the most provincial parts of Britain, and such movements as iatromathematics were replaced by the new empiricism.

What was really new about this so-called pragmatic medicine? Briefly, it was 'practical' in the attempt to collect as many data as possible about known diseases, their causes, symptoms, most commmon cures. If the seventeenth century – that 'wretched state of physic', as Dr Millar called it[36] – produced a few observational works collecting data, they were scanty in comparison to the number written after 1740. At about this time, a substantial number of works bringing together 'practical observations' on known diseases began to be printed: Richard Holland, *Practical Observations on the Small Pox* (1741); James Nihell, *New Observations Concerning the Pulse* (1741); Joseph Hurlock, *A Practical Treatise upon Dentition* (1742); Fielding Ould, *Treatise of Midwifery* (1742); Stephen Hales, *A Description of Ventilators* (1743); H. F. Ledran, *Practical Treatise on Gun-Shot Wounds* (1743); G. La Motte, *Treatise on Midwifery* (1746); Richard Manning, *The Symptoms, Nature, Causes, and Cure of the Febricula* (1746). Medical geography also appeared and physicians began describing diseases indigenous to remote parts of the world (the tropics, Greenland, Senegal, the American colonies) and to various military professions, the army, navy, miners. Dictionaries defining 'hard' and 'simple' medical words were produced, Robert James's *A Medicinal Dictionary* (3 vols., 1743–5) among the most important.[37] This was also the first era of 'case histories', as scrutiny of *The Catalogue of Printed Books in the Welcome Historical Medical Library: II: Books Printed from 1641 to 1850* (1966) makes abundantly evident – case histories reporting anomalies and typical cases from almost every region on earth and describing symptoms and manifestations. Doctors' guides (*Vade mecum*) and desk manuals were turned out almost annually, and more revised and expanded editions of the *London Pharmacopœia* appeared during the 1740s than in any previous decade. The 1740s also produced a greater number of medical titles than any earlier age and many were 'practical' rather than 'systematic' works.[38] Although it cannot be labelled a cause, the recent death in 1738 of Boerhaave was a contributing factor in the demise of rational medicine: he found no successor to his systematic rationalism. Even if he had, there is evidence that the tide had already turned. More consequential was the new conviction that medicine ought to uncover the *cause* of disease, especially by observing its symptoms. Dennis de Coetlogen, author of the *Universal Dictionary of Arts and Sciences* (2 vols., 1745), expressed the new notion in his definition of

'Physick': 'a perfect Knowledge of the Symptoms of Diseases depends more on *Practice* than on *Theory*.'

Combined with the new effort to gather many data was experimentation carried on especially in teaching hospitals and teaching schools. Between 1740 and 1780, seventy new teaching hospitals were founded in England and such schools as William Hunter's Great Windmill St 'anatomy theatre' proliferated. Again and again, the 1740s are a starting point and the student who surveys the century decade by decade will eventually sense it marks a historical division.[39] London and Edinburgh were no longer the only recognised medical centers. Bath because of her spa and Bristol because of the Hot Well had always been competitors of a sort as had been the two university towns, but now the Midlands and North, especially Birmingham, Peterborough, Manchester, York, and Newcastle, began to develop as well. Typical apothecaries like Simon Mason of Birmingham practised, wrote a few medical works, and exchanged information with other provincial apothecaries. 'The Practical Knowledge of our Hospitals', Mason wrote in *Practical Observations in Physick Wherein are Exhibited the Aetiology of Distempers* (1751,) 'is the greatest Advantage to any practicing medical man.'[40] The cumulative result of these communications was a belief that medicine was a genuine science. In the seventeenth century the question had been hotly debated, but the next century raised it again and convinced itself it had answered it once and forever. Medicine, buttressed by the discoveries of biology, was now deemed 'Queen of the Sciences', coronated by doctors and laymen because of her monumental utilitarian value to mankind. In conjunction with her new status, achieved by the forties, arose the preponderant view (especially among professionals) that this 'Science' was and would remain in her dawn, unlike astronomy or anatomy. Dale Ingram expressed it well:

> ... there are other Professions which must remain as it were in their Dawn, so long as this World exists: Among this last Class we may justly reckon *Mathematics, Physic* and *Surgery*. Tho' in each of these, vast Discoveries have been made, yet what an unbounded Abyss lies before their respective Professors? What they do know, is no more in Comparison of what they ought to know, than a single Atom is, in Comparison of the infinite Expanse.[41]

This was an entirely new conception of the potential of medical science, an idea engendered by structural changes in the medical profession (especially the resolution of long wars among physicians, surgeons, and apothecaries and the ensuing harmony) and by revolutionary technological advances: the building of hospitals, teaching schools, laboratories, and the compiling of data and facts not directly related to medicine but eventually imperative for its furtherance.

What had altered the course of empiricism in relation to medicine lay outside medical textbooks and treatises. Little change occurred in the philosophical theory of medicine, except for the developments surveyed in the section on 'Rationalism'. No philosopher of medicine writing in the 1740s was so concerned about the problem of cause and effect relationships as was Hume, but he had little, if any, influence on doctors of the eighteenth century, as Professor Laurens Laudan has shown.[42] Medical *theory* was still systematic, deductive, and logical. Few theorists existed whose empirical data radically affected their theorising and their hypotheses were rarely the result of what modern philosophers would call intuitive induction. The effect of the so-called empirical philosophers (Locke, Berkeley, Hume, Reid, the Scottish 'Common-sense School') on eighteenth-century medicine was practically nil, and the application of the term 'empirical' for both groups has produced more confusion than clarity. It is historically far more accurate to note that seventeenth-century doctors were as observational as eighteenth-century doctors, and that the only significant differences lay in the amount (i.e., the sheer bulk) of self-professed empiricism and in the bulk of accumulated data. Moreover, I have attempted to suggest that the reason for the eighteenth century's penchant for proclaiming its empirical aims so monolithically was precisely its own self-conscious feeling that it was not faithfully carrying out these intentions.

Such a thesis is difficult to prove and elusive to demonstrate but history bears it out. Some idea of the intermingling of rational systematic medicine and empirical writing may be gained by consulting the annual checklists of printed medical works in the Wellcome Institute of the History of Medicine. If we examine the years clustering around 1740, the transitional period, the point becomes clear.[43]

Year	Rational	Empirical
1738	J. Burton, *Treatise on the Non-Naturals Mechanically Accounted For*; John Marten, *Treatise of the Gout*.	S. Hales, *Statical Experiments on the Sap in Vegetables* (3rd ed.); D. Hartley, *Ten Cases of Persons who have taken Mrs. Stephens's Medicines for the Stone*; J. Hoofnail, *New Practical Improvements and Observations on Some of the Experiments and Considerations Touching Colours*.

The following two years saw a notable increase in the number of medical works printed and these demonstrate the way that self-conscious empiricism was catching up to and overtaking rational treatises:

Year	Rational	Empirical
1739	D. Bayne Kinneir, *Essay on the Doctrine of the Animal Spirits*; Feyjoo, *Uncertainty of Physick* (translated); P. Keir, *Enquiry into the Nature and Virtues of the Medicinal Waters of Bristol*; D. Turner, *Three Tracts* on medical subjects adopting dogmatic views and *Syphilis*.	S. Hales, *Philosophical Experiments*; J. Huxham, *Observationes de aëre*; T. Lobb, *Practical Treatise of Painful Distempers* and *A Treatise on Dissolvents of the Stone*; S. Sharp, *A Treatise on the Operations of Surgery*; G. Thomson, *Syllabus Pointing Out Every Part of the Human System*.
1740	R. Manningham, *Obstetrical Works in Latin*; George Martin, *Essays Medical and Philosophical*; T. Short, *Mineral Waters of Cumberland*; Thomson, *Discourse Concerning the Present State of Physick in Europe* (rational dogmatism); G. Cheyne, *Essays on Regimen* (3rd ed.); W. Stukeley, *Treatise of the Gout* (3rd ed.)	Four treatises on Mrs Stephens's Cures for the Stone; W. Beckett, *Chirurgical Tracts* and *Practical Surgery Illustrated*; W. Cheselden, *Anatomy of the Humane Body* (5th ed.); W. Hillary, *Practical Essay on the Small-Pox*; (anon.), *The Ladies Dispensatory*; Saviard's *Observations in Surgery* (translated).

If the bibliographical terrain of medicine were extended to 1800, one could see a clear pattern of empirical works overtaking others. By the 1770s sufficient numbers of case histories and practical observations of symptoms and cures of diseases had been written – or so it seemed to doctors commenting on this matter – and important theorists (in their own time often referred to as 'medical philosophers') thought the moment might be ripe once again to develop theories by intuitive deduction. But a vast change had occurred. No longer was the smugness of the iatromechanists possible, nor did doctors believe that the deepest questions in medicine really could be answered by quantification alone. It simply had not occurred to the iatromathematicians that their whole endeavour was suspect.[44] William Cullen Brown, summarising in 1804 his father John Brown's contributions to medicine, noted that his real achievement lay in overthrowing the 'hypotheses' (used derogatively) of previous doctors, especially by concluding his medical works 'with a masterly refutation of the once celebrated *hypothetical* doctrine, originally stated by Hoffman [sic], afterwards modified by the late Dr. Cullen, and then taught at the University of Edinburgh'.[45] The confidence of earlier medicine was swept away, as is evident when the methodological passages of medical works written 1760–1800 are examined. Scepticism about *all* hypotheses, regardless of their validity, was deeply ingrained. Doctors now exercised greater caution in promoting systems, as their collective writings show. Drs Cullen, Whytt, and Brown, if read in their entirety, demonstrate little of the confidence of Quincy, Morgan, and Cheyne. Quincy's 'Historical

Certainty, Moral Certainty, and Demonstration',[46] i.e., the earlier physicians' desiderata, were gone. By now a new age of uncertainty principles had arisen. No monumental discoveries, it is true, had been made – certainly none in comparison to Harvey's circulation of the blood or Willis's experiments on the nervous system. But a more rigorously critical type of inquiry was conducted that may not as yet have produced results but which laid the groundwork, the theoretical foundations as it were, for future progress – for nineteenth-century English clinicians, James Parkinson, Thomas Addison, Richard Bright, and a host of other important innovators.

Dr Brown, later known as the founder of 'Brunonian' medicine (i.e., the theory that animals and vegetables are endowed with a principle of *excitability*, which distinguishes living beings from inanimate matter), more or less summarised the changes occurring after 1740 when he declared that the sound medical scientist proceeds 'by a solid, cautious, and broad induction' and later 'ascends to a *fact* [italics mine] which unites them all, and which itself receives illustration and confirmation from each of them'.[47] The precise manner, Brown suggests, in which the scientist proceeds from data to 'facts' distinguishes rationalists from empiricists. The interesting point is that in his own research Brown did not actually execute this inductive plan but thought he had. It does not matter that Brown and his colleagues ignored what Whewell and Mill later highlighted: that there are philosophical difficulties involved in generalising inductively about some white swans, many white swans, and all white swans.[48] What counted was a new sense of the significance of induction in medical research. During the period 1740–80 doctors were not wholly able to put it into practice but they had successfully given it the stamp and seal of their authority. The next generation could continue.

CONTRIBUTING FACTORS

Social factors as well as unconscious assumptions and cultural presuppositions influence medicine in all ages. These assumptions are usually implicit and seldom questioned by the doctors of the age.[49] Together, they played a considerable role in the progress of eighteenth-century medicine, especially in its self-professed crusade toward furthering observation. Nine factors are noteworthy.[50]

1. *Professional feuds.* During the period 1660–1800, particularly in the earlier part of the period, the profession underwent greater social transformation than it had in the previous two centuries or than it would in the next. Feuds among physicians, surgeons, barbers, and apothecaries changed the guild structure of these professional groups and also consumed personal energy that could have been devoted to medical research. Medical

historians have studied these transformations and are correct to conclude that they were social rather than scientific.[51] Bitter conflicts carried out by eighteenth-century doctors in the name of professional standards and ethics are now seen to have been social and economic struggles. These gave rise to pamphlet wars, sometimes on the absurdest of topics, which drained the entire profession and curtailed its gathering of data.[52]

2. *Centres for medical research.* Despite increases in national wealth throughout most of the period, governmental support of medical experimentation (in England as well as Scotland) was virtually non-existent. The Royal Society (chartered in 1660) and the Royal College of Physicians of London continued to be centers for observation, but they operated on small sums when equated in the purchasing power of money.[53] Teaching hospitals opened during the middle and end of the period attempted to encourage observationist medicine – always in the name of medical education – but financial considerations, especially the waning of aristocratic patronage, limited experimenters to inferior laboratory facilities and instruments. Observational medicine made its most significant gains outside the teaching hospitals and in such private schools as William Hunter's in Great Windmill Street.[54] Surgery advanced more slowly than some doctors of the period anticipated for reasons given in items 3, 7, and 8.

3. *'Ancients' versus 'moderns'.* Medically speaking, the 'quarrel' was far from dead although it underwent radical transformation. Many theorists of different persuasions were distracted from a *magnum opus* to defend the ancients or moderns under the guise of a given problem: e.g., whether the moderns had truly improved upon Galen; whether medicine had become a true science in the recent past; whether Galen, Celsus, and Hippocrates had been surpassed by any combination of three modern physicians.[55] Typical among a plethora of medical books written to answer these and other similar questions was Dutens's *Enquiry into the Origin of the Discoveries Attributed to the Moderns. Wherein it is Demonstrated, that our most celebrated Philosophers have, for the most Part, taken what they advance from the Works of the Ancients* (1769),[56] a 459-page scholarly treatise written by a highly regarded physician attempting to prove conclusively that Newton's law of universal gravity was known to Plutarch and Lucretius, and his theory of colours by Pythagoras and Plato; that Linnaeus's discovery of the sexual system of plants was known to Claudian, Theophrastus, and Empedocles; and that even in the liberal arts the ancients had anticipated all modern architecture, painting, and mechanics (i.e., technological advances). Such works coupled to a belief held by many doctors that allegiances ought to be declared, caused polarisation into camps of ancients and moderns. Both groups – and there were several intermediary ones – tended toward antiquarianism and pedantry, especially in their submission to ancient medical works, but their antiquarianism also manifested itself in such

diverse circumstances as hostility to surgery and to the use of new instruments in obstetrics.[57] Medical historians are just beginning to comprehend the knotty problems of 'traditionalism' in medical theory during 1660–1800. If the demise of Galenic influence seemed complete by 1700, other ancients (especially Aristotle) still regarded as springs of veritable truth awaited toppling.

4. *Metaphysics*. The idea of an occult and metaphysical agency as a 'cause' of disease was disappearing but 'empirics' (quacks, mountebanks, charlatans in contrast to trained medical men) promoted it as their rationale,[58] especially when asked why their cures were universally applicable, and the clergy sustained the idea with much energy. That Nature is the best physician is as commonplace an idea among laymen throughout the epoch as the belief in 'general' and 'particular' Providence. By 1740, however, most licensed physicians, surgeons, and apothecaries had lost faith in metaphysical agencies as a cause of disease.[59]

5. *Satire of the Medical Profession*. No other professional group (lawyers, the clergy) was so vigorously and prolifically satirised in this age as medical men. The endeavour was indeed a national sport.[60] Satirists especially chose self-professed 'rationalists' as their targets, although virtually every attribute of doctors was lambasted: their pedantry, mercilessness, immodesty, public antics, bigotry, pretensions, panaceas. Personal and general medical satire flourished – indeed rose to the status of a genre in itself[61] – and this fact of English literary history cannot be explained away merely by arguing that in the eighteenth century *all* professional groups were excoriated. Medical men earned for themselves a low reputation because they deserved it. Among masses of extant medical ridicule is a preponderance of satire of 'systematists' – as may be seen in the *locus classicus*, Swift's *Tale of a Tub* (1704), written precisely at the time (1696–1704) when iatromechanists were at their zenith, although observationists, collectors of empirical data, and laboratory researchers were also criticised, sometimes just as bitterly.[62] Satirical poems and tracts in the British Museum and Wellcome Institute are sufficiently abundant to permit many generalisations about the exposure of particular groups within the medical profession, but no basis other than a statistical one exists to document the main point itself. Wrangling among medical men was fierce enough to give rise to a literature of this sort; and it is a bibliographical fact of eighteenth-century medicine that the historian with an axe to grind will discover enough evidence in satirical literature to buttress almost any point.

6. *Influence of other sciences and scientific subjects*. While biology and physiology probably influenced medical research, philosophy appears to have had no impact whatsoever.[63] It is perhaps not surprising that biological developments – especially in embryology but also in theoretical physi-

ology (i.e., mechanist-vitalist controversies over the definition of animal heat)[64] – should have very palpably influenced the direction of medical research. But philosophy's inability to penetrate the medical world (philosophers of medicine in particular) must be attributed to primitive modes of communication and the parochialism of theoretical physicians. While discoveries of eighteenth-century medicine influenced philosophers (e.g., Locke's theory of madness altered somewhat after reading medical literature on insanity),[65] the converse is untrue. Because they were not carefully read, the theories of Locke, Berkeley, Hume, and Reid barely reached the nitty-gritty of daily medical research, least of all Hume's powerful thesis that mankind can never be in a state of intimate contact with external physical reality. Locke's investigations, in *An Essay Concerning Human Understanding* (1690), of primary and secondary sense impressions associated into abstract ideas, went unnoticed by most 'rationalists', like Dr John Quincy, whose writings repeatedly approach Lockian methodology without any trace of awareness of so doing. 'When a Person sets out upon an Enquiry,' he wrote in *Medicina Statica* (1712), 'nothing can be of greater concern, than to be first well acquainted with the Powers and Capacities of his own Mind, and the Means by which only the Matter he has in Pursuit is attainable' (p. 3). Similarly, medical works written 1770–1800 echo the ideas in Reid's *Essays on the Intellectual Powers of Man* (1785).

7. *The argument of genius.* Although the Carlylean tendency to regard the history of human achievement as a succession of inexplicable geniuses arbitrarily bestowing knowledge upon mankind has finally been abandoned as simple-minded,[66] one cannot completely overlook the fact that there were no 'Newtons' in the medical world during the period 1660–1800; there was not even a Boyle or Galileo.[67] One rebuttal to this objection queries whether medicine could have accommodated the colossal talents of a Newton. Astronomy, physics, and mathematics were relatively 'advanced' sciences in 1700; medicine was in her infancy. Not until the middle of the nineteenth century did practical medicine really make monumental advances, although this observation should not detract from the solid achievements of some earlier medicine.

8. *'Empirics' and 'Empiricism'.* Quack doctors of every variety took advantage of disputes within the profession and, by the introduction of such pills and potions as fever powders, hungary water, mercury, tar water, drops, and electric therapy – to mention a few – persuaded the public at large of their competence. The number of empirics increased so sharply during the eighteenth century[68] – and here one must recall what a lucrative business it was in Britain[69] – that licensed physicians, already threatened by encroachment, fought for their economic survival by warning the public about such abuses. Together with 'ancients versus moderns' controversies, this added burden further deflected some serious medical

men from their research. Distraction of doctors by imposters was noted throughout the age[70] and is a refrain continuously mentioned in the literature of eighteenth-century England. Treatises like an anonymous work in the Bodleian Library entitled *An Examination into the Origin, and Meaning of the Words Empiricism, Empirick, Quack Doctor and Quack*. And, *An Exact Account of the Present State of Physick* (1749) defend the empiric's trial-and-error method and argue that his results are ultimately more useful than current deductive hypotheses of 'rationalists'. True as this may be, competition from quacks posed a serious threat to trained doctors, thereby adding to an already long list of reasons for becoming distracted from empirical research.

9. *Foibles of Medical Men*. Swift made Lemuel Gulliver's arch sin demented pride and Pope, in one of his foremost moments, commented that 'in Pride, in reas'ning Pride, our error lies'.[71] Eighteenth-century doctors would be surprised if they could borrow, just for a moment, the historian's perspective and ponder the number of charges of 'reas'ning Pride' made by other medical men in their own age. If these accusations had come from sources outside the profession we might discount them on several counts, and our own 'Peter Principle'[72] reinforces such discredit on sociological grounds; but they were most often uttered by insiders and therefore continue to carry weight.[73] A recurrent sentiment decrying the immodesty of medical researchers is so pervasive in this epoch that it can indeed be called a *leitmotif*. If medical men as a collective profession had not consolidated their 'rationalistic' and 'empirical' proclivities they also had not shaped for themselves an image of dignity and humility.

NOTES

1 'When are Statements Empirical?' *British Journal for the Philosophy of Science*, x (1959), 287. BHM = *Bulletin of the History of Medicine*.

2 'Empiricism and Rationalism in the Works of Thomas Sydenham', *BHM*, XLIV (1970), 9.

3 Modern logical positivists (Carnap and the Viennese School, Wittgenstein, Popper, Hempel, Ayer, etc.) have developed such criteria; see especially A. J. Ayer, *Language, Truth and Logic* (1946); Karl R. Popper, *The Logic of Scientific Discovery* (1959); and C. F. Hempel, 'Empirical Statements and Falsifiability', *Philosophy*, XXXIII (1958). But these criteria apply to theories of knowledge in philosophy, not to medical research. For the difference of empirical (i.e. verifiable) criteria in philosophy and science see Popper, 'The Demarcation between Science and Metaphysics', *Rudolph Carnap: Library of Living Philosophers*, ed. P. A. Schilp (1963), and P. B. Medawar, 'Hypothesis and Imagination', *The Art of the Soluble* (1967), 131–55. Historical approaches to empiricism are often the most satisfactory: for a brilliant sampling of this approach see D. J. Greene, 'Augustinianism and Empiricism: a Note on Eighteenth-Century English Intellectual History', *Eighteenth-Century Studies*, 1 (1967), 33–68.

4 Here Hume is a good example: no philosopher is more often called a 'British empiricist' and yet no thinker was so unrepresentative of his age. Although Hume

paid attention to some of the philosophical questions troubling Locke, he spent most of his energies elsewhere, in the theory of causation (cause and effect criteria); on strictly philosophical grounds he has little in common with Berkeley, Reid, and Dugald Stewart. These differences have now been demonstrated by Gerd Buchdahl in *Metaphysics and the Philosophy of Science: the Classical Origins, Descartes to Kant* (Oxford, 1969), 'Hume: the Critique of Causation', pp. 325–85.

5 'Harvey and Sir Francis Bacon', *The Life of William Harvey* (Oxford, 1966), 161.
6 Bodleian MSS. Aubrey 6, fol. 64, transcribed in Keynes, *Harvey* (1966), appendix i, 433.
7 *Ibid.*, 434.
8 Most dictionaries (Bayle, Quincy, Chambers, Robert James) offer no help. Johnson defines 'empiricism' briefly as 'dependence on experience without knowledge or art; quackery', and the *NED* ('empiricism') lists no uses of the term to denote an observational process until 1803. The nonce use of 'empiricism' in the *NED* is 1657, in George Starkey's *Natures Explication and Helmont's Vindication*, 245: 'The Chymistry of the Galenical Tribe is a ridiculous . . . and . . . dangerous Empiricism' (meaning quackery) and the term continued to be used as a synonym for quackish incompetence until the nineteenth century. Distinctions between trained physicians and unskilled 'empirics' abound during the period 1660–1800. See, for example, Dr William Hillary, *A Rational and Mechanical Essay on the Small Pox* (1735), preface, x: 'The distinguishing *Characteristic* of a true *Physician*, from an *Empiric*: [is that the former, unlike the latter, proceeds] . . . by a Method of reasoning from *Data*, founded upon observations and real Facts, that the *Healing Art* must be improved, and brought to a State of Perfection: for if we once quit our Reason for Mystery, and abandon a just Method of Mechanical and Geometrical Reasoning, for the unintelligible Terms of Occult Faculties and Qualities, with all such like Metaphysical and Chymical Jargon and Nonsense, heretofore too much used in the Schools; we must wander through endless Mazes, and dark Labyrinths, played at Hazard with Men's Lives, and suffer ourselves to ramble where ever conceited Imagination, or whimsical Hypotheses will lead us.'
9 From the Greek *iatros*, meaning doctor, i.e. mechanical doctor, mathematical doctor, chemical doctor.
10 'Rationalism in Early Eighteenth Century Medicine', *Journal of the History of Medicine*, XVIII (1963), 257–71; 'Some Problems of Causality in Eighteenth Century Medicine', *BHM*, XXXVII (1963), 15–24; 'Empiricism and Rationalism in the Works of Thomas Sydenham', *BHM*, XLIV (1970), 1–11. See also King's two books, *The Medical World of the Eighteenth Century* (Chicago, 1958) and *The Road to Medical Enlightenment 1650–1695* (1969), and Bernard Aschner, 'Empiricism and Rationalism in Past and Present Medicine', *BHM*, XVII (1945), 269–86.
11 See especially his conclusions (114–15) in *The Road to Medical Enlightenment*: 'The significant difference between 'rationalists' and 'empiricists' concerns the degree to which they relate with concrete observations. Both start with experience but the rationalist goes much more rapidly into abstractions, the abstractions are more rarefied, and the subsequent contact with experience much less frequent. The rationalist will end with first principles of highly abstract character, utterly remote from perpetual experience and deserving the term "speculative".' Buchdahl makes the same point (*Metaphysics and the Philosophy of Science* (Oxford, 1969), 5) although in a philosophical context: 'This difference [between Descartes and Locke] is one of a group that has traditionally been designated by the respective labels of "rationalism" and "empiricism," "schools" which have theirmost prominent representatives in Descartes and Locke respectively. But . . . what really divides these two philosophers is not a genuine "opposition" of viewpoints but rather differences of emphasis, and "descriptive appraisals" of certain aspects in the various fields of "knowledge." The fluidity of the language in which these appraisals appear is most pronounced in the

earliest of the thinkers of our classical period, though I do not think that this prevents them from representing types of the point of view here suggested.'

12 Quotations from *BHM*, XLIV (1970), 11, 5.
13 King, *The Road to Medical Enlightenment*, 193.
14 *Journal of the History of Medicine*, XLIII (1963), 264.
15 An analogy with 'blended genres' in literary theory suggests itself. Modern critics have only recently abandoned the notion of 'pure forms' for 'blended forms', as in the case of eighteenth-century georgic-didactic poems. See E. D. Hirsch, *Validity in Interpretation* (New Haven, 1966), 78–102 and John Chalker, *The English Georgic* (1970).
16 It is curious that Jones's pioneer study *Ancients and Moderns: a Study of the Rise of the Scientific Movement in Seventeenth-Century England*, rev. ed. (St Louis, 1961) has not been extended to the eighteenth century.
17 See J. G. Crowther, *Francis Bacon* (1960); Margery Purver, *The Royal Society: Concept and Creation* (1967); Paolo Rossi, *Francis Bacon* (1968). Cf. King, *The Road to Medical Enlightenment*, 130: 'Study of Sydenham indicates that the teachings of Francis Bacon simply had not reached the medical profession at that time, despite abundant lip service.'
18 For example, Dr William Hillary, an iatrochemical doctor opposed to the gathering of data. See his *Rational and Mechanical Essay* (1735).
19 A tradition strengthened by Joseph Glanvill in *The Vanity of Dogmatizing: Or Confidence in Opinions* (1661) and continued in the first half of the next century.
20 These medical writers include Drs Thomas Morgan, William Oliphant, Archibald Pitcairne, Peter Shaw, James Keill, George Cheyne, Nicholas Robinson, James Jurin, Richard Blackmore, Daniel Turner, Edward Strother, John Quincy, David Hartley, and dozens of others. For a bibliography of their writings, see Eric Gaskell, 'The Iatromathematicians: a Bibliographical Survey', *Medical History*, XVI (1972). The introduction and reception of iatromechanism has been studied by Theodore M. Brown, 'The College of Physicians and the Acceptance of Iatromechanism in England, 1665–1695', *BHM*, XLIV (1970), 12–30. During the next quarter of a century (1695–1720) the medico-mathematical works of Archibald Pitcairne began a controversy, especially among Drs Oliphant, Cheyne, Eizat, and Pitcairne. The most amusing and readable work in this pamphlet war is Eizat's *Appollo Mathematicus: or the Art of Curing Diseases by the Mathematicks, according to the Principles of Dr. Pitcairne . . . Never published in English before. To which is Subjoined a Discourse of Certainty* (1695). The most eloquent apologist for the group ('school' exaggerates their cohesiveness) during the 1720s was Dr John Quincy, who defended their methodology in the introduction to *Medicina Statica* (1712; 5th ed. 1737), in the 1730s Dr Hillary (see n. 8 above), and James Jurin, nicknamed 'Philalethes Cantabrigiensis' by his opponents, whose *Dissertationes Physico-Mathematicae* (1732) vigorously restated their position. David Hartley, the philosopher and author of *Observations on Man* (1749), was perhaps their last important spokesman.
21 See G. A. Lindeboom, 'Pitcairne's Leyden Interlude Described From the Documents', *Annals of Science*, XIX (1963), 273–84.
22 *Medicina Statica* (1712), 3.
23 See T. M. Brown, *BHM*, XLIV (1970), 12–30. Throughout the period there was also a marked antipathy to 'theories', as I demonstrate later. Dr John Woodward, the 'Fossile' of *Three Hours After Marriage*, and Thomas Baker (1656–1740) carried on a long correspondence, now in the Cambridge University Library, about the status of theories, from which I quote one passage from an undated letter by Woodward, probably written *c.* 1702–3 (Camb. Univ. Add. MS. 7647): 'I perceive the Author of the *Reflections upon Learning* [Baker published this work in 1699] is resolved, notwithstanding what I laid before him in my last, to have Dr. Woodward's Work

a *Theory*. The Dr. himself has very fitly intitled it *An Essay towards a Natural History of the Earth*: & I do not see why this ingenious Gent. should call it a *Theory*; especially when he blackens Theorys in the manner he does. He may as justly call it a Pastoral, an Elegy, or what else he pleases, as a Theory. Dr. W's Essay is as opposite to the Theoryes as it can well be. He has professedly shewn that one of them is utterly destitute of any manner of Foundation, or Countenance either from Nature or Scripture: & for the other, 'twas indeed wrote since the Dr's *Essay*, but any one who reads the *Essay* will meet with what will furnish him with Arguments that evince the apparent falsehood of the Theory. Would it not be pretty strange that a Man that has wrote against the Deists, bashed and defeated all their Arguments, & clearly made good the Cause of Christianity in opposition to them, should for this be stiled a Deist? as little reason is there to stile Dr. W. a *Theorist*.'

24 *A Treatise Concerning the Influence of the Sun and Moon upon Human Bodies, and the Diseases thereby Produced* (1748), xiv.

25 *Uncertainties of Physic* (1751), sect. vi, 172.

26 Camb. Univ. Add. MS. 7647, A.L.S., dated 28 Oct. 1723.

27 E. Stone, 'The Experimental Philosophers', *A New Mathematical Dictionary* (1726), n.p.

28 Ibid.

29 Preface, iv.

30 See G. S. Rousseau, 'Medical Prefaces of Mid Eighteenth-Century England', *BHM*, XLVII (1973).

31 *Influence of the Sun and Moon* (1748), preface, viii.

32 *The Nature and Laws of Motion of Fire* (1760), viii. Hillary (1697–1763), a Bath doctor who voyaged to the Barbados, wrote several books on diseases prevalent in England and the tropics.

33 *Practical Cases and Observations in Surgery* (1751), xi. Ingram's surgical practice, however, was based on new and innovative methods, and his case illustrates once again the dangers involved in classifying doctors.

34 'Observations on Former Systems of Medicine, and Outlines of the New Doctrine', *The Works of Dr. John Brown*, 2 vols. (1804), I, 6.

35 Preface, 3.

36 Ibid., 356. Elsewhere in the *Observations* Millar notes that the eighteenth century had not improved enough: 'the parade of scholastic learning is still too prevalent' (p. 3); 'visionary hypotheses must be exploded' (*ibid*.); 'the penury of useful observations' (4); 'there does not exist in the annals of physick a complete account of the popular diseases of any one country' (*ibid*.); 'it is become necessary to strip physick of that formal dress which hath rendered it contemptible, and to cultivate the science with that freedom, simplicity, and candour, which are at once the test and ornament of truth' (288).

37 Quincy's *Lexicon Physico-medicum* (1717) and John Barrow's *Dictionarium Medicum Universale* (1749), both in English, were also important.

38 See the Wellcome Library's unpublished 'Chronological Catalogue' and J. L. Thornton, 'Medical Books from 1701–1800', *Medical Books, Libraries and Collectors: a Study of Bibliography and the Book Trade in Relation to the Medical Sciences* (rev. ed. 1966), 108–45.

39 It is extremely difficult to document this point; but it does seem true after studying dozens of medical prefaces written from 1730 to 1750. The editors of the Oxford History of English Literature have intuited this in making 1740 a dividing year.

40 Preface, ii.

41 *Practical Cases and Observations* (1751), preface, xi.

42 See 'Peirce and the Trivialization of the Self-Correcting Thesis', in *Proceedings of the Indiana Symposium on Nineteenth Century Methodology* (Bloomington, 1971).

43 My purpose in listing books is to demonstrate the range of medical writing during these years. Titles themselves are admittedly misleading and I have classified these works as 'rational' or 'empirical' after thoroughly evaluating the author's statement of purpose, plan, and execution. An adequate history of medicine would take into account such extraneous influences as fashionable trends and the author's financial gain before generalising about works written in a given year.

44 See Quincy's defence of these rationalists in *Medicina Statica* (1737, 5th ed.), an *apologia* for the whole movement.

45 *Works of Dr. John Brown*, 2 vols. (1804), I, xv. Empirical doctors in France expounded similar beliefs in the 1740s and 1750s: see, for example, Quesnay's 'Préface' to his *Mémoires* (Paris, 1743), and F. Thiery, *La Médecine expérimentale* (Paris, 1755), 13-22.

46 *Medicina Statica* (1737), 8, in which Quincy argues that mathematical demonstration offers the surest type of scientific verification.

47 *Works of Dr. John Brown* (1804), I, 4-5.

48 J.S. Mill, *A System of Logic* (1843), sect. vi; Whewell, *Of Induction* (1849). See also Gerd Buchdahl, *Metaphysics and the Philosophy of Science* (Oxford, 1969), pp. 335-40. Dr Robert Jones, an obscure physician who favoured Brunonian medicine, anticipated Mill in *An Inquiry into the State of Medicine, on the Principles of Inductive Philosophy* (1781).

49 See Robert K. Merton, 'Puritanism, Pietism, and Science', in *Social Theory and Social Structure*, rev. ed. (New York, 1957), 586.

50 While no one would wish to suggest that any of these individually or even in combination produced (i.e., cause and effect) the intimate intermingling of rationalism and empiricism about which we have been speaking, only the most sceptical historian will deny that *cumulatively* they created the unique atmosphere and particular conditions for medicine to develop as it did. See Patrick Gardiner, Causal Connexion in History', *The Nature of Historical Explanation* (Oxford, 1952), 80-99.

51 See Sir George Clark, *A History of the Royal College of Physicians*, 2 vols. (Oxford, 1964-6); Cecil Wall and H. Cameron, *A History of the Worshipful Society of Apothecaries*, ed. and rev. by E. A. Underwood (1963), 130-5; Christopher Hill, *Puritanism and Revolution* (1963) and *Intellectual Origins of the English Revolution* (Oxford, 1965), 74-84.

52 The situation in France was different, although professional feuds also diverted medical men; see Toby Gelfand, 'Empiricism and Eighteenth-Century French Surgery', *BHM*, XLIV (1970), 40-53.

53 Some research on the matter is found in T. Sprat, *History of the Royal Society* (1667), 197; G. Clark, *A History of the Royal College of Physicians* (Oxford, 1964-6), II, 542-3; W. H. McMenemey, 'The Hospital Movement of the Eighteenth Century and its Development', in F. N. L. Poynter, *The Evolution of Hospitals in Britain* (1964); E. M. Sigsworth, 'A Provincial Hospital in the Eighteenth and Early Nineteenth Centuries', and J. H. Woodward, 'Before Bacteriology – Death in Hospitals', both in *The Royal College of General Practitioners' Yorkshire Faculty Journal* (June 1966), 24-31, and (autumn 1969), 1-12.

54 This is the opinion of recent historians of medicine, especially those experts (D'Arcy Power, Jane Oppenheimer, Jessie Dobson, C. Finch, Charles Singer, Lester King) knowledgeable of Hunter's career.

55 A vast and still unexplored literature, mostly in Latin, exists. See, for example: Richard Mead, *Mechanica exposito Venenorum Exercitationum Philosophicarum nonadecima* (Leyden, 1691); Conyers Middleton, *De Medicorum Apud Veteres Romanos Degentium Conditione Dissertatione* (Cambridge, 1726); *The London Medley . . . the Thesis Being*

Rationalism and empiricism 143

on a Parallel between the Ancients and the Moderns ([1731?]). The most amusing and ill-conceived of these treatises is Charles Lamotte's *An Essay Upon the State and Condition of Physicians Among the Ancients: Occasioned By a late Dissertation of the Reverend Dr. Middleton: Asserting that Physick was Servile and Dishonourable among the old Romans, and only practis'd by Slaves, and the meanest People* (1738). Lamotte, a Fellow of the Royal Society of Medicine and chaplain to the Earl of Montagu, produced biographies of ancient doctors but his methods were outdated and could have been written a half century earlier. Garbled statements abound: 'Homer so curiously describes the Wounds of his Heroes, and so nicely displays and discovers every Part, that no one can doubt but he had some Knowledge of Physick, or at least a competent Skill in anatomy' (pp. 10–11).

56 British Museum press mark 1135. i. 1.

57 An indication of the antiquarian tendencies of eighteenth-century doctors is gained by reading the Royal Society's *Philosophical Transactions*, especially during 1720–50.

58 'Empirics', echoing the arguments of second-century B.C. Alexandrians of the so-called 'School of Empiricists', maintained that trained physicians could no more explain why disease occurred when and where it did than they could account for the colour of the moon. The British Museum possesses the single largest corpus of writings by these quacks under the classification 'Medical Tracts' (BM. Tracts 301–498).

59 See Lester King, 'Problems of Causality in Eighteenth Century Medicine', *BHM*, XXXVII (1963), 17–18.

60 Some historical background for this development is found in G.S. Rousseau, 'Doctors and Medicine in the Novels of Tobias Smollett' (unpub. diss., Princeton Univ., 1966), chaps. i–ii.

61 See M. S. Day, 'Anstey and Anapestic Satire', *English Literary History*, XV (1948), 142–3.

62 They continued to be written throughout the century; see, for example, such anonymous poems as *The Prophetic Physician: An Heroi-Comic Poem, Address'd to the Physicians* (1737); *Physick's a Jest* (1739); *Various Ironic and Serious Discourses on the Subject of Physick* (1749).

63 See Laurens Laudan, 'Theories of Scientific Method', *History of Science*, VII (1968), 1–63 and Philip C. Ritterbush, *Overtures to Biology* (New Haven, 1966).

64 See June Goodfield, *The Growth of Scientific Physiology: Physiological Method and the Mechanist-Vitalist Controversy* (1960).

65 Kenneth Dewhurst, *John Locke, Physician and Philosopher: a Medical Biography* (1963).

66 See G. S. Rousseau, 'Science and the Discovery of the Imagination in Enlightened England', *Eighteenth-Century Studies*, III (1969), 132.

67 Eighteenth-century medical thinkers themselves recognised this or they would not have protested so vigorously for reform.

68 T. McKeown and R. G. Brown, 'Medical Evidence Related to English Population Changes in the Eighteenth Century', *Population Studies*, IX (1955–6), 119. See also McKeown's 'A Sociological Approach to the History of Medicine', *Medical History*, XIV (1970), 342–51, whose sober method offers cogent reasons to substantiate my referring to social factors that affected theoretical developments in medicine.

69 Popular medicines – those of Ward, Berkeley, Dover, James, Wesley, Hill, and others – enriched their sponsors. The clearest definition of an 'empirical' doctor I have found occurs in an anonymous pamphlet in the British Museum, *An Ernest Dissuasive from the Use of Empiricks* (1749), 26–7.

70 See, for example, *An Examination into the Origin, and Meaning of the Word Empiricism* (1749), 15–18; *Physick's a Jest* (1739); *Minutes of the Royal College of Physicians*.

71 *An Essay on Man*, 123.
72 Dr Laurence J. Peter and Raymond Hull, *The Peter Principle* (New York, 1969).
73 See, for example, Hillary, *Observations* (1735), preface, xix, 2; Dale Ingram, *Practical Cases* (1751), xi-xv; and John Millar, *Observations on Diseases in Great Britain* (1770), preface, vi, who indicted his colleagues as 'Men of a quick and lively fancy who have flattered themselves with hope of arriving, by their ingenuity, to that intimacy with the operations of nature, which is only to be obtained by careful, attentive, and unwearied observation'.

6

Praxis 1: Bishop Berkeley and tar-water

During the 1960s I collaborated at Princeton with Professor Marjorie Hope Nicolson, the distinguished historian of ideas who had been a student of A. O. Lovejoy and who moved from Columbia to the Institute for Advanced Study upon her retirement. We published essays and books about literature and science of the classical period, the best known of which is perhaps 'This Long Disease, My Life': Alexander Pope and the Sciences (Princeton, 1968). As our collaboration progressed, we found ourselves increasingly drawn to the medical realm, the territory I continued to believe had been imprudently under-represented by those who studied the muses. But I had not yet read Foucault, Althusser, Adorno, Bakhtin, etc. – had not yet been liberated by theory. All I knew were the French historians of medicine who, of course, harboured theoretical interests: Georges Canguilhem, Jacques Roger, the recently deceased Gaston Bachelard.

Nicolson strongly believed that writing of our 'serious kind' needed to be sprinkled, so to speak, with salt and pepper. Otherwise it became – in her words – 'too deadly dull'. I supposed her attitude to be all the truer when the subject for study was an Irish divine (Berkeley) who was a world-class philosopher. Why not sprinkle pepper generously over an assortment of nostrums and panaceas when quackery was running rampant in this way?

But Nicolson harboured positively no interest in the theory or institutional arrangements of the intersections we were developing: society, medicine, consumption. In a draft of this essay I originally included a brief but naive metacritique about the disciplinary implications of this kind of generated discourse, which she excised in red ink. As the senior partner, she won, of course. She was hospitalised in 1968–9 before I could explain to her the liberation I was gaining from theory. Just as well: as a purist historian of ideas in the school of Lovejoy, she would have been opposed to any preferment for social history or for an agenda that privileged ideological prose; opposed also to many aspects of race, class, and gender.

We wrote the essay in 1966 and published it in 1970 in a festschrift for the late distinguished Swift scholar Louis Landa (Oxford University Press).

George Berkeley, Dean of Derry – not yet Bishop of Cloyne – arrived in Rhode Island on 23 January 1729, and remained until shortly before he sailed home from Boston on 9 September 1731. He had come to the colonies in connection with his plan to establish in Bermuda a college for sons of English planters and for the education of Indians. During his residence on the mainland and in Newport he was marking time, waiting for confirmation of funds the government was supposedly assigning, which ultimately proved not forthcoming. In the late eighteenth and nineteenth centuries accounts of Berkeley's colonial visit came to be surrounded by various legends, one of which was that he arrived in Rhode Island by accident, because 'the captain of the ship in which he sailed could not find the island of Bermuda, and having given up the search after it, steered northward until they discovered land unknown to them, and which they supposed to be inhabited only by savages'.[1]

For the purposes of this study a more important legend which has been questioned is that Berkeley's acquaintance with tar-water came from the Narragansett Indians, on the mainland of Rhode Island; it may rather have come from South Carolinians. Such is the contention of Miss Alice Brayton,[2] who tends to discount Berkeley's interest in the Narragansett Indians as part of a sentimentalisation of the 'noble savage' tradition, in which the distinguished Irish philosopher turns to the lowly Red Man for medical knowledge. According to Miss Brayton, Berkeley first heard of tar-water in the spring of 1730, when a violent epidemic of smallpox broke out in Boston and a stranger of Newport died of it. Among the Newport residents were several self-exiles from South Carolina, friends of Mrs Berkeley, who used tar-water for their own protection and that of their slaves; they had learned the principle from Indians in South Carolina.[3]

Miss Brayton may be correct in her theory that Berkeley's knowledge of tar-water came from the South Carolinian *émigrés*, but she is probably not correct in discounting his interest in the Narragansett Indians. There is abundant evidence in Updike and other early historians of Berkeley's study of the Narragansetts both early in his stay on the mainland and later when, with the elder Updike, McSparran, and others, 'he repeatedly visited Narragansett . . . to examine into the condition and character of the Narragansett Indians'.[4] Since the proposed college in the Bermudas was in part for the education of Indians, it seems inevitable that Berkeley should have studied with care what he could learn of the mentality and other characteristics of the first Indians he had known.

Berkeley's knowledge of the Narragansett Indians is further shown in the sermon he preached on 18 February 1731, shortly after his return to London, before the Society for the Propagation of the Gospel in Foreign Parts.[5] Here he developed in some detail his observations on the condition of both Indians and Negroes in Rhode Island. He referred to the diminution in number of Indians, a result in part of wars and smallpox, but chiefly through the 'slow poison' of 'the Use of strong Liquors' taught them by their supposed masters. But we are here chiefly concerned with Berkeley's interest in tar-water.

The work in which Berkeley treated the subject was *Siris: a Chain of Philosophical Reflexions and Inquiries concerning the Virtues of Tar Water*, published in 1744. Until recently it was believed that *Siris* was not the original title but was added in later editions.[6] We now known that the first edition was that published in Dublin in March 1744, in which the title appeared. Berkeley considered *Siris* a contribution to philosophy, but the publisher of the first London editions realised that the introductory portion – the first 119–odd sections – describing a simple home-remedy specific for the prevention or cure of many diseases, would greatly attract general readers, particularly if the unfamiliar word 'Siris' were omitted and the stress seemed to be upon 'the Virtues of Tar Water'. Another indication of the publisher's lowering the work to the capacity of the general reader was the addition to the second Dublin edition (September 1744) of a 'Vocabulary of certain words not commonly understood'. Correct in their surmise, the publishers had a 'best-seller'. Six editions were sold out within a few months.[7]

For many years *Siris* continued to be regarded as a contribution to medicine rather than philosophy. With the exception of a few references considered below, it first began to assume philosophical importance in 1871 in Alexander Campbell Fraser's *The Works of George Berkeley*, D. D.[8] Fraser called it 'the curious and beautiful work of speculation in which he celebrated the new medicine'. Berkely said that *Siris* had cost him more thought and research than any other work. Fraser adds: 'No one who examines its contents can be surprised to hear this. The book is full of fruit gathered in the remoter by-ways of science and philosophy.'

Fraser's estimate is shared by Professor John Wild in one of the most important twentieth-century treatments of Berkeley's philosophy.[9] Wild went even further than Fraser, challenging 'many of his modern commentators [who] have dismissed the *Siris* altogether as a senile aberration'. He devoted some fifty pages to a study of its ideas and their sources in early philosophy. Yet many modern readers have been perplexed by *Siris*, in part because of the first lengthy section, with its stress upon a simple household remedy for man and beast, the very part of the work with which most eighteenth-century readers stopped their reading. As recently as 1960 John Linnell declared:[10] 'If Berkeley had not written the *Siris*, his significant place in the history of philosophy would be what it actually is; if he had written only *Siris*, he would very likely have had little or no place in that history.'

An interesting example of a change in attitude may be seen in Professor A. A. Luce, Berkeley's most recent biographer, and co-editor with Professor T. E. Jessup of his *Works*. In 1936 he raised the question, 'Is there a Berkeleian Philosophy?'[11] and spoke of *Siris* as 'an old man's ramble through quack remedies to Elysian fields'. But in his biography,[12] published in 1949, his treatment of *Siris* is sympathetic and understanding. He recognises various reasons for the banishment from serious philosophy it had encountered:

> Tar is a black and sticky substance with none too good a name in letters, and the very idea of a bishop discarding his white lawn sleeves and handling it and extracting a nasty medicine from it is too much for our sense of gravity, and Berkeley's tar-water has become a jest. . . . No one objects to a laugh or two about it; Berkeley could see the funny side of it himself, and jokes his tar-drinking friends. But the joke has been carried too far; the whole affair has been treated as a craze, as a proof of unbalanced temperament and failing faculties, if not of a disordered mind. Serious biography must leave the comic aspect aside.

Professor Luce then shows how different the sections on tar-water seem in their historical perspective. Berkeley's tar-water, he rightly emphasises, was no jest in his own day, except to a limited number. 'Thousands of sick and sufferers blessed his name. The apostles of old could heal the sick; a bishop in the eighteenth century could with perfect propriety make the attempt, and situated as he was Berkeley simply *had* to do so.' Luce's justification of the Bishop's activities reminds us of a quatrain, written in an eighteenth-century hand in the Princeton University copy of Berkeley's *Miscellany*:

> Who dare deride what pious Cloyne has done.
> The Church shall rise and vindicate her son.
> She tells us all her Bishops shepherds are
> And shepherds heal their rotten sheep with tar.

Luce calls our attention to the situation Berkeley faced when he began to administer tar-water to the Cloyne sick. The winter of 1739–40 was one of the most severe ever experienced, causing a famine. In Dublin there were nurses and doctors, even some hospitals, but in country districts there was no provision for the sick and poor, who inevitably turned to the clergy for aid. During the great frost Berkeley gave twenty pounds in gold to the poor each Monday. He and his wife ministered to their physical ailments with any medicines they could make. Sickness and plague followed upon the famine, and a severe epidemic of dysentery raged through Ireland and descended on Cloyne. Berkeley wrote to Thomas Prior on 8 February 1741[13] of cures he had been trying for the epidemic, and added: 'I believe tar-water might be useful to prevent (or to perfect the cure of) such an evil; there being, so far as I can judge, no more powerful corrector of putrid humours.' Some time during this period Berkeley began to experiment with the tar-water which in Rhode

Island had proved effectual against smallpox. For a time he became a doctor on the one hand, a chemist on the other. A room in the episcopal residence was set apart for preparation of tar-water, and here he experimented on such matters as the stirring-time and the clarifying-time.[14] He had already read all that had been written about tar by the ancients, who knew its virtues though they did not use tar-water. Undoubtedly he kept himself abreast of anything germane to tar-water in the natural history of his own time.[15] When the violence of the plague and dysentery subsided and crops began to grow again, Berkeley had time to devote himself to the writing of *Siris*, and a little later to the tracts that followed it.

Among the controversies caused by even the earlier editions of *Siris*, the most immediate and the one that was to be of longest duration was that due to the opposition aroused among conservative members of the medical profession, to whom the new 'cure-all' seemed in a class with Joshua Ward's 'Pill and Drop', the most popular nostrums of the preceding decade.[16] Of such opposition we shall hear more as we continue. Involved with this in many minds, not limited to the medical faculty, was the question whether the Bishop considered that he had found what alchemists had sought so diligently, the universal panacea. On this Berkeley spoke a number of times, most vigorously in *A Letter to Thomas Prior, Esq.*,[17] the sequel to *Siris*, published in Dublin in July 1744, while editions of *Siris* were still coming from the press:

> The great objection I find made to this medicine is that it promises too much. What, say the objectors, do you pretend to a *panacea*, a thing strange, chimerical, and contrary to the opinion and experience of all mankind? Now, to speak out, and give this objection or question a plain and direct answer – I freely own that I suspect tar-water is a panacea. I may be mistaken, but it is worth trial: for the chance of so great and general benefit, I am willing to stand the ridicule of proposing it. . . .
>
> . . . by a panacea is not meant a medicine which cures all individuals (this consists not with mortality), but a medicine that cures or relieves all the different species of distempers. . . .
>
> After all that can be said, it is most certain that a panacea sound odd, and conveys somewhat shocking to the ear and sense of most men, who are wont to rank the Universal Medicine with the philosopher's stone, and the squaring of the circle. . . . I do not say it is a panacea, I only suspect it to be so – time and trial will shew.

Other letters to Thomas Prior were published later, but we may leave them for a time to consider Prior's own contribution to the *Siris* controversy, *An Authentic Narrative of the Success of Tar-Water, in curing a great Number and Variety of Distempers*.[18]

Thomas Prior of Rathdowney had been a fellow student at Kilkenny College with Berkeley, who entered in his eleventh year; they remained close friends all their lives. Prior did not enter a profession, though he seems to have read law, but devoted himself to the remedy of public and social evils. He acted as Berkeley's legal agent, and was the founder and the Secretary of the Dublin Society. The *Authentic Narrative* was dedicated to Philip Stanhope, the fourth Earl of Chesterfield, in the hope that he would find it a contribution to the public good. The work contained 'An Alphabetical Index of the several Distempers' which had been alleviated or cured by tar-water, and a remarkable list it is, of approximately 120 'Distempers', ranging alphabetically from 'ague' to 'wind', a few certified in only one instance, others many times. There are twenty-seven instances of the curing of coughs, twenty-one of scurvy, nineteen of asthma, eleven of the king's evil. Prior begins his pamphlet by discussing an advertisement that had appeared in the *Dublin Journal* for 3 July 1744, asserting that many patients in Stephen's Hospital had not benefited from tar-water. Upon investigation he found only six affidavits, and went on to point out in his narrative that the treatment had not been continued long enough. Most of the *Authentic Narrative* is devoted to brief case-histories of cure or relief by tar-water. These Prior had secured largely through advertisements in various periodicals. The most interesting one, which attracted widespread interest, as references to it show, was an affidavit, sworn to before the Mayor of Liverpool, by Captain Drape,[19] master of the ship *Little Foster* of Liverpool. On a voyage from Guinea to Jamaica he had carried over 200 Negro slaves, among whom 170 contracted smallpox. A passenger advised the master to give them tar-water. One who refused to drink it died; all the others recovered.

In addition to 120 testimonials, Prior included 'A Pindarique by the Right Honourable, L. C. J. M., inscribed to the author of *Siris*', and two anonymous sets of verses, which as we shall hear, were written by Berkeley. There is also an unsigned letter to the editor of the *General Evening Post* for 4 June 1744,[20] paying high tribute to *Siris*. This we mention because the author is one of the few eighteenth-century writers we have found who seems to have read *Siris* from beginning to end, and to have realised that it was much more than a disquisition on tar-water. Berkeley is praised as chemist, physician, philosopher, and divine. 'While he gradually leads me on from the simplest operations of nature, thro' the animal and vegetable world, up to the great author of both, I am charmed with my progress, and think I see in this *chain* of his, that golden one, which hung down to earth from heaven, as this by several links carries us up thither.' In the system of *Siris* the correspondent finds 'a principle of pure light', while in other philosophical systems he feels 'nothing but gravitation'.

Among those who seemed to forget that *Siris* was a philosophical rather than a medical work was its author. Most of the short remainder of the Bishop's life was devoted to tar-water. In 1744 he published *A*

Letter to Thomas Prior, and in 1746 *A Second Letter from the author of Siris to Thomas Prior*.[21] Both are practical advice on the preparation and administration of the remedy, with a 'suspicion' that this is a panacea. Berkeley's last publication was *Farther Thoughts on Tar-water*, which appeared in 1752.[22] He looks back happily on the remarkable increase in the use of tar-water: from Ireland and England it had spread throughout Europe. 'Many barrels of tar-water' were shipped from Amsterdam to Batavia. It was successfully used in the East and West Indies, and Berkeley had had accounts, by post and from travellers, of its widespread use in the colonies, 'particularly by those who possess great Numbers of Slaves'. For all practical purposes, tar-water had proved a panacea to man, woman, child, and beast. Other case-histories are quoted. So obsessed had the Bishop become with his American remedy, that he believed that tar-water always succeeds where Nature fails, offering among its successes such extreme examples as these: 'A Gentleman with a wither'd Arm had it restored by drinking Tar-Water. Another who, by running his Head against a Post, had a Concussion of the Brain attended with very bad Symptoms, recovered by drinking Tar-Water after other Medicines had failed. In my own Neighbourhood, one had lost the use of his Limbs by Poison, another had been bitten by a mad Ass; these Persons drank Tar-Water, and their Cure was attributed to it.' Bishop Berkeley had indeed found the universal panacea!

The effect of *Siris* was almost instantaneous. Tar-water warehouses were established in London and elsewhere. Advertisements for tar and tar-water appeared widely in periodicals. On 10 June 1744 William Duncombe wrote to Archbishop Herring:[23] 'It is impossible to write a letter now without tincturing the ink with tar-water. This is the common topic of discourse both among the rich and poor, high and low; and the Bishop of Cloyne has made it as fashionable as going to Vauxhall or Ranelagh.' Luce says: 'A dispensary was opened in St. James's Street, London, by the "Proprietors of the tar-water warehouse", who published a tract professing to explain Berkeley's terms and giving instructions for making tar-water well. Lady Egmont took the remedy, and the Earl seems to have introduced it into Court circles. The Princess Caroline tried, and so did the Duke of Newcastle.'[24]

Some indication how quickly news of tar-water crossed the sea may be seen in a rare pamphlet, which, so far as we have been able to determine, is to be found only in the New York Public Library: *An Abstract from Dr. Berkeley's Treatise on Tar Water, with Some Reflexions therein. Adapted to Diseases Frequent in America. By a friend to the Country* (New York, 1745). Dr. Saul Jarcho has shown that the author was Cadwallader Colden (1688–1776), 'the distinguished colonist physician, savant, and government official',[25] who contributed the papers anonymously to the *New York*

Weekly Post-Boy, in which they constitute numbers 109–14, 18 February to 25 March 1745. In the articles Colden quoted a section of *Siris*, and followed it with twenty-seven paragraphs of discussion, on the whole decidedly finding for Berkeley. Dr Jarcho says that in a fashion typical of the period, 'Colden's comments contain a great deal of reasoning and a very small measure of observation'. There is little question that the essays in the New York paper would have attracted a good deal of attention to the remedy that had proved so effective in Ireland. Within a year tar-water had come home to America.

Evidence of the popular interest aroused in England by tar-water may be seen by following some of the many references in the *Gentleman's Magazine*. We deliberately choose this periodical rather than the *Dublin Journal*, still richer in materials, because that might be considered a prejudiced witness, since its printer and editor was George Faulkner, a close friend of Berkeley's, associated with him in various movements for the public good. The first mention of tar-water in the *Gentleman's Magazine* is in the number for January 1739 (IX, 36), remarkably early, since *Siris* was not to be published for five years. This is a short notice, 'A Receipt to make and use Tar-Water, for preventing Infection by the Small-Pox, and for a consumptive Habit'. The recipe is followed by a statement: 'By this Remedy several Persons in Charles Town, South-Carolina, where the Small Pox was lately very mortal, escap'd the Infection, though conversant with the Infected. . . .' The entry is referred to again in the issue for April 1744 (XIV, 193–6), in connection with a notice of the first London edition of *Siris*, which had recently appeared. 'The Virtues of [Tar-Water] have lately been set in so strong a Light, by an Author of the greatest Character for Learning, Penetration, and Veracity, that we should be negligent of our Duty to the Public, should we omit his judicious Observations.' The lengthy review is made up of extracts from the first third of *Siris*, sufficient to tell the reader how to prepare the water and to give him some idea of the many ills for which it was specific. One of the earliest reviews gives the reader no idea that *Siris* is a philosophical work, but limits its comments and excerpts to the medical section.

The issue for January 1745 (XV, 34) suggests that the publisher of the early London editions was correct in believing that the title *Siris* was not a good one, since the general reader would not understand its significance. An anonymous correspondent, writing to 'Mr. Urban', the editor, says that, because several of his acquaintance had been at a loss to understand the title, he will explain to the public that, according to 'Dionysius', Siris is the name given by Ethiopians to the Nile because of the darkness of the waters. 'Mr. Urban' adds an editorial note to the effect that Berkeley intended an allusion to the Greek word signifying 'chain'.

The next number of the periodical contains a communication to the effect that members of the medical profession give warning that tar-water is dangerous in inflammatory cases, but the Bishop of Cloyne has written to one of the correspondent's acquaintance, insisting that the water had recently cured the Bishop's young son of a fever. In the issue of March 1745 (xv, 163), the editors quote from the *Dublin Journal* an account of the affidavit made by Captain Drape about the Negroes on his ship who recovered from smallpox, thanks to tar-water.

Beginning with the issue of June 1745 (xv, 317–19), we find testimonials sent to Thomas Prior, who was collecting them for his *Authentic Narrative*. One account that must have interested readers was that of a woman, twice married, who never became pregnant until she drank tar-water; the inquiring correspondent learned that the husband also drank it. Many of the reports on cures were picked up and used as advertising by the Tar-Water Warehouse in Painter's Court, Bury Street, St. James's, which announced 'Remarkable Cures perform'd by Tar-Water; collected out of the "Gentleman's Magazine" to be had of the Proprietor'. In issues of 1746 and 1747 readers were told that cures are not limited to human beings; tar-water also cures various distempers in animals. The magazine was obliging enough to publish a recipe showing how to prepare it for cattle (xvii, 22). In March 1748 (xviii, 120), when there was 'cause to be apprehensive' about the plague, the editors included an extract from Berkeley's treatise indicating 'the usefulness of Tar-Water in the Plague'. In November 1748 (xviii, 485–6), 'Mr. Urban' is told that although 'our college of physicians' has not yet seen fit to introduce tar-water into the British Pharmacopoeia, the French have adopted it in the *Formules de Pharmacie*.[26] Opposition of the medical profession to tar-water was continuing, as is evidenced by a quotation sent to the magazine in June 1749 (xix, 247) from *Reflections on Catholicons or Universal Medicines*, by Thomas Knight, M.D., a lengthy tract upon Berkeley's claims for his cure-all of which we shall hear later.

Occasionally correspondents of the *Gentleman's Magazine* launched into verse so far as *Siris* was concerned. The issue for March 1745 (xv, 160), included 'SiRis. A Vision'. In an elysian garden, near a crystal fountain and a mossy grotto, the poet in vision saw a nymph whose name proved to be 'Siris', the daughter of Phoebus and Hygeia, whose 'dotal wealth' was her father's skill and her mother's health. With water from the limpid stream and tar from 'the wounded fir's disclosing side', she mixed a draught which she bade the poet drink: 'her orders I obeyed', / And quick as thought, felt all my anguish fled'. In more serious mood is a Latin ode 'To the Author of Siris' (xxii, 472–3), together with a translation, in which Berkeley is hailed among the great of all time:[27]

> There Newton, studious with extensive aim,
> And Boyle, the friend of man, my rev'rence claim:
> There great Hippocrates, with Syd'nham join'd,
> Share the sweet friendship of a kindred mind.

Most amusing among the verses are lines by one of the few contributors who seems to have read more of *Siris* than the introductory section: 'On the Bishop of Cloyne's SIRIS, which, after treating of the Virtues of Tar, enters upon the sublime Mystery of the Trinity.' Since Tar and Trinity are indissolubly united, modern Arians are urged to drink 'the juice of pines', because

> The *Irish* prelate's *Terebinthian* draughts
> Delude old *Antitrinitarian* t houghts.
> Swallow the julep of the *Norway tree*,
> You'll find the *three in one*, and *one in three*.
> How *orthodox* a *soup!* how glorious *pitch*,
> That cures *coughs, scurvy, heresy* and *itch!* . . .
> Ye *surgeons*, arm'd with lancets, cease to bleed;
> Ye *readers*, drop the *Athanasian* creed;
> Plain *tar*, by bishop *bless'd*, all art controuls;
> It *purifies* your *blood*, your *faith*, and *souls*.

Although 'Mr. Urban' was not aware of the fact, one group of verses published by the *Gentleman's Magazine* in October 1744 had been written by Berkeley himself. On 3 September he had sent them to Thomas Prior with the comment: 'The doctors, it seems, are grown very abusive. To silence them, I send you the above scrap of poetry, which I would by no means have known or suspected for mine. You will therefore burn the original, and send a copy to be printed in a newspaper, or the *Gentleman's Magazine*.' The verses 'On *Siris* and its Enemies. By a Drinker of Tar-Water' begin:

> How can devoted Siris stand
> Such dire attacks? The licens'd band,
> With upcast eyes and visage sad
> Proclaim, 'Alas! the world's run mad.
> The prelate's book has turn'd their brains,
> To set them right will cost us pains.
> His drug too makes our patients sick;
> And this doth vex us to the quick.'
> And vex'd they must be, to be sure,
> To find tar-water cannot cure,
> But makes men sicker still and sicker,
> And fees come thicker still and thicker.[28]

A scarcer set of Berkeleian verses, which appeared only in some copies of *Siris*[29] in the second Dublin edition, suggest the scope of *Siris* as a whole rather than that merely of the medical sections:

> Hail vulgar juice of never-fading pine!
> Cheap as thou art, thy virtues are divine.
> To shew them and explain (such is thy store)
> There needs much modern and much ancient lore.
> While with slow pains we search the healing spell,
> Those sparks of life, that in thy balsam dwell,
> From lowest earth by gentle steps we rise
> Through air, fire, aether to the highest skies. . . .
> But soon as intellect's bright sun displays
> O'er the benighted orb his fulgent rays,
> Delusive phantoms fly before the light,
> Nature and truth lie open to the sight:
> Causes connected with effects supply
> A golden chain, whose radiant links on high
> Fix'd to the sovereign throne from thence depend
> And reach e'en down to tar the nether end.

As another example of various poetic rhapsodies evoked by *Siris*, we quote from 'Tar Water, A Ballad',[30] referred to but not quoted by the *Gentleman's Magazine* for January 1747. The verses probably appeared anonymously that month and sold for sixpence. They were written by a reader of Thomas Prior's *Authentic Narrative*, dedicated to Lord Chesterfield, and begin:

> Since good Master Prior,
> The War Water Squire,
>   Without being counted to blame,
> Vulgar Patrons hath scorn'd,
> And his Treatise adorn'd
>   With the Lustre of Chesterfield's Name.

The author also dedicates to Chesterfield his ballad, then passes to Berkeley and the virtues of the panacea:

> Then come, let us sing;
> Death, a Fig for thy Sting!
>   I think we shall serve thee a Trick;
> For the Bishop of *Cloyne*
> Has at last laid a Mine,
> That will blow up both thee and Old Nick.
>
> Have but Faith in his Treatise
> Tho' you've Stone, Diabetes,

> Gout or Fever, Tar Water's specifick;
> If you're costive, 'twill work;
> If you purge, 'tis a Cork,
> And if old, it will make you prolifick.

It would seem that the ballad was later claimed by Sir Charles Hanbury Williams, ambassador to Russia and Saxony, and included in his *Works*.[31]

Hardly had *Siris* come from the press than pamphlets – replies and defences – began to appear. At least nine appeared during the first year, 1744.[33] Some of these were written by medical men, largely inveighing against the entrance into their lists of a complete amateur, who, like John Wesley in *Primitive Physick* (1747), ventured to prescribe for many ills.[34] One was the work of an apothecary,[35] welcoming the new remedy – as apothecaries well might, with money flowing into their coffers – but suggesting changes in the recipe. At the opposite extreme is the most important scientific paper, that of Stephen Hales, which we shall consider presently. In addition is a small but interesting group of more 'literary' replies, naturally appealing more than the others to the literary historian. These we shall consider in some detail.

Most of the 'replies' of doctors we shall treat only briefly since they are largely repetitious. The first was chronologically the sixth paper in the war of words, appearing in July 1744: *A Cure for the Epidemical Madness of Drinking Tar-Water*, by T. R. M.D. This sixty-six-page pamphlet was the work of Dr Thomas Reeve, whose chief claim to fame is that he was later President of the Royal College of Physicians for some ten years.[36] Reeve's basic position is that *Siris* is filled with errors because the author was ignorant of chemistry. He takes exception to many things, one of which is Berkeley's use of medical terminology. This section of his work may well have been responsible for the fact that the Dublin publisher of *Siris* added a glossary of terms to the edition that appeared in September 1744 and is found in some other editions. Even more than his errors in terminology, Reeve criticises Berkeley's logic. Among the many virtues of tar-water, the Bishop had emphasised its cheapness in contrast with many expensive medicines. In that case, asks the physician, why did not Berkeley go the whole way with John Hancock and recommend the use of cold water only? This pamphlet was answered by one of the few doctors who ever spoke in Berkeley's defence, *The Bishop of Clyne Defended; Or, Tar-Water Proved Useful, By Theory and Experiments. In Answer to T. R. M.D.*, which Fraser dated August 1744. The author identifies himself only as 'Philanthropos', but he was clearly a physician of some sort, since he uses medical terminology throughout and refers to various subjects in the history of medicine. He defends Berkeley's theory of the value of tar-water, though he criticises him for overstatement and exaggeration of its virtues. Cannily he suggests that, while its use was undoubtedly justified in Cloyne, where the villagers were poor and underfed, better remedies are available to the Enhish, who live more comfortably.

We shall omit discussion of a later pamphlet by a doctor,[37] and devote ourselves to the most interesting of the medical pamphlets, and the only one to which we know Berkeley's reaction. In June 1744 appeared an anonymous *Letter to the Right Reverend and the Bishop of Cloyne*,[38] rightly, we believe, attributed to James Jurin, M.D., Secretary of the Royal Society, later President of the Royal College of Physicians, important in the spread and development of inoculation techniques. Jurin had earlier written two replies to Berkeley on another subject.[39] The *Letter* begins with a doffing of the hat in the presence of a Lord Bishop, but soon becomes an attack on Berkeley's attempt to bring 'again the whole Knowledge of Medicine to its Primitive Darkness to specific Remedies, and occult Qualities'. Jurin slyly makes clear that he is speaking of Berkeley's own opinions, 'not what you have copy'd from other Authors (which makes more than two Thirds of your Treatise)'. Jurin takes the Bishop's medical theories and opinions to be 'perfectly exalted above all Sense and Understanding'. Clearly the medical army is drawn up on one side of the field, and close to them in the Battle of the Books (there are vague echoes of Swift in some of Jurin's pages) are 'pernicious Wits', who have 'unmercifully *roasted* your Lordship'. 'I profess', says Jurin, 'I never hear a Horse-Laugh in a Corner of a Coffee-house, but I guess your Lordship is the subject of it.' We are told that a group of apothecaries, when their business was bad, had prepared a petition for the building of hospitals and alms houses, but suddenly finding their business 'prodigiously encrease', they turned their petition into an 'Address of Thanks' to 'your Lordship'. One wit after another is quoted[40] to indicate the laughter evoked by tar-water, but a single sample may suffice: 'Rot me, says another, and cocks his feather'd Hat with an uncommon Smartness, but the Author has serv'd his Medicine just as *Pope* serv'd him, by giving it every Virtue under Heaven, he had made all the World conclude it never had any.' To the paraphrase of Alexander Pope's line on Berkeley, Jurin adds a note querying whether Pope wrote sincerely, or whether it was true, as some thought, 'that he never praised any Body but with an Intent that the World should construe it into something more bitter than his severest Raillery'.

Jurin inveighs against Berkeley's misunderstanding and misinterpretation of the medical authorities to whom he has referred. The Bishop, he says, shows awareness that soap, opium, and mercury come closest to justifying the term 'universal medicine', though each has its limitations, but he puts in place of all of them his own universal panacea. The tone of the conclusion is bitter:

> Why would a Man of the Bishop of Cloyne's Character throw by the Reverence and Good Will the World had to him and make himself so egregiously a Jest, by endeavouring to revive, against the Dictates of Sense, Science and right Reason, a Medicine which he knows was laugh'd out of the World more than Forty Years ago? ... And to conclude in your Lordships own Way of Speaking: As Bishop of *Cloyne*, I *honour* and *respect*, but as a Physician, I *despise* and *pity* you.[41]

With the wits' gallery joining its forces to those of the medical fraternity, the *Letter*, like so much pamphleteering of the age, is a *mélange*, but it is far more interesting to the literary student than any other of the pamphlets written by doctors.

To this tract Berkeley made no public reply. However, on 19 June 1744 he wrote to Prior, enclosing a poem: 'Last night being unable to sleep for the heat, I fell into a reverie on my pillow, which produced the foregoing lines; and it is all the answer I intend for Dr. Jurin's letter.[42] The verses follow:

> To drink or not to drink! that is the doubt,
> With *pro* and *con* the learn'd would make it out.
> *Britons, drink on!* the jolly prelate cries:
> What the prelate persuades the doctor denies.
> But why need the parties so learnedly fight,
> Or choleric *Jurin* so fiercely indite?
> Sure our senses can tell if the liquor be right.
> What agrees with his stomach, and what with his head,
> The drinker may feel, though he can't write or read.
> The authority's nothing: the doctors are men:
> And *who drinks tar-water will drink it again.*

From the point of view of the history of science, the most significant paper in the Berkeley controversy is that of Stephen Hales, *An Account of some Experiments and Observations on Tar-Water*, which was read before the Royal Society and published in December 1744.[43] Unlike the pamphlets discussed so far, this was not prepared as either attack on or defence of Berkeley, though it proved a most important defence. Hales was not concerned with the practical problems discussed by Berkeley of the making of tar-water and its administration to cure or alleviate many diseases. His was an attempt to analyse tar-water chemically, in order to learn wherein its undoubted curative properties lay. He discusses, among other things, experiments he had made upon tar from America, Norway, and Barbados. The paper is so highly technical that it has not the interest for the layman of some of Hales's other publications or of the man himself. Stephen Hales had read for holy orders at Cambridge and spent nearly all his life as 'the perpetual curate of Teddington', ministering to his parishioners. Actually he spent most of his time in scientific experimentation. His *Vegetable Staticks* of 1727 became a basic text in plant physiology, his *Haemastaticks* of 1733 a milestone in animal anatomy. Both were read by many laymen. He had become well enough known to the scientific world to be elected Fellow of the Royal Society in 1717. The Society awarded him its Copley Medal in 1739. He is known to many literary students through his acquaintance with Alexander Pope, since Teddington was a neighbour of Twickenham. The relation between pastor and poet was close enough for Hales to be one of the two witnesses to Pope's will. Pope referred to him twice in poetry, on one occasion as if he had recently been sketching 'parson Hales'. However, the Popean reference to Hales most frequently repeated is that in Spence's *Anecdotes*, in which Pope regretted that the good parson cut up animals, particularly dogs.

Berkeley replied to Hales in 'A Letter to Dr Hales', which appeared in the *Gentleman's Magazine* for February 1747, and was also published by Berkeley, with a letter to Prior, as *Two Letters from the Right Reverend Dr. George Berkeley . . . to Thomas Prior . . . to the Rev. Dr. Hales*.[44] To the second edition of Hales's *Account* was added *A Letter to Dr. Hales Concerning the Nature of Tar*, by A. Reid, Esq.[45] Writing in 1747 he indicated that it had become the custom for the sick, particularly consumptives, 'to repair to the *Red-house* at Deptford as their last Resort' to drink 'the clear Liquor from the barrelled Tar, and be cured'. Reid warmly agreed with Hales on the importance of tar-water, but he offered some correction to Berkeley's recipes. Among the great of remote or recent past who, he thinks, would have approved tar-water, he numbered Pliny the Elder, Glauber, Van Helmont, Boyle, and Boerhaave. Berkeley had good reason to appreciate the agreement he found in Stephen Hales and his correspondent, in opposition to the various medical men who expressed their adverse views about tar-water.

By all odds, the most elusive of the pamphlets listed by Alexander Campbell Fraser proved to be *Reflections upon Catholicons, or Universal Medicines* (1749), by Thomas Knight, M.D. It is not in the British Musuem nor in libraries in Edinburgh, not in the Berkeley collection in Trinity College, Dublin, nor, so far as the good offices of Professor Lester Conner could determine for us, elsewhere in Dublin. We finally found a copy in the Bodleian Library. Because of its rarity we shall treat it in slightly more detail than some others. The author, a Fellow of the Royal College of Physicians, had earlier written another scientific-medical work, *An Essay of the Transmutation of Blood* (1725), which stirred a controversy in the 1730s about the definitions and chemical composition of blood. In *Reflec-*

tions upon Catholicons Knight shows his familiarity with earlier controversialists, Prior, Hales, Reid, and 'Risorius', to whom he frequently refers. The main point of the pamphlet is that Berkeley's 'universal Catholicon', although beneficial at times, is not the panacea that its author and the apothecaries claimed it to be; for this reason the physician personally refuses to try it. 'Every Body', he says 'takes Tar-Water, but that is not a sufficient Reason for me to take it; Custom is not a sovereign Law to judicious Persons, the Deluge of bad Example drowns the whole World; for not to follow the Fashion is look'd upon to be the worst Character a Man can have in this World.'[46] His basic criticism is expressed early in the work:

> Hence it appears by experimental Observations and demonstrable Principles, that the Philosophical Reflections, and the Theory so artfully handled, setting forth the virtues of Tar-Water, is no more than a bare Hypothesis, being not founded upon real but imaginary Principles; the fugacious, fine, active, acid, volatile Spirit, so bland and temperate, and the subtile aetherial Oil, which were to do the Feats, are lost, there remaining only in the Tar the grosser and less active Principles, the caustic Oil or Sulphur, and Salt corrupted and render'd alkaline by burning the Wood close cover'd up to make Tar.

Much of the pamphlet is devoted to a discussion of 'the acid Spirit' in tar. Here he shows himself abreast of recent trends in chemistry, since the 'universal acid' in the air was among the most frequently discussed chemical topics in the 1740s and 1750s.[47] As indications of this widespread interest we may note that Tobias Smollett mentioned the 'vague acid' (*acidum vagum*) briefly in his novel *Ferdinand Count Fathom* (1753), and that Henry Fielding made Tom Jones ask whether Sophia's illness midway in the chase was not caused by some 'unknown universal acid in the air'. According to Knight, a main reason why tar-water is not so efficacious as Berkeley claimed is that 'it is but reasonable to imagine, that if the acid Spirit in Tar-Water renders it cool and unnoxious', tar-water actually does nothing to the body but heat it, thus 'effecting its opposite'. 'It is a *Proteus*, which changes its Form every Moment, in assuming contrary Qualities answerable to the Occasion of those who take it.'

More readable than the pamphlets of medical men – with the exception of Dr. Jurin's – are the ones to which we now turn. On 24 July 1744 appeared *Remarks on the Bishop of Cloyne's Book Entitled Siris*, by Risorius, M.A. Oxon. The author has not been identified. This can hardly be called a 'literary' pamphlet, but it was written by an amateur, not a doctor or scientist, since 'Risorius' shows little knowledge of medicine and frequently derogates doctors. He condemns the medical sections of *Siris* as 'a very rash and hasty Performance . . . by no means worthy of its learned, and reverend, Author'. He takes exception to Berkeley's implication that tar-water is a Catholicon, a universal panacea. As we continue through the six-part critique, tedious except when lightened by irony, we incline to believe that his basic criticism is that Berkeley is Irish, while 'Risorius' is a true-born Englishman. Tar-water, he declares, may conceivably be an excellent remedy when administered by 'the Bishop's masterly Hand', but may prove dangerous 'under the Management of the more Illiterate'. It is probably 'better adapted to the Climate of *Ireland* than to *English* Constitutions; in which Case, the prudent Part would be, to leave it entirely to the *Irish*, those Neighbours of the Bishop, who have the opportunity of applying immediately to his Lordship for advice. I am indeed persuaded, 'tis not *calcualted to do good* to the *English*.'

Anti-Siris,[48] an anonymous pamphlet of some sixty pages, seems to have been the earliest of the replies to *Siris*, since it was published in May 1744. The epistolary style, while less literary, suggests the *Persian Letters* or Goldsmith's *A Citizen of the World*. A 'Foreign Gentleman', resident in London, wrote to his 'Friend abroad' that the 'continual fluctuations' and faddism he found in London in clothes and diversions, and in politics, was now paralleled in 'Drugs and Physick'. When he had first arrived in London, the current medical fad was '*Sugar:* and the Public swallowed it voraciously, till it had rotted half the Teeth in the Nation, and was supplanted by *Water*'. These were followed by Joshua Ward's 'Pill and Drop'.[49] But the popularity of all these pales into insignificance in contrast with the '*Nostrum of Tar Water*, lately come into general Vogue and Use here on the bare Assertion of a Spiritual Q—k'. The 'Pill and Drop' were sold for much more than their Weight in Gold, 'while the B-p's Specific costs scarce anything'. Ward jealously kept his recipe secret; it is to the credit of the Reverend Quack that he publishes his. 'Anti-Siris' has his suspicions of the Bishop's reasons for his generosity, as we shall see.

The writer knows some true details of the Bishop's discovery of tar-water, although he is wrong about others. He is aware that Berkeley married just before crossing the sea and that his wife had taken a woman friend as her companion. He takes for granted, however, that Berkeley went to Bermuda and understands incorrectly that he returned to England in a year. He adds still another to the various legends that came to surround Berkeley's American sojourn:

> [Joshua Ward] travell'd no farther and *Paris* in quest of Fame and Wealth, whereas his L——p went half Way to the *Antipodes* to seek for the Virtues of *Tar*; and 'tis not improbable but he would have gone all the Way, so intent was he to become famous and talk'd of, had he not luckily contracted an Intimacy with an *Indian*, whom he intended to convert to Christianity, who initiated the modern *Apostle* into all the *American* Arts and Sciences,

and particularly into that Knowledge of *Tar*, which he now so generously communicates to his Fellow Subjects.[50]

The Anti-Sirian suggests that the Bishop's real purpose in publishing *Siris* was less to cure the bodies of men than to save their souls. His suspicions of Berkeley's motives go further:[51] 'It is pretty extraordinary too that he chose to publish his latent Poison, at so very critical a Conjuncture of the present, at a time when England is at war with France and that the *Pretender's* Son is within call.' *Anti-Siris* does not doubt that *Siris* will have political repercussions: 'The Call for this Commodity here must raise the Price in *Norway*, which will not only endear our Prince and Nation to the *Danes*, but will enable their King to help us against *France* without a *Subsidy*.' Indeed, the pine-growing Norwegians would reap a rich harvest, since the author computes that, if the demand for tar continues at its present rate, the consumption will amount to 29,784,569 pounds and 3/7/8/ ounces. What Bishop Berkeley had initiated, he could not control.

By all means the liveliest and most interesting of the pamphlets is *Siris in the Shades*, with a sub-title, *A Dialogue concerning Tar Water; Between Mr. Benjamin Smith, lately deceased, Dr. Hancock, and Dr. Garth at their Meeting upon the Banks of the River Styx* (July 1744). No author is known. He was probably not a scientist, since there is nothing technical in the paper, and not a physician, since again the medical discussions are not in technical language. He may well have been a journalist; the style is lively and informal. Two of the characters may be readily identified, but the lately deceased 'Benjamin Smith' remains a mystery.[52] Dr Garth is the well-known Samuel Garth, author of *The Dispensary* (1699), who had died in 1719. Less familiar to literary students, but known in the history of medicine, is John Hancock, D.D., Prebendary of Canterbury, whose name was mentioned in an earlier pamphlet. He was probably chosen for two reasons: like Berkeley, he was of the cloth; like him, too, he had come into fame and infamy because of his advocacy of a cure-all, cold water, the widespread use of which he recommended in his *Febrifugium Magnum* in 1722, at a time when there was fear of the plague. Hancock was mentioned by name in Defoe's *Journal of the Plague Year*, and two papers in the Hancock controversy were formerly attributed to Defoe.[53]

As the title implies, *Siris in the Shades* is a 'dialogue of the dead'. On the banks of the Styx Hancock meets Smith, who had died of 'a Burning Fever'. He blames his physician for not using the remedy recently discovered 'to cure all Diseases, and prolong a Man's Life to the Age of Methuselah'. If the medicine continues to be as popular as it was when Smith died, Charon may lay up his boat and take a holiday, since there will be no work for him. Smith praises tar-water, Hancock his own remedy, until they are at an impasse. When Dr Garth appears, they refer the issue to him because he, they find, has read *Siris*. Smith learns to his surprise that books had been common in Hades until Pluto recently made an edict forbidding further traffic in them, since so many are now the work of free-thinkers. Pluto had been so impressed by all he had heard of *Siris* that he made an exception, and also released Tantalus from his punishment and ordered Dr Garth to make tar-water and administer it to him. Hungry and thirsty, Tantalus drank it eagerly, but in three days it made him so horribly sick that he begged for his lake again.

With the arrival of Garth, Smith has more and more difficulty in his defence of Berkeley. The doctor tears his argument to pieces, showing how impossible it is – medically and logically – that the same medicine should be effective upon opposite conditions of hot, cold, moist, dry. Whoever wrote *Siris in the Shades* had read more of *Siris* than many of the readers we have encountered. Smith, newly arrived in the place of darkness, defends tar-water as embodying the '*luminous Spirit*, or the *solar Light*.'[54] Garth fails to understand how so many errors of fact could be made by a 'Person of his Lordship's Learning and Abilities, one that soars so high in Spirituals, and who can demonstrate the doctrine of the Trinity, by a necessary Chain of Reasoning, from the Virtue of *Tar-Water*'. He finally concludes that the sections on tar-water had been deliberately introduced by the Bishop to attract the attention of general readers to philosophy. ''Tis past a Doubt to me, that the Bishop's Book was writ with no other View, but to make Converts to his Philosophy; and that Tar-water was intended only for a Bait to draw worldly-minded People in to read it; or, as one may say, a Ladder, by which you may mount up to the Trinity.' His Lordship is not a quack, because 'the Book is not in Reality, a Treatise of *Physics*, but of *Metaphysics*'. It is conceivable that the author was the anonymous correspondent to the *General Evening Post*, mentioned above, one of the few who realised that *Siris* was a philosophical work. The letter was published on 4 June 1744, *Siris in the Shades* in July. Garth's last interpretation of *Siris* is somewhat similar to that in the letter.

We have largely concerned ourselves with pamphlets of the year of publication of *Siris*, mentioning a few later ones in notes, but the controversy continued for some years. Indeed, we find echoes of it thirty years later in a different connection. In 1776 William Hawes, a London apothecary wrote *An Examination of Mr. John Wesley's Primitive Physick* (1776), blasting Wesley's remedies in general, but particularly his excessive prescription of tar-water in various ills. When Wesley prescribed it for 'St Anthony's Fire', then a common disease, Hawes denounced the prescription on the grounds that tar-water is 'a very heating medicine'; later, when Wesley recommended the taking of tar-water as a cure for 'Old Age',

Hawes again denounced him, as he did repeatedly in the paper. But we shall leave the pamphlet war, and turn in the final section of this study to consider the interest in tar-water of men – and women – of letters.

The almost instantaneous effect of *Siris* may be seen in a letter of Thomas Gray's. The publication of the first edition in Dublin had been 20–4 March 1744; the first London edition was 'promised in a few days' on 30 March. Gray wrote to Thomas Wharton on 26 April:[55] 'oh Lord! I forgot to tell you, that Mr Trollope[56] & I are in a course of Tar-Water, he for his Present, and I for my future Distempers: if you think it will kill me, send away a Man & Horse directly, for I drink like a Fish.' Horace Walpole[57] wrote to Sir Horace Mann on 29 May 1744: 'We are now mad about tar-water, on the publication of a book that I will send you, written by Dr Berkeley Bishop of Cloyne. The book contains every subject from tar-water to the Trinity; however all the women read it, and understand it no more than they would if it were intelligible. A man came into an apothecary's shop t'other day, "Do you sell tar-water?" "Tar-water!" replied the apothecary, "why I sell nothing else!" ' Edmund Burke, a student at Trinity College, wrote to Richard Shackleton in July 1744:[58] 'I am sure Tar is the universal Medicine here notwithstanding the opposition of its Enemy's the physicians. Does anyone in your Villa, or Academy use it?'

Siris seems to have been the last book Alexander Pope ever read. His comment on it, to which we shall return, marked the approaching end of a friendship between Pope and Berkeley that had continued for over thirty years. Berkeley was not a Scriblerian, but he had been associated with the group during the period that he was a 'London Wit'. Swift he had probably known in Ireland.[59] It was through Swift that Berkeley received his appointment as Chaplain to the Earl of Peterborough, with whom he travelled on the Continent; probably through Swift that he began his association with Dr John Arbuthnot, who rendered various services, and to whom it was due that one of Berkeley's letters found a place in the *Philosophical Transactions of the Royal Society*. Writing to Arbuthnot from Naples on 17 April 1717, Berkeley graphically described several ascents when Vesuvius was in eruption. Arbuthnot read the letter to the Society, and it was published in the records.[60]

When the younger Scriblerians, Alexander Pope and John Gay, were drawn into the charmed circle, Berkeley may well have felt even closer to them than to the seniors, since Berkeley and Gay were the same age, Pope three years younger. The group kept in touch with each other throughout their lives. Relationships among them have been discussed by their various biographers. We limit ourselves to Pope, since he was the only one who lived to read *Siris*.[61] His acquaintance with Berkeley had begun at least as early as 1713, since on 7 March of that year Berkeley mentioned in a letter to Sir John Percival that Pope had given him a copy of *Windsor Forest*.[62] Various letters to Pope from Berkeley, particularly during his Continental travels (1713–14, 1716–20), are extant.

All the Scriblerians followed with interest Berkeley's Bermuda project. Gay wrote to Pope in September 1725 that he had just seen their friend Dean Berkeley, who was most solicitous about Pope's health. 'He is now so full of his Bermuda's project that he hath printed his Proposal, and hath been with the Bishop of London about it.'[63] Joseph Spence included in his *Anecdotes*[64] a tale Lord Bathurst told Joseph Warton

> that all the members of the Scriblerus Club, being met at his house at dinner, they agreed to rally Berkeley, who was also his guest, on his scheme at Bermudas. Berkeley having listened to all the lively things they had to say begged to be heard in his turn; and displayed his plan with such astonishing and animated force of eloquence and enthusiasm, that they were struck dumb, and after some pause, rose up all together with earnestness, exclaiming, 'Let us all set out with him immediately.'

In the same connection Spence also says that Berkeley liked to apply to the Bermudas Horace's description of the Fortunate Isles, in epode xvi, 'and was so fond of this epode on that account, that he got Mr. Pope to translate it into English'. In October 1725 Pope lightly warned his friend Robert Digby that in another year he might 'carry you all with me to the *Bermudas*, the seat of all Earthly Happiness, and the new Jerusalem of the Righteous'.[65]

Berkeley went to America and returned, his scheme a failure, but tar-water in his mind. For a short time he remained in England, but when he was elevated to the see of Cloyne he returned to Ireland, there to spend almost all the rest of his life, there to develop tar-water and weave it into a philosophy. In the spring of 1734 Pope mentioned his 'strong inclination' to see the Bishop of Cloyne before his departure for Ireland. During the stay in London Pope presumably showed him the manuscript of *An Essay on Man*, since – again according to Spence[66] – Pope said: 'In the Moral Poem I had written an address to our Saviour, imitated from Lucretius' compliment to Epicurus, but omitted it by the advice of Dean Berkley.' In 1738 Pope published in his *Epilogue to the Satires*[67] the line of tribute to Berkeley we have heard paraphrased:

> Ev'n in a Bishop I can spy Desert;
> *Secker* is decent, *Rundel* has a Heart,
> Manners with Candour are to *Benson* giv'n,
> To *Berkley*, ev'ry Virtue under Heav'n.

Pope's final comment on Berkeley was in a letter to Hugh Bethel, characteristically undated, probably written in April 1744. 'I have had the bishop's book as a present', he wrote,[68] 'and have read it with a good deal of pleasure.' Undoubtedly Pope recognised *Siris* – of which he had probably received the first edition – as the work of philosophy it was, but there is little question that the medical sections would most have attracted his attention, since at that time his various physicians were desperately seeking relief for the condition they had diagnosed as dropsical asthma. Pope may have consulted them about the possibility of his taking tar-water, since he wrote to Bethel, who also suffered from asthma: 'my own doctors [have] disagreed with your Yorkshire Dr. Thomson, on the use of water in a dropsical asthma'.[69] Berkeley must deeply have regretted their decision, since Prior's alphabetical list included reports on the cure or alleviation of nineteen cases of asthma, three of dropsy, by the use of tar-water. Pope died without the universal panacea.

From Pope we pass to his once-adored, later-detested Lady Mary Wortley Montagu, who mentioned tar-water in her letters on three occasions, though one is a mere passing reference. On 24 April [1748] she wrote from Brescia to her husband, thanking him for volumes of Sir Charles Hanbury Williams's poetry, in one of which was the 'Ballad on Tar Water' that has been mentioned. Bishop Berkeley's cure, she suggests, has taken the place of Joshua Ward's 'pill and drop' (which Wiliams had also satirised). She continued:[70]

> Tis possible by this time that some other Quackery has taken the place of that. The English are easyer than any other Nation infatuated by the prospect of universal medicines, nor is there any Country in the World where the Doctors raise such immense Fortunes. I attribute it to the Fund of Credulity which is in all Mankind. We have no longer faith in Miracles and Reliques, and therefore with the same Fury run after receits and Physicians. The same Money which 300 years ago was given for the Health of the Soul is now given for the Health of the Body, and by the same sort of People: Women and halfe witted Men. In the countries where they have shrines and Images, Quacks are despis'd, and Monks and Confessors find their account in manageing the Fear and hope which rule the actions of the Multitude.

Three years later she again mentioned the remedy to her husband:

> Tar Water is arriv'd in Italy, I have been ask'd several Questions concerning the Use of it in England. I do not find it makes any great progress here. The Doctors confine it to a possibility of being usefull in the Case of inward ulcers, and allow it no farther merit. I told you sometime ago the method in this Country of making it the Interest of the Physician to keep the Town in good Health.

Those very satisfactory correspondents, Mrs Elizabeth Carter and Miss Catherine Talbot, who seem to have read nearly all books published, mentioned tar-water to each other over a period of nearly a dozen years. Mrs Carter first referred to it on 20 July 1744, some three months after *Siris* first appeared, experiencing some of the confusion of various later readers:[71]

> I make no doubt but you have read *Siris*, as I have to no great purpose you will think, as I fairly confess I have no clear idea what one half of it means: what I can understand of it extremely pleases me, but possibly its being beyond the reach of my comprehension is the cause that some parts of the book appear entirely visionary, and more like the glittering confusion of a lively imagination, than any regular system of distinct reasoning.

She added: 'Pray what is your opinion of tar-water?' Miss Talbot replied on 7 September that, although she was considered the local 'quack', she had not yet ventured to try it on any of her neighbours, 'though by what I can learn of it, it is very good if properly applied'. The next sentence indicates that she was aware of some of the 'replies' to Berkeley we have considered: ''Tis very hard I think the good man, who published his opinion of it from no other notice than a general benevolence, should be so vilely abused for it, as he has been by various paltry scribblers.'

Two years later, on 1 November 1746, Mrs. Carter indicated that she had begun a course of tar-water:[72]

> Have people utterly left off writing books? I have not heard of a new one this century, excepting one on the wonders of Tar-water. I thought the strong appetite to this medicine had been greatly worn off, and that folks now were universally agreed in the fashionable fury of drinking up the sea, an experiment perhaps much the less safe of the two. Tar-water being thus again in high repute, several of my acquaintance have persuaded me into a consent to drink it, though I depend but little upon its efficacy with regard to myself; however, as one ought to give a medicine fair play, I intend to persevere as far as a hogshead will go, before I pronounce that it does me no good.

Probably to her own surprise, Mrs. Carter found that the remedy suited her very well after she had persevered for some time 'with great resolution', since on 8 December 1746 she reported: 'I really think it has done me some good, for the first effect I perceived was that I could bear the sight of beef and pudding, and the next that I arrantly eat it, and upon the whole I am better.'[73]

That tar-water was considered a standard remedy, long after its novelty had ceased, is indicated by passing remarks between the two friends. Mrs Carter, who had been away from her home in Deal, wrote on 21 May 1750 that she 'had a most formidable idea of being sick in a land overrun with physicians, and not like Deal flowing with tar-water'. As late as 1755 she teased her friend, to whom, she declared, tar-water had become one of her two 'specifics'.

Henry Fielding tried every 'specific' known to his period in a vain attempt to regain health. He was attended for a time by the notoriously sceptical Dr Thomas Thompson, the last of Pope's physicians, and then asked to be treated by tar-water, this move offering the measure of his desperation, Fielding's basic trouble, like Pope's, was diagnosed as dropsical asthma. In the last year of his life (1754) he was treated by Joshua Ward of 'pill and drop' fame, to whom he paid high tribute, although the treatments were ineffectual in what Fielding described as his 'weak and deplorable condition, with no fewer or less diseases than a jaundice, a dropsy, and an asthma, altogether uniting their forces in the destruction of a body so entirely emaciated, that it had lost all its muscular flesh'.[74] Defending Ward against the imputation of failure, Fielding went on to say that no one medicine could be a panacea for all ills, and continued:

> But even such a panacea one of the greatest scholars and best of men did lately apprehend he had discovered. It is true, indeed, he was no physician; that is, he had not by the forms of his education acquired a right of applying his skill in the art of physic to his own private advantage; and yet, perhaps, it may be truly asserted, that no other modern hath contributed so much to make his physical skill useful to the public; at least, that none hath undergone the pains of communicating this discovery in writing to the world. The reader, I think, will scarce need to be informed that the writer I mean is the late bishop of Cloyne, in Ireland, and the discovery that of the virtues of tar-water.
>
> I then happened to recollect, upon a hint given my by the inimitable author of the Female Quixote, that I had many years before, from curiosity only, taken a cursory view of bishop Berkeley's treatise on the virtues of tar-water[75] which I had formerly observed he strongly contends to be the real panacea which Sydenham supposes to be in existence in nature, tho' it yet remains undiscovered, and, perhaps, will always remain so.

Evidently Fielding re-read *Siris* – at least the medical sections – to find that Berkeley considered tar-water effectual in dropsy. After a brief period on a milk diet, he betook himself to tar-water, dosing himself morning and evening with a half-pint. He had earlier been tapped by surgical trochar for his condition. We may let him continue the account:

> It was no more than three weeks since my last tapping, and my belly and limbs were distended with water. This did not give me the worse opinion of tar-water: for I never supposed there could be any such virtue in tar-water, as immediately to carry off a quantity of water already collected. For

> my delivery from this, I well knew I must be again obliged to the trochar; and that if the tar-water did me any good at all, it must be only by the slowest degrees; and that if it ever should get the better of my distemper, it must be by the tedious operation of undermining, and not by a sudden attack and storm.
>
> Some visible effects, however, and far beyond what my most sanguine hopes could with any modesty expect, I very soon experienced; the tar-water having, from the very first, lessened my illness, increased my appetite, and added, though in a very slow proportion, to my bodily strength.

Temporarily he showed improvement; when he was tapped again the surgeon withdrew three quarts less than the previous time, and Fielding bore the ordeal better than before, without faintness. But his health had gone too far for improvement by any medicine, and he started on his last journey, the voyage to Lisbon.

Throughout the eigthteenth century – both before and after the author's death – *Siris* remained largely a medical rather than a philosophical work. From it William Cowper drew a figure that has come to be associated with him rather than with Berkeley, who wrote in *Siris* § 217: 'the luminous spirit lodged and detained in the native balsam of pines and firs is of a nature so mild, and benign, and proportioned to the human constitution, as to warm without heating, to cheer but not inebriate'. Cowper versified this idea in *The Task* (iv, 38–40):

> And while the bubbling and loud hissing urn
> Throws up a steamy column, and the cups
> That cheer but not inebriate wait on each . . .

Coleridge, who had gone through a period of Berkeley-enthusiasm sufficient to cause him to name his second son 'Berkeley', showed little admiration for *Siris* in a passage in his notebooks in which he said:[76]

> Much injury has been done to society by the naked *assertion* of Truths which have been repeated till at least they have [been] treated with contempt as old Paradoxes. Ex. gr. If Berkley instead of asserting that England could sustain three times its number, tho' it were encompassed by a brass Wall 50 cubits high, had written a treatise as long & eloquent as that which he squandered upon Tar Water & layed open the whole of the good, & all of the evil & delusion of Commerce, & artificial Wealth – my God, what a difference –

Professor D. J. Greene, in 'Smart, Berkeley, the Scientists and the Poets',[77] finds in Christopher Smart striking parallels to Berkeley, some of them in passages from *Siris* – though they are, of course, from the philosophical rather than the medical sections. The philosopher and the poet, he believes, were close together in their dismissal of both Newton and Locke. Whether the similarities between the poet and Berkeley were coincidental, or the result of 'influence', Mr Greene does not know: 'What Smart's acquaintance was with Berkeley and his writings, I do not pretend to determine; I have found no evidence on the point. It may be that Smart arrived at all this independently or through some intermediary source. But the parallel in thought is striking.' William Blake owned a copy of *Siris* – one of the editions published at Dublin in 1744 – and annotated it with marginalia.[78] It is significant that these do not begin until page 203 – well after the long medical passages.

Since we have made no pretence of treating *Siris* as a work of philosophy, we shall not deal with writers of the nineteenth century, during which it ceased being a medical treatise. If, however, we may believe Pip in *Great Expectations* (1861), tar-water continued to be used. 'Some medical beast', he says, 'had revived tar-water as a fine medicine', and Pip is so frequently dosed with it by his sister that he is 'conscious of going about, smelling like a new fence'. We conclude with that modern descendant of Berkeley, Smart, and Blake – William Butler Yeats. In his Introduction to J. M. Hone and M. M. Rossi's *Bishop Berkeley*, Yeats wrote:

> ... did he think that if he could stop all thought with his Utopian drug – what thinker has not felt the temptation – the mask might become real: he that cannot live must dream. Did tar-water, a cure-all learnt from American-Indians, suggest that though he could not quiet men's minds he might give their bodies quiet, and so bring to life that incredible benign image, the dream of a time that after the anarchy of the religious wars, the spiritual torture of Donne, of El Greco and Spinoza, longed to be protected and flattered. The first great imaginative wave had sunk, the second had not yet risen.[79]

The third sage in 'The Seven Sages' remarks: 'My great-grandfather's father talked of music, / Drank tar-water with the Bishop of Cloyne.' In 'Blood and the Moon', in *The Winding Stair and Other Poems*, Yeats described his tower, with its 'winding, gyring, spiring treadmill of a stair', a figure that well describes *Siris*'s 'chain', leading the reader from tar-water to the Trinity:[80]

> I declare this tower is my symbol; I declare
> This winding, gyring, spiring treadmill of a stair is my ancestral stair;
> That Goldsmith and the Dean, Berkeley and Burke have travelled there ...
>
> And God-appointed Berkeley that proved all things a dream,
> That this pragmatical, preposterous pig of a world, its farrow that so solid seem,

Must vanish on the instant if the mind but change its theme . . .
Everything that is not God consumed with intellectual fire.

NOTES

1. This legend, which seems to have been persistent for many years, was repeated in these words by the grandson of one of Berkeley's closest Rhode Island associates, Wilkins Updike, *History of the Episcopal Church in Narragansett* (New York, 1847), 395. Alexander Campbell Fraser, *The Works of George Berkeley* D.D. (Oxford, 1871), IV, 154–8, disproved it from Berkeley's correspondence and other historical evidence. Rhode Island had been settled for a century, and Newport was an important cultural and trade centre, well known in England. Berkeley had chosen it deliberately. This visit was described in a letter from William Byrd to Sir John Percival. See Benjamin Rand, *Berkeley and Percival* (Cambridge, 1914), 238, 244; other material about the period is in Rand, *Berkeley's American Sojourn* (Cambridge, 1932). In the most recent treatment of the American period, Alice Brayton, *George Berkeley in America* (Newport, 1954), 4, suggests that the stop in Virginia may have been made because two of Berkeley's travelling companions, Richard Dalton and John (later Sir John) James, both gentlemen of substance (who, she suggests, might have paid for the ship that made the crossing), wished to make a protracted visit there. Throughout this article we refer to the Luce and Jessop edition of *The Works of George Berkeley*, 9 vols. (1948–57), as 'Luce and Jessop, *Works*'.

2. *Op cit.*, xiv–xv, 99–100.

3. In the passage referred to above Miss Brayton says – without reference – that Berkeley mentioned South Carolinian Indians in connection with tar-water. We have not been able to locate the passage. In *Siris* (1744), 4, he says: 'In certain parts of America tar-water is made. . . . This cold infusion of tar hath been used in some of our colonies, as a preservative or preparation against the small-pox.' On p. 9 he mentions 'the method used by our colonies in America, for making tar and pitch'. A. A. Luce, *The Life of George Berkeley, Bishop of Cloyne* (1949), 200, says, in connection with the smallpox epidemic, that Berkeley 'heard that the Indians used tar-water as a specific preventive, though he did not actually meet it in use'. He does not indicate which Indians. One sentence in *Siris* (§ 115) leads us to wonder whether Berkeley had used tar-water in Rhode Island, perhaps at the time of the smallpox epidemic. He refers to 'my own manner of making it, and not the American; that sometimes makes it too strong, and sometimes too weak'. Unless otherwise stated, all references to *Siris* are to the reprinted London edition of 1744.

4. Updike, *History*, 176. Alexander Fraser in his *Life* takes for granted Berkeley's interest in the Narragansetts, as does Luce, *Life*, 122, *et passim*.

5. This, the only one of his sermons published during his lifetime, was printed at London in 1732. It may be found in *A Miscellany containing Several Tracts on Various Subjects by the Bishop of Cloyne* (1752), 215–35.

6. This theory was generally accepted by Berkeley scholars until 1955, when E. J. Furlong and W. V. Denard published 'The Dating of the Editions of Berkeley's *Siris* and of his first letter to Thomas Prior', *Hermathena*, LXXXVI (Nov. 1955), 66–76. Since six editions were published in 1744, the problem was inevitably complicated. Furlong and Denard show that the first edition was the Dublin one announced 20–4 Mar. 1744.

7. Perhaps at Berkeley's insistence, the title *Siris* was restored in the London edition announced on 30 Apr.

8. *Op. cit.*, IV, 293.

9. *George Berkeley: a Study of his Life and Philosophy* (Cambridge, 1936), 411.

10 'Berkeley's *Siris*', *The Personalist*, XLI (winter 1960), 5.
11 In *Hermathena* L (1936), 197.
12 *Life*, 197.
13 *Letters* in Luce and Jessop, *Works*, VIII, 249; see also 272–3.
14 E. J. Furlong in part determined the order of the various editions of *Siris* upon such internal evidence as these matters. See *Hermathena*, LXXXVII (1956), 37–48.
15 We had hoped to be able to determine more accurately than we can Berkeley's probable reading in contemporary natural science. A catalogue of the sale of his library by Leigh and Sotheby in 1796 is in the British Museum, but R. I. Aaron believes that many books Berkeley once owned are now missing. See his 'Catalogue of Berkeley's Library', *Mind*, XLI (1932), 465–75. Berkeley had left in America many of the books he had brought with him, which he had used in the writing of *Alciphron*. Of the forty-seven works to which he referred there, Aaron found only one in the catalogue. Among the approximately 1,000 books he gave or sent to Yale, ten were upon natural philosophy. These included the *Philosophical Transactions of the Royal Society* from its beginning to 1720. Twenty-five volumes deal with 'Anatomy, Physick, and Chyrurgery'. On this see Henry M. Fuller, 'Bishop Berkeley as a Benefactor of Yale', *Yale University Library Gazette*, XXVIII (1953), 1–18; Andrew Keogh, *ibid.*, IX (1934), 1–25. J. M. Hone and M. M. Rossi, *Bishop Berkeley: His Life, Writings and Philosophy* (1931), 236, describing Berkeley's involvement with tar-water, says: 'The library shelves were filled with books on medicine, mostly ancient and mediaeval.' They give no authority for the statement.
16 See Marjorie Nicolson, 'Joshua Ward's "Pill and Drop", and Men of Letters', *Journal of the History of Ideas*, XXIX (1968), 177–96.
17 In Alexander Campbell Fraser (ed.), *Works* (Oxford, 1871), III, 465–70.
18 We have used the 'New Edition', London, 1746, published in Dublin and reprinted in London. The first edition had appeared earlier that year.
19 He is so called by both Prior and Berkeley. In the *Gentleman's Magazine* his name is given as Draper.
20 *Op. cit.*, 16.
21 In the British Museum copy these are attached to the 1746 *Authentic Narrative*.
22 In *A Miscellany, containing Several Tracts on Various Subjects* (1752), 9–28. The most recent edition is in Luce and Jessop, *Works*, V.
23 Quoted A. A. Luce, *Life*, 201.
24 *Ibid.*
25 'The Therapeutic Use of Resin and of Tar Water of Bishop George Berkeley and Cadwallader Colden', *New York State Journal of Medicine*, IV (1955), 834–40. On pp. 834–5 Dr Jarcho points out a fact Berkeley's biographers have not stressed, that in 1711 Berkeley had effectively used resin for the treatment of diarrhoea.
26 Tar-water was later admitted to the British Pharmacopoeia, and remained in the Dublin list well on in the nineteenth century. In 'Bishop Berkeley and his Use of Tar-Water', *Annals of Medical History*, 3rd ser., IV (1942), 463–4, Burton Chance says: 'In the latest practical United States Dispensatory two or three columns are given to the chemical composition of the water, and directions for its occasional employment.' Dr Saul Jarcho, in his article on Cadwallader Colden (840), mentions its presence in 'the remarkable pharmacopacia which was prepared for the French forces in the American Revolution, and published at Newport in 1780'.
27 There is also a set of 'Verses sent to a Gentleman, on the Use of Tar-Water' in the issue of Oct. 1757 (XXVII, 471).
28 *Letters* in Luce and Jessop, *Works*, VIII, 273–4.

Bishop Berkeley and tar-water

29 Luce and Jessop, *Works*, v, 225–6. Another poem of Berkeley's will be found later.
30 *Tar Water, a Ballad, Inscribed to the Right Honourable Philip Earl of Chesterfield: Occasioned by reading a Narrative on the Success of Tar Water, Dedicated to his Lordship by Thomas Prior, Esquire.*
31 *Works*, Horace Walpole, Earl of Oxford (ed.) (1822), II, 21–4. That volume also includes a short poem, 'On Charles Stanhope, Esq. Drinking Tar-Water'.
33 Alexander Campbell Fraser, *Works of George Berkeley*, II, 355–7, listed twelve tracts in what he considered their chronological order. Earlier than that, an anonymous writer in the *Restrospective Review*, I (1853), 20–35, discussed a number of them, including *Anti-Siris, Siris in the Shades*, and the pamphlets by 'Risorious', Jackson the chemist, 'T. R.', 'Philanthropic', and Stephen Hales. Three others are mentioned, but so briefly that it is impossible to identify them. These are all largely passing comments and offer nothing in dating or identification.
34 See G. S. Rousseau, 'John Wesley's *Primitive Physick* (1747)', *Harvard Library Bulletin*, XVI (1968), 242–56.
35 *Reflections concerning the Virtues of Tar-water*, published in June 1744, ascribed to H. Jackson, who is obviously an apothecary.
36 Sir George Clark, *A History of the Royal College of Physicians of London* (Oxford, 1966), II, 550–1, merely mentions that as President he proposed a candidate in 1754, and again mentions a resolution of thanks to him on his retirement as President in 1764. He seems to have had little stature ten years earlier when he wrote the pamphlet.
37 Malcolm Flemyng, M.D., *A Proposal for the Improvement of the Practice of Medicine* . . . (Hull, 1748). The work is dedicated to Dr Richard Mead. Flemyng was a well-known English physician, who later took part in the animal-spirits controversy with his *Nature of the Nervous Fluid; or, Animal Spirits Demonstrated* (1751); he also wrote a poem about hypochondria, *Neuropathia; sive de Morbis Hypochondriacis et Hystericis* . . . (Hull, 1740).
38 The rest of the title is: *Occasion'd by his Lordship's Treatise on the Virtues of Tar-Water. Impartially Examining How Far that Medicine deserves the Character His Lordship has given of it* (June 1744).
39 In 1734 Berkeley had published *The Analyst: Or, a Discourse Addressed to an Infidel Mathematician*. Addressed to Edmund Halley, this was an attack upon mathematics as Isaac Newton propounded it, and an analogical vindication of the mysteries of religious faith. Jurin replied, under the pseudonym of 'Philalethes Cantabrigiensis', in *Geometry no Friend to Infidelity* (1734). Berkeley answered in *A Defence of Free-Thinking in Mathematics* (Dublin, 1735), and Jurin again in *The Minute Mathematician . . . a Defense of Sir Isaac Newton*, in the same year under the same pseudonym. In the *Letter*, discussed in our text, p. 8, Jurin asks ironically, 'What, I would ask him, may they not expect of one who can . . . improve upon Sir Isaac Newton?' The controversy was discussed by Alexander Campbell Fraser in his edition of Berkeley's *Works*, III, 301–2.
40 The quotations thus far have been from pp. 8–9; those that follow are from pp. 10–11.
41 In the Houghton Library, Harvard College, is a group of pamphlets on the tar-water controversy, a number of which have never been catalogued. The index is in a nineteenth-century hand. The group includes all but three of the pamphlets listed by Alexander Campbell Fraser. In addition there is one bound with the Jurin letter, beginning at p. 17 of that work. This is an anonymous *Remarks on a Letter to the Right Reverend the Bishop of Cloyne* (July 1744). Intending a defence of Berkeley, the author answers Jurin point by point. This is a very dull piece of work. A 'justification' for a non-medical man's venturing to write on medicine is found on p. 19: 'The two great Names of Bacon and Boyle will be lasting Monuments, that the Knowledge of Physick is not confined to the Profession only.'

42 A. A. Luce published the lines in *Letter* (Luce and Jessop, *Works*, VIII, 271). He says (IX, 117) that he had been unable to find the verses in the *Dublin Journal*. Prior published them in his *Authentic Narrative*, 16.

43 Several of Hales's papers and references to others appear in the *Philosophical Transactions*, but there is no mention of this one. Since it is long and highly technical, Hales apparently preferred to publish it as he did, *in toto*, rather than to have a digest made. It is very curious that there seems to be no reference to tar-water in the *Philosophical Transactions*. Two brief papers on tar were published at the end of the preceding century, XIX (1695–7), 544; XX (1698), 291. These had to do only with methods of extracting tar.

44 The entry in the *Gentleman's Magazine* is XVII, 64f. *Two Letters* was published in London, 1747.

45 Andrew Reid, who migrated to London from the Scottish provinces in 1720, edited a journal, *The Present State of the Republic of Letters*, and was a compiler for scientific subjects. He was responsible for the abridgement of the *Philosophical Transactions* of the Royal Society from 1720 to 1732, although he was not a member of the Society. He was also in charge of the abridgement of Newton's *Chronology*.

46 Our quotations are from pp. 106–7, 34–5, 66.

47 See G. S. Rousseau, 'Smollett and the *Acidum Vagum*', *Isis*, LVIII (1967), 244–5.

48 *Anti-Siris: Or, English Wisdom Exemplify'd by various Examples, But, Particularly, The present general Demand for Tar Water, On so unexceptionable Authority as that of a R–t R–d Itinerant Schemist, and Graduate in Divinity and Metaphysicks. In a Letter From a Foreign Gentleman at London, To his Friend Abroad.*

49 The fashion for sugar is attributed in a note on p. 19 to 'Doctor *Slayer*, who wrote a learned Treatise on the Usefulness of *Sugar*'. The vogue of water is attributed on p. 22 to the Revd Mr Hancock, mentioned in a medical treatise above. Ward's nostrums are alluded to on p. 28.

50 *Ibid.*, 30, 34.

51 *Ibid.*, 50–1.

52 Various bibliographical aids we have consulted include only one work by a Benjamin Smith, *Method of Raising a Loaded Cart, When the Horse in the Shaft has fallen* (c. 1730). The only clue in the work itself is that Smith is 'a great mercury man'. Since the other speakers are given their own names, there seems no reason for disguising one of several 'mercury men' of the period.

53 *Remarks upon Febrifugium Magnum; Flagellum, or a dry answer to Dr. Hancocke's wonderfully comic liquid book.* Both were published in 1722.

54 The quotations in this section are from pp. 12, 30, 33–5.

55 Paget Toynbee and Leonard Whibley (eds.), *Correspondence of Thomas Gray* (Oxford, 1935), I, 225. This is not a Warton of the poetic family, but Thomas Wharton, Pensioner of Pembroke Hall and student of medicine at Cambridge.

56 Gray was Fellow of Peterhouse, Cambridge; William Trollope, Fellow of Pembroke, seems, from various of Gray's references, to have been a semi-invalid. He died in 1749.

57 Lewis, Smith, and Lam (eds.), *Correspondence with Sir Horace Mann* (New Haven, 1954), 452.

58 Thomas W. Copeland (ed.), *The Correspondence of Edmund Burke*, I (Chicago, 1958), 26.

59 A. A. Luce in his *Life* of Berkeley discusses the relationship between Swift and Berkeley in detail, including that in appendix ii, 232–3. The unexpected bequest of 'Vanessa' to Berkeley lies beyond the scope of this study.

60 XXX No. 354, pp. 709–13. It is also in Fraser, *Life and Letters*, and more recently in

Bishop Berkeley and tar-water

Jessop and Luce, *Works*, IV, 78–81. In *Philosophical Transactions* Berkeley's name is given as 'Edw.'.

61 Swift was still alive, but his mental faculties were clouded.

62 Benjamin Rand, *Berkeley and Percival* (Cambridge, 1914), 110. The Berkeley-Pope letters are in George Sherburn (ed.), *The Correspondence of Alexander Pope* (Oxford, 1956). Professor Sherburn quotes (II, 104n.) a letter of Berkeley's written in 1721, when the plague threatened England, 'a preservative against the plague . . . the Jesuits bark taken as against the ague'. This prescription, Berkeley says, he had from Dr Arbuthnot.

63 *Correspondence*, II, 324.

64 In S. W. Singer's edition of the *Anecdotes* (1964) the Bermuda accounts are on pp. 154–5. They are not included in James M. Osborn (ed.), *Observations, Anecdotes, and Characters of Books and Men* (Oxford, 1966). The source was Joseph Warton, *Essay on Pope* (1782), II, 204n., where the same story is given verbatim, except that Warton said 'the members', rather than 'all'. If the story was true – Lord Bathurst often exaggerated in his stories – and if 'all' the members were present, it might have been during the summer of 1726, when Swift was back in England and when Berkeley's enthusiasm was at a peak, although Harley was by this time dead.

65 *Correspondence*, II, 330. In the 1735 edition of his Letters Pope added a note, calling attention to the fact that, just at the time he wrote, 'Dean Berkley' had developed his project of 'erecting a Settlement in Bermuda for the . . . Propagation of the Christian Faith, and of Sciences of America'.

66 In Singer's edition of Spence (1964), 103; in Osborn's, I, 135.

67 John Butt (ed.), *Imitations of Horace* (London, 1961), 316–17.

68 Dr Thompson (Thomson) is discussed at length in Marjorie Nicolson and G. S. Rousseau, *'This Long Disease, My Life': Alexander Pope and the Sciences* (Princeton, 1968), 295–305.

69 *Correspondence*, IV, 514.

70 Robert Halsband (ed.), *The Complete Letters of Lady Mary Wortley Montagu* (Oxford, 1966), II, 397; 486–7; 20 June [1751].

71 *A Series of Letters between Mrs. Elizabeth Carter and Miss Catherine Talbot, from the year 1741 to 1770 . . . in Two Volumes* (1809). The two first references are I, 44, 46. It is interesting to find that during the 1740s and 1750s, Berkeley's son was one of Mrs Carter's correspondents.

72 *Ibid.*, I, 114–15; Miss Talbot mentioned her friend's new course briefly (I, 116).

73 *Ibid.*, I, 119.

74 *The Journal of a Voyage to Lisbon* by Henry Fielding, ed. Austin Dobson (1902), 27, and 36–7 for the following passage.

75 Fielding had in his library a copy of one edition of *Siris*, listed as 'Berkeley and Prior on Tar Water', 1744. See the 'Sale Catalogue' of his library in Ethel M. Thornbury, *Henry Fielding's Theory of the Comic Prose Epic* (Madison, 1931), 177.

76 Kathleen Coburn (ed.), *The Notebooks of Samuel Taylor Coleridge* (New York, 1957), I, 893. The note indicates that this was a 'garbled reference' to a passage in *The Querist*.

77 *Journal of the History of Ideas*, XIV (June 1953), 327–52. The passage quoted is on p. 343.

78 David V. Erdman (ed.), *Poetry and Prose of William Blake* (New York, 1965), 562–4.

79 P. XVIII.

80 Peter Allt and Russell K. Alspach (eds.), *The Variorum Editionaa of the Poems of W. B. Yeats* (New York, 1957), 480–1.

7

Praxis 2: Pineapples, pregnancy, pica and Peregrine Pickle

After Nicolson's illness incapacitated her, I began to write in the same vein on my own – so much practical work relating literature to medicine remained to be completed, and we seemed to have been among the few scholars willing to undertake it.

I remember the delight I experienced when I first realised the intentions of Tobias Smollett (the eighteenth-century novelist) at the opening of his playful second novel, Peregrine Pickle. *This very long book narrates the story of its hero, the naughty and wilful Peregrine, whose mother despises him. Smollett, like his fellow novelist Laurence Sterne in* Tristram Shandy, *spends a great deal of time narrating the circumstances of his hero's birth. But it was only as a consequence of extensive reading in the medicine of his day that I came to understand the terms of Smollett's wit.*

Writing of the sort found in this essay – as Nicolson would have been the first to remark – was fun: fun for the original author (Smollett), for his interpreter (the critic), for his reader (all of us). I also remember the fun I had writing it, although I am certain that many theorists today would claim it is utterly devoid of theoretical profile and therefore of little consequence.

I published the essay in 1971 in an anthology of critical essays about Smollett (Oxford).

I

'This young lady, who wanted neither slyness nor penetration . . . replied with seeming unconcern, that for her own part she should never repine, if there was not a pine-apple in the universe, provided she could indulge herself with the fruits of her own country.'[1] Mrs Pickle's remark is calculated and represents the basis of a plan to rid herself of the 'teizing and disagreeable Mrs. Grizzle'. In fact, Smollett tells us that if a certain 'gentleman happening to dine with Mr. Pickle' had not mentioned pineapples, Peregrine might not have seen the light of day: the ridiculous Mrs Grizzle would have remained at the side of her sister-in-law throughout gestation, teasing and torturing her with obsequiousness, dementing her imagination, which at this period seemed to be strangely diseased, and 'marking' her still unborn child.[2] A 'diseased imagination' in the mother, Mrs Grizzle would have argued, produced inferior progeny (Peregrine born with the image of a pineapple clearly defined on his body!), but a 'dish of pineapples' could produce no progeny at all. Mrs. Grizzle's search for pineapples for 'three whole days and nights' is thus futile: Peregrine, our hero, must be born. But then, a pregnant woman's desires must not be balked – Mrs Grizzle would have continued – and Mrs Pickle had told her that 'she had eaten a most delicious pine-apple in her sleep'. Unaware of the consequences of these alternatives, Mrs Grizzle makes the wrong choice, furnishing her sister-in-law with two ripe pineapples, 'as fine as ever were seen in England'. Propitiously, however, Mrs Pickle did not partake of the fruit – at least, Smollett never tells us that she did. Her swooning at the dinner table, the ensuing hysteria, the nocturnal dream – all are devices to encourage Mrs Grizzle to leave the Pickle household.

It is curious that these chapters (v-vi) describing the incubation and birth of Peregrine have never received any attention. This fact is further puzzling when one recalls that they are among the most amusing in the novel. Smollett's eighteenth-century readers would have been amused by Mrs Grizzle's obstetric handbooks and pious medical beliefs, and perhaps entertained by Mrs Pickle's intense dislike of her 'sister'. But they must have been doubly amused by the elder's 'researches within the country' (Smollett's hyperbole) for pineapples: such peregrinations for fruits 'which were altogether unnatural productions, extorted by the force of artificial fire, out of filthy manure!'[3] For her distaste for pineapples and her pretentious medical learning – Aristotle and Nicholas Culpepper[4] – eighteenth-century audiences could have forgiven Mrs Grizzle; they could not forgive her for an inability to put learning into practice. Her immense concern for pineapples, however, must have struck them as particularly topical, for readers of *Peregrine Pickle* would have understood Mrs Grizzle's new-fangled theories of the effects of this fruit on pregnant women in the context of eighteenth-century medicine. We, as modern readers, view her actions and statements as part of Smollett's use of learning for the purposes of wit. A clue to the extensiveness of Smollett's knowledge of the subject can be observed in the first full-length book on pineapples in English, John Giles's *Ananas, a Treatise on the Pine Apple* (London, 1767), a book which Smollett may or may not have read. This technical handbook for scientists and expert gardeners confirmed Mrs Grizzle's belief that pineapples were an unlucky sign for pregnant mothers. Much of Smollett's comic irony in these chapters derives from the audience's familiarity with contemporary ideas about pineapples. It is not surprising, however, to discover medical learning used for comic purposes in the novels of an author who was a physician by profession and whose works abound with a variety of scientific references.

For our purposes chapters v and vi of *Peregrine Pickle* contain the key passages in which Mrs. Pickle describes her unexplainable desires: for a fricassee of frogs, for a porcelain chamber-pot, for three black hairs from

Mr Trunnion's beard – and for pineapples. Then we are told that only in the case of exotic foods did Mrs. Grizzle interfere:

> She restricted her [Mrs Pickle] from eating roots, pot-herbs, fruit, and all sort of vegetables; and one day when Mrs. Pickle had plucked a peach with her own hand, and was in the very act of putting it between her teeth, Mrs. Grizzle perceived the rash attempt, and running up to her, fell upon her knees in the garden, intreating her, with tears in her eyes, to resist such a pernicious appetite. Her request was no sooner complied with, than recollecting that if her sister's longing was baulked, the child might be affected with some disagreeable mark, or deplorable disease, she begged as earnestly that she would swallow the fruit, and in the mean time ran for some cordial water of her own composing, which she forced upon her sister, as an antidote to the poison she had received.[5]

The witty context of this description and the ridiculous actions of the women it involves should not lead us to conclude that Smollett was distorting or ridiculing contemporary theories of embryology. Actually Smollett's accounts of the treatment of pregnant women (such as Mrs Pickle) were based upon experiments with foetuses performed in the third and fourth decades of the eighteenth century and speculations of respected scientists of the Royal College of Physicians such as James Augustus Blondel, Daniel Turner, and William Smellie, a leading obstetrician to whom Smollett was apprenticed and in whose medical library he educated himself. It is essential to note biographically that at the very same time – 1750 – Smollett was composing *Peregrine Pickle*, he was also editing, annotating, and preparing for the press Smellie's *Treatise on the Theory and Practice of Midwifery*, which was published the same year as Smollett's novel. In 1750, then, Smollett was deep in the study of obstetric medicine and, particularly, in abnormal pregnancy, for that subject occupies the largest portion of Smellie's book. Smollett's novel was the first in English (a decade before *Tristram Shandy*)[6] to refer to heated controversies about the role of imagination in abnormal pregnancy, a subject smellie treated at length and which abundantly stimulated Smollett's own imagination. The humorous episode built upon Mrs Grizzle's extreme distrust and dislike of pineapples also reflects the popularity of this exotic fruit in England and in Scotland in the 1740s and 1750s, its alleged medicinal qualities, and its being a forbidden fruit to pregnant women. Although Smollett did not live long enough to see the appearance of Edward Topham's widely read *Letters from Edinburgh written in the years 1774, and 1775*, he would have taken Scottish pride in Topham's observation that in the 1740s and 1750s Scotland, no less than England, was the garden of Europe – at least as far as exotic *ananas* were concerned:

> . . . if the Scotch are deprived, by the nature of their situation, of enjoyment of natural fruit, they have the opportunity of furnishing themselves with hot-houses . . . and, in this respect, have the advantage of the rest of Great Britain. There are few gentlemen of any consequence that are not supplied with fruit by this means; and indeed, melons, pineapples, grapes . . . are produced here with great success.[7]

'Mrs. Pickle's longings', Smollett tells us, 'were not restricted to the demands of the palate and stomach, but also affected all the other organs of sense, and even invaded her imagination, which at this period seemed to be strangely diseased'. The notion that a mother's 'imagination' influenced her foetus and subsequently, her child, was as old as Aristotle. In his treatise *De Generatione et Corruptione*, the Greek philosopher had discussed the matter at length in 'Rules for the First Two Months of Pregnancy', 'Let none present any strange, or unwholesome thing to her, not so much as name it, lest she should desire it, and not be able to get it, and so either cause her to miscarry, or the child have some deformity on that account.'[8] The sections in Aristotle's book dealing with gestation and pregnancy were extracted in the seventeenth century and bound together, under the title *The Experienced Midwife*. Galenic medicine of the seventeenth century had little to add to Aristotle's precepts. Occasionally an author such as Shakespeare capitalised upon Aristotle's admonishment: thus Pompey, the jester in *Measure for Measure*, comes on stage bellowing that his Mistress Elbow 'came in, great with child, and longing for stewed prunes; Sir, we had but two in the house, which at that very distant time stood in a fruit dish, a dish of some threepence'. Distinguished physicians like Thomas Sydenham and Thomas Willis adhered to the ideas stated in *The Experienced Midwife*, and less illustrious doctors were equally obeisant.

Throughout the seventeenth and early eighteenth centuries this work was a standard textbook for physicians and midwives, and was the major source for Nicholas Culpepper's *Directory for Midwives* (first published in 1651), a book which Smollett must have read at least by the time he prepared Smellie's *Treatise on Midwifery* for the press. Culpepper's *Directory* had become so popular by the 1730s – years during which Smollett was in medical school – that it evoked in 1735 this enthusiastic rhapsody from an anonymous commentator:

> And if he [any young doctor] applies himself to the Obstetrical Act, let him turn over Culpepper's *Midwife enlarg'd* night and day. That little Book is worth a whole Library. All that is possible to be known in the Art is there treasur'd up in a small *Duodecimo*. Blessed, yea for ever blessed, be the memory of the inimitable Author, who, and who alone, had the *curious happiness* to mix the profound learning of Aristotle with the facetious Humour of Plautus.[9]

A physician himself, Culpepper repeated Aristotle's advice to pregnant women who, like Mrs Trunnion, wished to give birth to unscarred infants. 'Sometimes there is an extraordinary cause, as imagination, when the Mother is frightened, or imagineth strange things, or longeth vehemently for some meat which if she have not, the child hath a mark of the colour or shape of what she desired, of which there are many examples.'[10]

It was not the pregnant mother's 'strange longings' – a condition known as 'pica'[11] – that disconcerted physicians and midwives so much as the ill effects of these yearnings on the foetus. According to Culpepper even a single instance of bizarre desire would produce 'Hermaphrodites, Dwarfs and Gyants', and this idea was repeated again and again in medical works of the period. Jon Maubray discussed it at length in *The Female Physician, Containing all the Diseases Incident to that Sex, in Virgins, Wives and Widows* (1724). In a chapter entitled 'Of Monsters' he complained that too few authors were 'ready to discuss the proper Causes of *Monstrous* BIRTHS', and continued to give his explanation for the occurrences:[12]

> First then, I take the Imagination to have the most prevalent *Power* in Conception; which I hope may be readily granted, considering how common a Thing it is, for the *Mother* to mark her child with *Pears, Plums, Milk, Wine*, or any *thing else*, upon the least trifling *Accident* happening to her from thence; and *that* even in the latter ripening *Months*, after the Infant is entirely formed, by the *Strength of her Imagination* only, as has been already manifestly set forth at large.

Maubray extended his case to the male as well, noting that 'a Foetus with a *Calf's, Lamb's, Dog's, Cat's-Head*, or the Effigy of any other thing whatsoever,' might be the result of '*a copulating* Man, if he should imprudently set his Mind on such Objects, or employ his perverted *Imagination* that way'. The powerful and lasting effects of the imagination, Maubray contended, were not limited to humankind, but extended also to lower species. 'This absurd *Imagination* takes place even among the very *Brutes*, as Lemnius relates of a Sheep with a Seal's or *Sea* Calf's-Head, having no doubt seen that Animal in the critical Time of *Conjunction* or *Conception*.' Like many doctors in the previous century, Maubray enjoined would-be mothers to suppress their 'absurd Imaginations', lest they bring into the world no children but 'Monsters formed in the Womb.'

An incident late in 1726 which Smollett probably had heard about, contributed much to the popular fear that women with 'absurd Imaginations' during pregnancy would bring forth monsters: this was the extraordinary case of the pregnant Mary Tofts of Godalming who insisted that she had eventually given birth to at least seventeen rabbits and other curious progeny. Unable to afford rabbits, she nevertheless craved them throughout her pregnancy, and one day, while working in the fields, she actually saw one who may or may not have frightened her. The case itself has recently been discussed in such detail that I do not pretend to add new discoveries or theories,[13] but will suggest that its widespread fame and the satires it provoked – for example, Hogarth's 'Credulity, Superstition, and Fanaticism' – added further fuel to existing fears. On 5 December 1726, Pope wrote to John Caryll who lived not far from Godalming and might be expected to have heard more details than Londoners: 'I want to know what faith you have in the miracle at Guildford; not doubting but as you past thro' that town, you went as a philosopher to investigate, if not as a curious anatomist to inspect, that wonderful phenomenon. All London is now upon this occasion, as it generally is upon all others divided into factions about it.'[14] If Caryll replied to Pope's inquiry, as he may well have, the letter is not extant, but numerous Scriblerian satires are. One of these was a series of verses by Pope, 'The Discovery: Or, *The* Squire turn'd Ferret. An Excellent New Ballad. To the Tune of *High Boys! up go we; Chevy Chase;* Or what you please', published in December 1726, again in January, and several times thereafter. Here Pope turned his light artillery particularly upon two scientists, Nathaniel St André, a Swiss anatomist and medical attendant on the King, and Samuel Molyneux. Mary Tofts had first been attended by John Howard, a Guildford surgeon and male midwife who had not known her until he was called in on her case. After having devoted most of his time to her for several days, delivering nine rabbits, he moved her to Guildford, to which he invited anyone who doubted the veracity of the reports he had been giving. St André, unfortunately for himself, accepted the invitation and made the trip to Guildford, taking with him Samuel Molyneux, secretary to the Prince of Wales, a scientist of great distinction, particularly important for his work in developing the reflecting telescope. Molyneux was not a medical man and made no pretence to knowledge of midwifery; St André, on the other hand, although he had taken no degree, had been apprenticed to a surgeon and had held the post of local surgeon to the Westminster Hospital Dispensary. It was the unoffending Moyneux, however, who bore the brunt of Pope's satire – perhaps because Pope knew more about telescopes (he had grown interested in the optical effects of the reflecting telescope) than about midwifery:[15]

> But hold! says Molly, first let's try
>   Now that her legs are ope,
> If ought within we may descry
>   By help of Telescope.
> The Instrument himself did make,
>   He rais'd and level'd right.
> But all about was so opake,
>   It could not aid his Sight . . .

> Why has the Proverb falsely said,
>   Better two Heads than one;
> Could *Molly* hide this *Rabbit's* Head,
>   He still might show his own.

Pope's satire on the 'Rabbit Breeder' was one among many. He himself may or may not have contributed another 'ballad' on the Tofts case to the *Flying Post*, published on 19 December 1726. In the *Flying Post* it was 'Said to be Written by Mr. Pope to Dr. Arbuthnot'; the published title is simple 'Mr P—— to Dr A——t'. Here the satire is chiefly directed at Sir Richard Manningham, son of the Bishop of Chester and godson of Sir Hans Sloane, 'society's most distinguished man-midwife', whose attention to Mary Tofts had been ordered by King George himself. Many others may be found in the Library of the Royal Society of Medicine in an apparently unique scrapbook, 'A Collection of 10 Tracts' on 'Mary Tofts, the celebrated pretended Breeder of Rabbit'. Among these is a one-page set of verses, 'The Rabbit-Man-Midwife', inscribed in an eighteenth-century pencilled hand, 'by John Arbuthnot'. Another is a tract of ten pages, *The Opinion of the Rev'd Mr. William Whiston concerning the Affair of Mary Tofts, ascribing it to the Completion of a Prophecy of Esdras*, written by William Whiston, formerly Lucasian Professor of Astronomy at Cambridge, Isaac Newton's successor. In the apocryphal book of Esdras, said Whiston, ''Tis here foretold that there should by "Signs in the Women" or more particularly that Menstrous [*sic*] Women should bring forth Monsters'.[16] Presumably writing years after the Mary Tofts affair, when the story had been 'long laughed out of Countenance', Whiston insisted that he believed it to be true 'as the fulfiling of this Ancient Prophecy before us'. Still another satire in the Royal Society of Medicine scrapbook is a pamphlet of 1727, purportedly written by 'Lemuel Gulliver, Surgeon and Anatomist to the Kings of Lilliput and Blefuscu, and Fellow of the Academy of Sciences in Balnibarbi': *The Anatomist Dissected: or the Man-Midwife finely brought to Bed*, 1727.[17] If one of the Scriblerians was responsible for this thirty-five-page pamphlet, it was Dr Arbuthnot, who need not have concealed authorship since *The Anatomist Dissected* was far above the average tract on the rabbit-woman. It professes to be chiefly 'An Examination of the Conduct of Mr. St André, Touching the late pretended Rabbit-bearer', based on St André's defence of himself, *A Short Narrative of an Extraordinary Delivery of Rabbits*. But scrutiny reveals that it is the work of a physician with extensive knowledge of anatomy and experience in childbirth. He points out inconsistencies in St André's account, pausing over matters of the temperature and pulse of a woman in labour, and the influence of her demented imagination on the foetus.

I have strayed afield and treated the Tofts case at length because it stirred a controversy among physicians and other scientists that was to last in England more than forty years. Less than a month after the episode of the 'rabbit breeder', James Augustus Blondel, a distinguished member of the College of Physicians of London, brought out a treatise denying the possibility of such an occurrence: this work was later advertised by Dr Blondel as 'My first Dissertation, *The Strength of Imagination in pregnant Women examin'd*, published upon the Occasion of the Cheat of *God-alming*, hastily, and without Name, as coming from one, who neither designed to be known nor to meddle any more in this Controversy'. Trained in Leyden by Boerhaave, Blondel maintained that deformities in birth were caused by other factors – such as actual delivery – than the mother's imagination. He was challenged by Daniel Turner, another physician and Fellow of the Royal College of Surgeons, who dogmatically asserted the opposite.[18] The two physicians, both distinguished and both fellows of the same society, stirred considerable debate within the College of Physicians. Since Blondel felt he had been attacked personally by Turner, he responded with a 155-page defence of himself: *The Power of the Mother's Imagination Over the Foetus Examin'd. In Answer to Dr. Daniel Turner's Book, Intitled a Defence of the XIIth Chapter of the First Part of a Treatise, De Morbis Cutaneis* (1729). In his preface Blondel stated his purpose: 'My Design is to attack a vulgar Error, which has been prevailing for many Years, in Opposition to Experience, sound Reason, and Anatomy: I mean the common Opinion, that Marks and Deformities, which Children are born with, are the sad Effect of the Mother's irregular Fancy and Imagination.' Without providing any historical survey of the controversy Blondel noted 'that the Doctrine of Imagination, relating to the Foetus, had gone through several Revolutions', and continued to indict the 'Imaginationists'.[19]

> 'Tis silly and absurd; for what can be more ridiculous, than to make of Imagination a Knife, a Hammer, a Pastry-Cook, a Thief, a Painter, a Jack of all Trades, a Juggler, Doctor Faustus, the Devil and all?
> 'Tis saucy and scandalous, in supposing that those, whom God Almighty has endowed, not only with so many charms, but also, with an extraordinary Love and Tenderness for their children, instead of answering the End they are made for, do breed Monsters by the Wantonness of their Imagination.
> 'Tis mischievous and cruel; it disturbs whole Families, distracts the Brains of credulous People, and puts them in continual Fears, and in Danger of their Lives: In short 'tis such a publick Nuisance, that 'tis the Interest of every Body to join together against such a Monster, and to root it entirely out of the World.

In less than six months the 'Imaginationists' led by Turner retorted with a reassertion of the influence of the mother on her foetus. Parodying Blondel's title, Turner called his treatise *The Force of the Mother's Imagination upon her Foetus* (1730). This work was over 200 pages and was a definitive defense of the majority view of the time. One of Turner's arguments was 'the authority of Antiquity': he had culled hundreds of ancient and medieval medical writings and prepared a list of monstrous births in which the mother's demented imagination was the apparent cause. Most common among these, Turner affirmed, was her craving for exotic fruits – plums, cherries, grapes, prunes, and now pineapples. Unlike Blondel, Turner was not concerned with the physiological processes by which the foetus was actually 'marked'. He abstained from proving the truth of his argument by appealing to 'sensation, the nerves, and the circulation of the foetus'. Instead, he cited numerous ancient and modern authorities, known and obscure – Hesiod, Heliodorus, Jacobus Horstius, Ambroise Paré, Johann Schenkius, Thomas Bartholin, Charles Cyprianus, Robert Boyle, Sir Kenelm Digby – who had reported instances of deformed children; moreover, Mary Tofts herself, although 'a cheat', had been frightened by rabbits while sowing in the fields, and in Turner's estimation there was a definite connection between the imagination and the foetus. Learned though his argument seemed, Turner was scurrilously satirised by his opponents in several pamphlets and poems, perhaps the most witty and scrofulous among them a burlesque set of verses written in Butlerian octosyllables entitled 'The Porter Turn'd Physician', and published in 1731.

Although Blondel and Turner after 1730 did not publish works concerning the mother's imagination, the controversy over which they differed continued to be a topic of concern in medical and lay circles for at least three decades. Fellow of the Royal Society, many of whom were Smollett's personal friends, and other scientifically inclined gentlemen, hesitated to drop so controversial an issue. Doctors Hunter and Monro spent considerable time on the topic of their anatomical lectures, and we know that Smollett not only heard these but was in 1748–50 in medical dialogues with these men. As late as 1747, John Henry Mauclerc, an M.D. of no great distinction, published a lengthy treatisè entitled *Dr. Blondel confuted: or, the Ladies Vindicated, with Regard to the Power of Imagination in Pregnant Women: Together with a Circular and General Address to the Ladies on this Occasion* (London).[20] Odd as it may appear, Mauclerc's book approved rather than confuted – his own term – Blondel's theory that imagination alone was unable to harm the foetus. 'The Design of the Dissertation', he wrote in the preface of his book intended especially for women, 'is to prove that the Opinion, which has long prevail'd, that the Marks and Deformities, Children bring into the World, are the sad Effect of the Mother's irregular Fancy and Imagination, is nothing else but a vulgar Error, contrary to sound Reason and Anatomy.' Later in his preface Mauclerc writes as if the controversy were still inflaming the hearts of medical men, twenty years after the fact: 'I don't despair of Success: Interest alone should prevail, upon the Party, which is chiefly concerned in this Controversy.' Dr Mauclerc disbelieving in old wives' tales and other odd superstitions, presents – to quote his own words – 'a Sketch of the true Cause of Monsters – I hope, 'tis sufficient for the present, to give a general, and yet a clear Solution of those strange Phenomena [monsters]'. If Blondel had his supporters, so did Turner, although both men had been dead for many years. In 1765, Isaac Bellet, a French physician residing in London, wrote *Letters on the Force of the Imagination in Pregnant Women*, in which he denied the possibility that pregnant women could mark their children, but was compelled to agree rather with Turner's explanations than Blondel's. The result was a second-rate medical work fraught with contradiction, but nevertheless one showing how very much alive the matter still was. Dr Smollett, probably the author of the review in the *Critical Review*,[21] found the book very appealing: 'We declare upon the whole', he wrote, 'that he [Bellet] has fulfilled his scope, and executed his undertaking with great precision, and that he was clearly demonstrated the impossibility of a pregnant woman's marking her child with the figure of any object for which she has longed, or which may have made a deep impression upon the imagination.' It is difficult, if at all possible, to state accurately what Smollett's views on the subject were almost twenty years earlier, when he was writing the early chapters of *Peregrine Pickle*, but there is every reason to believe that even then the probability of a mother marking her child seemed remote to him. Hunter, Monro, and Smellie doubted the possibility, and medically speaking, they exerted much influence on his thought.

Throughout the 1740s cases of extraordinary childbirth of every sort continued to interest the English public, especially physicians and scientists. One could compile with ease a long list of works written in the decade about strange childbirths. Nor was the subject treated in books and tracts only. Popular periodicals that enjoyed large amateur audiences devoted much space in their issues to these freaks of nature. *The Gentleman's Magazine*, for example, contained no fewer than ninety-two articles (essays, reviews, and letters) on the question of extraordinary childbirth.[22] Curiosity was especially aroused by a woman who never became pregnant until she drank Bishop Berkeley's tar-war in 1745. More specialised in its reading audience, the *Philosophical Transactions* of the Royal Society (to which Smollett never was elected but many of whose Fellows he knew) was flooded with communications about bizarre births attesting to the interests of its fellow members in the subject. Professor Knapp's biography makes it clear that Smollett was familiar with some of these publications.

An idea of the range and diversity of such cases may be gained by listing a few of the titles of these articles:[23] 'Account of a monstrous boy'; 'Account of a monstrous child born of a woman under sentence of transportation'; 'An Account of a monstrous foetus resembling an hooded monkey'; 'Case of a child turned upside down'; 'A remarkable conformation, or lusis naturae in a child'; 'Part of a letter concerning a child of monstrouse size'; 'Account of a child's being taken out of the abdomen after having lain there upwards of 16 years'; 'a letter concerning a child born with an extraordinary tumor near the anus, containing some rudiments of an embryo in it'; 'An account of a praeternatural conjunction of two female children'; 'Part of a letter concerning a child born with the jaundice upon it, received from its father's imagination, and of the mother taking the same distemper from her husband the next time of being with child'; 'An account of a monstrous foetus without any mark of sex'; 'An account of a double child born at Hebus, near Middletown in Lancashire'. So interesting to laymen and amateur scientists were many of these cases that they were abstracted from the *Philosophical Transactions* and reported in abbreviated form as news items in *The Gentleman's Magazine*.

The controversy originally stirred by Blondel and Turner, and about which Smollett must have heard, also provoked considerable commentary in the 1740s in books (essays, novels, and poems). Fielding Ould, among the most famous male-midwives of his age and best known for *A Treatise of Midwifery* (1742)[24] – frequently called by historians of medicine the first important text on midwifery in English – considered the mother's imagination important in the health of the foetus, although he seems to have doubted the validity of Turner's views. Not infrequently the subject appeared in novels. Sir John Hill, an arch-enemy of Smollett in 1751, created a marvellous female character in *The Adventures of George Edwards, A Creole*,[25] who touched a robin redbreast during pregnancy (her fancy having led her to this curious action!) and, thereafter, bore a child with a red breast – or red chest. In the same year that *Peregrine Pickle* was published, 1751, he also wrote a satire on the subject of curious births entitled *Lucina sine concubitu*. Here the reader finds the theories of preformation (according to which organisms are already fully formed in their seeds) and panspermism (minute organisms developed in fluids owing to the presence of germs) satirically treated, as well as those of the influence of the mother's imagination on her child. Although Hill was unable to attain membership in the Royal Society because of his disagreeable personality, his hoax *Lucina* was nevertheless widely read by such professional medical men as Dr Smollett, who had good reason to take note of Hill in 1751.[26] The famous case,[27] two years earlier, of a woman 'who carried with child 16 years', also helped to create the background for Smollett's witty treatment of the state of the mother's imagination.

Peregrine Pickle was published twelve years too early to bear any traces of awareness of George Alexander Stevens's *Dramatic History of Master Edward* (London, 1763). But this collection of extraordinary occurrences in 1730–1760 – written, as the title page indicates, by the 'Author of the celebrated Lecture upon Heads' – demonstrates how fully formed a type of writing (most accurately described as a *leitmotif*) about abnormal births had emerged by the 1760s. In the opening pages (7–13), Stevens has Thomas recount to David a series of histories, all dealing with abnormalities during pregnancy and shortly after birth. The stories, culled from authors in different countries in Europe, are as bizarre and grotesque as scenes from gothic romances (then coming into vogue). Two histories in particular appeal to David, so much so that Stevens included illustrations of them in his revised edition of 1785. In the first, '*Aldrovandus* [a seventeenth-century naturalist] relates, that a woman in Sicily observing a lobster taken by a fisherman, and being moved by an ernest longing for it, brought forth a lobster, altogether like what she had seen and longed for'. Stevens's 'lobster woman' is not very different from Mary Tofts, the 'rabbit woman'. The second history richer in complexity and more touching relates, as Stevens writes, 'something singular beyond all these':

> . . . [it] is the tale of *Languis*, of a woman longing to bite the naked shoulder of a baker passing by her; which, rather than she should lose her longing, the good-natured husband hired the baker at a certain price. Accordingly, when the big-bellied woman had bit twice, the baker's wife broke away from the people who held her, would not suffer her to bite her husband again; for want of which, she bore one dead child, with two living ones.

Smollett himself had shown interest in cases of extraordinary childbirth five years before the composition of *Peregrine Pickle*. In *Advice: a Satire* (1746), the character Poet refers to the strange conception, or near-conception, of a hermaphrodite:[28]

> But one thing more – how loud must I repeat,
> To rouse th' engag'd attention of the great,
> Amus'd, perhaps, with C——'s prolific bum,
> Or rapt amidst the transports of a drum.

Here Smollett's own note reads: 'This alludes to a phenomenon not more strange than true; the person here meant, having actually laid upwards of forty eggs, as several physicians and fellows of the Royal Society can attest, one of whom, we hear, has undertaken the incubation, and will, no doubt, favour the world with an account of his success. Some virtuosi affirm, that such productions must be the effect of a certain intercourse of organs not fit to be named.' Smollett's source for 'C–'[29] remains a mystery, although his satiric habit of mind was unlikely to fabricate a source. London newspapers in 1745–6 were filled with reports of such odd occurrences and the populace seemed to be diverted, if not instructed, by these accounts. In 1750, Smollett, while preparing for publication Dr William Smellie's *Midwifery*, probably read about cases of unnatural birth in the hours he spent annotating. During that year he may have seen 'Michael Anne Drouvert', the much talked about Parisian hermaphrodite who was displayed in London and written up as a case history in the *Philosophical Transactions;* may have read James Parsons's *Inquiry into the Nature of Hermaphrodites* (1741) or George Arnaud's new book entitled *Dissertation on Hermaphrodites* (1750); or may even have heard that John Hill's forthcoming book, *A History of Animals* (1750), contained a modern epitome of the subject. At any rate, Smollett's extensive reading in obstetrics in the library of Smellie, and earlier in Dr James Douglas's library, would have revealed a wealth of real cases from which to create fictional characters and episodes relating to pregnancy. Smollett's own observations, printed in Smellie's *Treatise* (II, 4–5), 'On the Separation of the Public Joint in Pregnancy', gives ample testimony to and palpable evidence of his interest in cases of abnormal birth.

In fact, Smollett never lost interest in the subject and continued from 1750 to 1764 to edit and revise all Smellie's obstetrical works. Smollett's extensive medical reading coupled with his knowledge of the extraordinary case of Sarah Last,[30] who in 1748 underwent normal pregnancy without ever giving birth to her foetus, must have inspired him to draw a parallel case in Mrs Trunnion in *Peregrine Pickle*. Readers will recall the bizarrely constructed chapters in which the pregnant lady is found to have been swelled with air! Smollett was reflecting contemporary fears about strange childbirth when he described the ultimate chagrin of the Trunnions:

> At length she and her husband became the standing joke of the parish; and this infatuated couple could scarce be prevailed upon to part with their hopes, even when she appeared as lank as a greyhound, and they were furnished with other unquestionable proofs of their having been deceived. But they could not forever remain under the influence of this sweet delusion, which at last faded away, and was succeeded by a paroxism of shame and confusion, that kept the husband within doors for the space of a whole fortnight, and confined his lady to her bed for a series of weeks, during which she suffered all the anguish of the most intense mortification.[31]

The first pineapple grown in a hothouse in England may well have been planted during the Restoration. A well-known extant painting, bearing the inscription *Rose, The Royal Gardener, presenting to Charles II the first pine-apple grown in England*, is ascribed to the Dutch artist Danckerts and the gardener is John Rose.³² Just how long before Monsieur Le Cour, a Frenchman residing in Leyden, Holland, 'hit upon a proper Degree of Heat and Management so as to produce pine-apples equally as good as those which are produced in the West Indies'³³ is not known. Even in Leyden, where winters were less brutal than in England, Le Cour used stoves to grow the tropical fruit. Chambers's compendious *Cyclopaedia* (1728) reports that the gardens of England were supplied with pineapples by Le Cour himself, and John Evelyn wrote in his *Diary* on 9 August 1661: 'I first saw the famous Queene-pine brought from Barbados presented to his *Majestie*, but the first that were ever seen here in England, were those sent to Cromwell, four-years since' (1658).³⁴ Evelyn continued several days later with a description of the 'rare fruite called the King-Pine', the first he had seen: he tasted it and found it not to his liking.

Whether or not those described by Evelyn were the first grown – it is at least plausible that an occasional fruit had been grown earlier – a more likely possibility is that pineapples were first artificially produced in quantity at any rate, in the hothouses of Sir Matthew Decker's famous garden in Richmond. He was a well-known London merchant (president of the East India Company) of Dutch origin who apparently enjoyed a 'truly Dutch passion for gardening'.³⁵ Richard Bradley, an authority on gardening in the early decades of the eighteenth century and Professor of Botany at the University of Cambridge, wrote that Sir Matthew's gardener, Henry Tellende, grew the first pineapples for his master 'circa 1723'.³⁶ Seven years before this, Lady Mary Wortley Montagu had eaten pineapples at the table of the Elector of Hanover. She wrote to Lady Mar about '2 ripe Ananas's, which to my taste are a fruit perfectly delicious', and continued to note surprise that pineapples had not as yet been cultivated in her native land. 'You know they are naturally the Growth of Brasil, and I could not imagine how they could come there but by Enchantment. Upon Enquiry I learnt that they have brought their Stoves to such perfection, they lengthen the Summer as long as they please, giveing to every plant the degree of heat it would receive from the Sun in its native Soil. The Effect is very near the same. I am surpriz'd we do not practise in England so usefull an Invention.'³⁷ From the lenghty discussion about pineapples that Horatio and Cleomenes have in Mandeville's *Fable of the Bees* (1714),³⁸ it may be assumed that Tellende and his lord, Decker, had raised the delicious and exotic fruit approximately in 1720–3. Mandeville's characters – not dissimilar to several enthusiasts in *Peregrine Pickle* – comment upon a new and 'fine Invention' as well as on the intrinsic attributes of pineapples: Horatio says, 'I was thinking of the Man, to whom we are in a great measure obliged for the Production and Culture of the *Exotick*, we were speaking of in this Kingdom; Sir Matthew Decker: the first *Ananas*, or Pineapple, that was brought to Perfection in *England*, grew in his Garden at Richmond.' That garden was still viewed in the 1730s and Smollett, who had come to London in 1739, may have visited it.

As Richard Bradley had explained in his essay 'A particular easy Method of managing Pine-Apples' (1726),[39] the difficulty of cultivation was due to poor hothouses and stovs. Pineapples required a full three years for growth, an exact temperature, ideal moisture conditions, and correctly constructed stoves. As soon this was achieved, the fruit could be grown in domestic gardens, even if at great financial expenditure. Such was the case and it applied not only to pineapples but to other exotic fruits, limes, papayas, guavas, bananas, and even grapes. The cultivation of pineapples in the third and particularly the fourth decades of the eighteenth century became a hobby – not quite a popular sport – among expert gardeners and aristocrats. Prominent families who could afford the expense sent their head gardeners to Decker's hothouses to observe the new method and educate themselves. Among the first to display home-grown pineapples on their tables were the opulent Earls of Bathurst, Portland, and Gainsborough. The Duke of Chandos, long incorrectly identified as 'Timon' in Pope's *Epistle to Burlington*, not only grew pineapples on his estate at Shaw Hall in Berkshire, but he also sold them 'at a half a guinea a time',[40] a price even Mrs Grizzle would have been willing to pay for her sister-in-law!

Smollett may not have known first-hand the early history of pineapples in England, but he was old enough to be familiar with its more recent peregrinations. Few people were more excited about the new art of growing pineapples than Alexander Pope. Together with his gardener John Serle, whom the poet employed in 1724, Pope was growing 'ananas' by 1734. He had, however, tasted the fruit long before this. On 8 October 1731, Pope wrote Maratha Blount, 'I'm going in haste to plant Jamaica Strawberries, which are to be almost as good as Pineapples.'[41] In the spring of 1735, he wrote to William Fortescue that he was improving and expanding his garden, 'making two new ovens and stoves, and a hothouse for anana's, of which I hope you will taste this year'. Two or three pineapples grown in the Twickenham hothouse and sent to an intimate friend was perhaps the greatest honour Pope could confer. During this period he was continually experimenting for cheaper and better ways to raise the fruit. In August 1738, Pope and Serle 'borrowed' Henry Scott, Lord Burlington's gardener who was an expert in growing pineapples, to consult with him 'about a Stove I am building'. It is possible that Pope was also reading modern handbooks on the subject. Whatever the case, by 1741 Pope thought he had discovered with the aid of Scott the long-sought method, and attempted to make it known to his friends. How well-circulated among the London *literati* Pope's 'discovery' was, it is now

Pineapples and Peregrine Pickle 191

impossible to tell; but by 1741, the magisterial poet was too conspicuous among men to veil any of his activities, even his pastimes and hobbies. Smollett, then young and still an *ingénue* among the 'wits' in London, kept his ears and eyes open and possibly may have heard about the new pineapple method of Mr Pope, his favourite author among all authors and a poet whose influence was to rub off considerably on his own writings. In any case, Pope soon wrote to Ralph Allen (to whom he occasionally sent a pineapple or two): 'In a Week or two, Mr. Scot will make you a Visit, he is going to Set up for himself in the Art of Gardening, in which he has great Experience, & particularly has a design which I think a very good one to make Pineapples cheaper in a year or two.'[42] Scott and Allen's gardener, Isaac Dodsley, were apparently successful in building the new type of hothouse with new stoves, for Pope wrote next year to Allen: 'I would fain have it succeed, for two particular reasons; one because I saw it was Mrs. Allen's desire to have that fruit, & the other because it is the only piece of Service I have been able to do you, or to help you in.'[43]

Poets and prose writers varied in their response to the fashionable king of fruits, some equating it with luxury and viewing it as a symbol of evil, others seeing in its beautiful colos and exotic shape an expression of the beauty of Nature and God. Pineapples were, as James Thomson wrote in 'Summer' (a poem which Smollett singled out for praise in the preface of *Ferdinand Count Fathom*), the fruits of the Gods in the Primitive Ages of the world:[44]

> . . . thou best Anana, thou the pride
> of vegetable life, beyond whate'er
> The poets imaged in the golden age:
> Quick let me strip thee of thy tufty coat,
> Spread thy ambrosial stores, and feast with Jove.'

Still other authors wrote about the medical properties of pineapples. In his didactic poem *The Art of Preserving Health* (1744), John Armstrong, with whom Smollett was on intimate terms throughout his life, chose the fruit as an example of a product raised that exhibited the differences and the extremes of cold and heat in diet:[45]

> . . . in horrid mail
> The crisp ananas wraps its poignant sweets.
> Earth's vaunted progeny: in ruder air
> Too coy to flourish, even too proud to live;
> Or hardly rais'd by artificial fire
> To vapid life. Here with a mother's smile
> Glad Amalthea pours her copious horn.

Smollett knew his friend's poem very well, had read it numerous times, and called it in *The Present State of All Nations* (1768, II, 227), 'an excellent didactic poem.'

Long before the prose encyclopaedists (Ephraim Chambers, John Harris, Robert James) discussed the fruit, medical authors had commented upon it. From the time of Nicholas Culpepper's popular handbook, *A Directory for Midwives*, which Mrs Grizzle had studied so assiduously, pineapples were strictly forbidden to expectant mothers as one of the 'Summer Fruits nought for her and all her Pulse'. In *The English Physitian* (1674) Culpepper had devoted an entire section to the benefits and ill effects of the fruit: 'It marvelously helpeth all the Diseases of the Mother used inwardly, or applied outwardly, procuring Women's Courses, and expelling the dead Child and After-birth, yea, it is so powerful upon those Feminine parts that it is utterly forbidden for Women with Child, and that it will cause abortment or delivery before the time . . . Let Women forbear it if they be with Child, for it works violently upon the Feminine Part.'[46] Seventeenth-century herbalists like Thomas Parkinson also warned their readers not to eat the artificial food. But it was not until pineapples were actually grown in English gardens that physicians and obstetricians became alarmed and abandoned superstition for medical science. Observation had revealed that pregnant women who ate this food miscarried again and again. Dr Robert James, inventor of the famed 'fever powders', writing in the London *Pharmacopaeia Universalis* (1742), commented: 'This Fruit is esteemed cordial, and analeptic; and is said to raise and exhilarate the Spirits, to cure a Nausea, and provoke Urine. But 'tis subject to cause a Miscarriage, for which Reason Women with Child should abstain from it.'[47] One year later James (whose *Medicinal Dictionary* Smollett knew well) was even more precautionary, stating the pineapples definitely caused miscarriage. Similarly, dietitians and other authors on nutrition warned the pregnant woman to refrain from the pineapple. M. L. Lemery, a prolific author on diet whose works were translated into English because of their popularity, wrote in *A Treatise of All Sorts of Goods:*[48] 'Ananas is a delicious fruit, that grows in the West Indies, whose juice the *Indians* extract, and make excellent Wine of it, which will intoxicate. Women with Child dare not drink of it, because they say, it will make them miscarry.' Francis Spilsbury, the author of *Free Thoughts on Quacks* (London, 1777), a treatise explaining the circumstances of Oliver Goldsmith's death, compared pineapples to gout (a strange comparison even for an eighteenth-century apothecary!) since he found a 'universal comprehensiveness' in both. In his words, just as 'the Ananas (vulgarly known under the name of *Pine-Apple*) is considered as containing the taste and flavour of many different fruits, so a great many disorders of the body are, under different appellations to be found in the gout'.[49]

Philosophers as well as medical thinkers pointed to the pineapple as a rare fruit with strange qualities and an exotic taste. Less concerned than physicians with the medicinal aspects of pineapples, they frequently referred to the fruit when discussing the taste. As early as 1690, John Locke singled out pineapples as the best obtainable example of a food whose taste could not be comprehended without actually partaking of it. In a well-known passage in *An Essay Concerning Human Understanding* on 'the Blind Man', to which Fielding referred several times in *Tom Jones*, Locke wrote of the impossibility of words replacing direct sensory experience:[50]

> He that thinks otherwise, let him try if any words can give him a taste of a pine apple, and make him have the true idea of the relish of that celebrated delicious fruit. So far as he is told it has a resemblance with any tastes whereof he has the ideas already in his memory, imprinted there by sensible objects, not strangers to his palate, so far may he approach that resemblance in his mind. But this is not giving us that idea by a definition, but exciting in us other simple ideas by their known names; which will be still very different from the true taste of that fruit itself.

Also speaking of the origin of ideas and the fact that they are grounded in sensory experience (i.e. direct sense experience), Smollett's countryman David Hume noted in the opening paragraph of his *Treatise of Human Nature* (1739) that 'we cannot form to ourselves a just idea of the taste of a pineapple, without actually having taste it'.[51] That is, the rare and uncommon pineapple affords the student of philosophy a splendid opportunity to observe that 'all our simple ideas in their first appearance are derived from simple impressions, which are correspondent to them, and which they exactly represent'. And David Hartley, writing two years before the publication of *Peregrine Pickle*, may not have commented upon pineapples but he increased speculative interest in abnormal pregnancy by the inclusion in his *Observations on Man* of a chapter entitled, 'To Examine How Far the Longings of Pregnant Women are agreeable to the Doctrines of Vibrations and Associations'. Hartley, a physician by profession, underplayed the variety of longings found in pregnant women – an impressive range, as Smollett's female figures in the novels show – and demonstrated instead that abnormal cravings are caused in the first place by means of 'nervous Communications between the Uterus and the Stomach'. Both, Hartley maintained, are in 'a State of great Sensibility and Irritability' during pregnancy, a view Smollett himself had taken in writing about the public joint. Smollett may not have read Locke, Hume, and Hartley – although that is highly unlikely – but he was certainly aware of popular references to pineapples and pregnancies in their works, ideas then so common that they probably required little documentation to a literate eighteenth-century man.

Thus, a decade before the publication of *Peregrine Pickle*, physicians, scientists, gardeners, philosophers, and literary men had all reacted in various ways to the new 'King Fruit' which by 1751 had become much more popular than in John Evelyn's day; all had seen in the body of beliefs and superstitions embracing the fruit something different. If gardeners found it their delight and joy, philosophers were not far behind in using it as an emblem of singular sensory experience. If physicians, especially obstetricians, called it the bane of their pregnant women patients, other scientists (biologists, botanists, physiologists) were equally ominous in their belief that pineapples contained strange and unknown chemical properties.

It was therefore left to a literary man, who was a physician as well as a novelist, to see the comic possibilities in all these prevailing theories. It may also be that Smollett, himself editing the obstetric volumes of William Smellie at the time he was composing his novel, saw that the fantastic (indeed absurd) theories of the mother's imagination together with the many muddles and mysteries that had grown up about pineapples could be wedded into one episode. The early chapters of *Peregrine Pickle* illustrate once again how adeptly Smollett used science, particularly medical learning, for the purposes of wit.[52] His satiric portraitures of characters such as Mrs Pickle, Mrs Grizzle, and Mrs Trunnion place great demands on the modern reader who wishes to comprehend the author's powerful wit. But his contemporary readers would have felt much more at home than we do in viewing his comic spectacle: they would have realised that he was using medical and scientific learning for pure levity and genial farce, and in this sense would have read his works as they were reading those of his great contemporary, Laurence Sterne.

We should not be surprised to observe a process of carry-over in Smollett's novels: from his medical writings to the novels and vice-versa. Although he was never a successful physician, if daily practice is a yardstick of measurement, his entire life demonstrates a continuing interest in medical theory. It is, therefore, to be expected that Smollett's medical works – short essays, unsigned medical tracts written pseudonymously for financial purposes,[53] and virtually all the reviews of medical books in the *Critical Review* 1756–60 and possibly later – would have rubbed off on his fiction. Indeed, the sensibility pervading both worlds, medical and fictive, was one, and Smollett was at their center. Such interaction serves to remind us, that the place occupied by medicine and by the social aspects of that science which daily seemed to take on ever greater consequence in the eighteenth century, is something of which we have yet to take account in our criticism and biography of Smollett.

NOTES

1. *Peregrine Pickle*, I, 32.
2. *Peregrine Pickle*, I, 31. Passages quoted in this paragraph are from *Peregrine Pickle*, I, 32–6.
3. *Peregrine Pickle*, I, 32. Grown in elephantine stoves in specially-built hothouses, pineapples were considered an exotic and artifical fruit in the eighteenth century. I discuss the fruit more fully below.
4. *Peregrine Pickle*, I, 31: 'She purchased Culpepper's midwifery, which, with that sagacious performance dignified with Aristotle's name, she studied with indefatigable care, and diligently perused the Compleat House-wife, together with Quincy's dispensatory, culling every jelly, marmalade and conserve which these authors recommend as either salutory or toothsome, for the benefit and comfort of her sister-in-law, during her gestation.'
5. *Peregrine Pickle*, I, 31.
6. Some attention to embryological theory of the eighteenth century and *Tristram Shandy* is given in Louis A. Landa, 'The Shandean Homunculus: The Background of Sterne's "Little Gentleman" ', in Caroll Camden (ed.), *Restoration and Eighteenth Century Literature* (Chicago, 1963), 49–68. I have discussed Smollett's novels and medicine in 'Doctors and Medicine in the Novels of Tobias Smollett' (Princeton University dissertation, 1966).
7. *Letters from Edinburgh* (Edinburgh, 1776), 'On . . . Gardening', 229.
8. *Aristotle's Compleat Master-Piece: Displaying the Secrets of Nature in the Generation of Man*(32nd ed., 1782), 33–4. See Fielding H. Garrison, *An Introduction to the History of Medicine* (4th ed. rev., 1929), 101ff.
9. *An Essay for Abridging the Study of Physick* (1735), 17.
10. *A Directory for Midwives* (London, 1684), p. 145.
11. The name given to the condition by Ancient Greek physicians, and also called 'citta' or 'malatia'. See Hermann Heinrich Ploss, Max and Paul Bartel, 'The Longings of Pregnancy' in *Woman: an Historical and Anthropological Compendium* (1935), II, 455–60. A recent study of the medical aspects of *pica* is by M. Cooper, *Pica* (Springfield, 1957).
12. *The Female Physician*, 368. See also part ii, chap. 7, which discusses numerous cases of foetuses that have been marked. Maubray was one of the first physicians in London to offer private instruction for midwives. See F. H. Garrison, *History of Medicine*, 399. A great believer in monsters, he earned notoriety in 1723 by assisting in the delivery of a Dutch woman, who produced a monstrous manikin called *de Suyger*, with 'a hooked snout, fiery sparkling eyes, a long neck and an acuminated, sharp tail'. Maubray called it a moldy-warp (mole) or sooterkin.
13. The most complete account is by S. A. Seligman, 'Mary Tofts: the Rabit Breeder', *Medical History*, V (1961), 349–60, which is based upon extant contemporary accounts. Another less extensive treatment is by K. Bryn Thomas, *James Douglas of the Pouch and his pupil William Hunter* (1964), 60–8.
14. George Sherburn (ed.), *Correspondence of Alexander Pope* (Oxford, 1956), II, 418–19.
15. I follow the text given by Norman Ault in *Minor Poems of Alexander Pope, The Twickenham Pope: vol. vi* (1964), 259–64. St André was a Fellow of the Royal Society and contributed papers to the *Philosophical Transactions*. His appointment as Surgeon and Anatomist to the Court seems to have been made rather for his linguistic ability than his medical ability. Mary Tofts's confession put an end to his Court position, and he was never again to attain medical recognition.
16. According to the title page of the tract, these pages were copied from the second

17 edition of Whiston's *Memoirs*, published in London in 1753. The interpretation of the Tofts case does not appear in the first edition of 1749, and, so far as I can determine, was not published separately. See K. Bryn Thomas, *James Douglas* (1964), 65.

17 See Marjorie Nicolson and G. S. Rousseau, *'This Long Disease, My Life': Alexander Pope and the Sciences* (Princeton, 1968), 114: 'Early in 1727 when the small talk of London seems to have been divided between Mary Tofts and Lemuel Gulliver – *Gulliver's Travels* had appeared the preceding autumn and provoked almost universal applause – it was inevitable that at least one pamphlet on the rabbit woman should be attributed to Jonathan Swift. Those who have done him that dubious honor have failed to notice that Swift had returned to Ireland a month before the Tofts affair, and while he probably heard of it in letters, he had no such background for parody as Scriblerians in London.'

18 The chronology of works in the controversy was as follows: Turner, *De Morbis Cutaneis* (1726); first published in 1714); Blondel, *The Strength of Imagination in Pregnant Women Examin'd* (1727); Turner, *A Discourse concerning Gleets . . . to which is added A Defence of . . . the 12th Chapter of . . . De Morbis Cutaneis, in respect of the Spots and Marks impress'd upon the Skin of the Foetus* (1729); Blondel, *The Power of the Mother's Imagination Over the Foetus Examin'd* (1729); Turner, *The Force of the Mother's Imagination upon her Foetus in Utero . . . in the Way of A Reply to Dr. Blondel's Last Bok* (1730). Turner's first work, *De Morbis Cutaneis*, was written in part as a defense of Malebranche's theory that the mother marks her child. See *Father Malebranche's Treatise concerning the Search after Truth*, trans. T. Taylor (Oxford, 1694), 'Book the Second Concerning the Imagination'. As a conclusion to this book, Malebranche wrote: 'When the Imagination of the Mother is disordered and some tempestuous passion changes the Disposition of her Brain . . . then . . . this Communication alters the natural Formation of the Infant's Body, and the Mother proves Abortive sometimes of her foetus' (60). When the Tofts case revived the issue among medical men, Turner turned from Malebranche to a then real-life example. In this and other footnotes, I have dealt at length with these medical tracts because they were clearly read by the masses in their time and now are so little known.

19 *The Power of the Mother's Imagination*, XI. Blondel singled out from medical literature the six most common causes of spotted children: '1. A strong Longing for something particular, in which Desire the Mother is either gratified, or disappointed. 2. A sudden Surprise. 3. The Sight and Abhorrence of an ugly and frightful Object. 4. The Pleasure of Looking on, and Contemplating, even for a long Time, a Picture or Whatever is delightful to the Fancy. 5. Fear, and Consternation, and great Apprehension of Dangers. 6. And lastly, an Excess of Anger, of Grief, or of Joy' (2). Later (4) Blondel notes that item (1) of the list above was the most common of the six, especially in the case of certain fruits, 'the strong Desire of *Peaches*, or *Cherries*'. Presumably he would have included in this list pineapples!

20 An earlier version of this work appeared in 1740 with the title page, *The Power of Imagination in Pregnant Women Discussed: with an Address to the Ladies in Reply to J. A. Blondel*. So far as I can learn Mauclerc published no other works.

21 *Critical Review*, XX (July 1765), 63–5. On pp. 125–33, Bellet provides a fair estimate – in his opinion – and a history of the controversy from the time of Malebranche. There is reason to believe that Smollett and Bellet had met and that Smollett was impressed by his knowledge of medical history.

22 A list of references is too long to be given here. The two cases that attracted the most attention werte 'A Foetus of Thirteen Years', *The Gentleman's Magazine*, XIX (1749), 415, and in the same publication, 'Fatal Accident: Woman carry'd a child sixteen years', XIX (1749), 211.

23 Respectively *Philosophical Transactions*, XLI (1740), 137; XLI (1741), 341; XLI (1741), 764; XLI (1741), 776; XLII (1742), 152; XLII (1743), 627; XLIV (1747), 617; XLV (1748),

24 325; XLV (1748), 526; XLVI (1749), 205; XLVII (1750), 360. Some of these cases were abridged and printed in popular monthlies such as *The Gentleman's Magazine* and *The Monthly Review*.

24 First published in Dublin and numerous times thereafter in London. For Ould's contributions to midwifery see John R. Brown, 'A Chronology of Major Events in Obstetrics and Gynaecology', *The Journal of Obstetrics and Gynaecology*, LXXI (1964), 303; and Fielding H. Garrison, *An Introduction to the History of Medicine* (4th ed. rev., 1929), 338–40.

25 First published in 1751 and reprinted in *The Novelists' Magazine*, XXIII (1788). One episode is discussed by William Scott, 'Smollett, John Hill, and *Peregrine Pickle*', *Notes and Queries*, CC (1955), 389–92.

26 Two weeks before the publication of *Peregrine Pickle*, Dr Hill anticipated Smollett's novel by bringing out *The History of a Woman of Quality: or the Adventures of Lady Frail*. While Smollett believed the presence in his novel of Lady Vane's 'Memoirs' would enhance sales, Hill's earlier publication greatly diminished sales.

27 *The Gentleman's Magazine*, XIX (1749), 211.

28 James P. Browne (ed.), *The Works of Tobias Smollett, M.D.*, (1872), I, 294.

29 There is no mention of this case in the *Philosophical Transactions* or other scientific literature I have examined. Perhaps 'C——' was the famous 'Charing Cross hermaphrodite', about whom Dr William Cheselden had written in the *Anatomy of the Humane Body* (1713) and about whom Dr James Douglas wrote many medical fragments in the 1720s. Smollett was too young to have seen that curious organism. By 1751 this hermaphrodite may have become too stale a subject for satire, although K. Bryn Thomas, *James Douglas and his pupil William Hunter* (1964), 190, does not think so. I am inclined to believe it was a more recent occurrence, about which Smollett was informed, as the tone of his note indicates.

30 Among numerous accounts of her case the most interesting I have found is in the *The Gentleman's Magazine*, XXI (1751), 214–15; 'About the beginning of August 1748, Sarah Last, a poor woman in Suffolk, had the usual Symptoms of pregnancy, which succeeded each other pretty regularly thro' the usual period, at times she as seiz'd with pains . . . the child did not advance in birth . . . after the pains were gone off, the woman grew better . . . her menses return'd at proper seasons as if she had been deliver'd of a child, and continued to do so for several months . . . the poor woman recover'd, and is now perfectly well.' The editor commented that 'the foregoing case is not singular; we see two of the same kind recorded in the Memoirs of the Royal Academy of Surgery at Paris for the last year . . . and [one] communicated to the French academy'.

31 *Peregrine Pickle*, I, 72. Smollett's description of Mrs Trunnion's expectant state tallies well with observations made in John Pechey's *Complete Midwife's Practice Enlarged* (5th ed., 1698), especially the section 'Of False Conception' (57–62). According to a manuscript in the Hunterian Museum, Pechey was included among required authors to be read by students at the Glasgow Medical School, which Smollett attended 1736–9.

32 Without pretending to summarise the vast literature dealing with the date and author of this painting, I mention the following: George W. Johnson, *A History of English Gardening* (1829), 72–81; Alicia Amherst, *A History of Gardening in England* (1910), 238ff.; Miles Hadfield, *Gardening in Britain* (1960), 126; J. L. Collins, *The Pineapple* (Honolulu, 1960), 70–86; William Gardener, 'Botany and the Americas', *History Today*, XVI (Dec. 1966), 849–55, where the picture was most recently reprinted. In his *DNB* life of John Tradescant the younger, gardener to Charles I, G. S. Boulger writes: 'There is a tradition that the younger Tradescant first planted the pineapple in England in the garden of Sir James Palmer at Dorney House, Windsor, where a large stone cut in the shape of a pineapple by way of extant . . . The pineapple pits were therefore pre-Charles II. Surely then John Tradescant the younger grew pine-

apples here for Charles I. The fact that there is no painting of John the elder or John the younger presenting a home-grown pineapple to Charles I does not disprove the possibility. The Tradescants would have had ready access to pineapples thanks to Sir William Courteen who was one of their principal benefactors. Sir William took out the first settlers to Barbadoes in 1625. The West Indies were one of his regular trade routes.' This theory is supported by M. Allan in *The Tradescants: Their Plants, Gardens and Museum 1570–1662* (1964), 143–5. Tradescant was known for his exotic fruits, as is seen in Tom Brown's *Amusements Serious and Comical*, particularly the section entitled 'The Philosophical or Virtuosi Country'.

33 Robert James, M.D., *A Medicinal Dictionary* (1743–5), article entitled 'Ananas'.

34 E. S. de Beer (ed.), *The Diary of John Evelyn* (Oxford 1955), III, 293, 513.

35 Hadfield, *Gardening in Britain*, 126.

36 *Dictionarium Botanicum* . . . (1728), article entitled 'Ananas', and 'A Particular Easy Method of Managing Pine-Apples' in *New Improvements of Planting and Gardening* (1726), 605. Bradley's assertion was challenged in 1780 by Horace Walpole, who wrote to the Reverend William Cole: 'There is another assertion in Gough [*British Topography*, 1768], which I can authentically contradict. He says Sir Matthew Decker first introduced ananas. My curious picture of Rose, the royal gardener, presenting the first ananas to Charles II proves the culture here earlier by several years' (W. S. Lewis (ed.), *Letters to the Reverend William Cole* (New Haven, 1937), II, 239). Walpole had acquired the painting from William Pennicott in 1780. In his popular handbook, *The Gardener's Dictionary* (1724), Philip Miller attributed the first pineapple to Tellende. See also E. S. Rohde, *The Story of the Garden* (1932), 178.

37 R. Halsband (ed.), *The Complete Letters of Mary Wortley Montagu* (Oxford, 1965–7), I, 290.

38 Edited by F. B. Kaye (Oxford, 1924), II, 193–5.

39 Pages 605–6. Bradley, among other authors, notes that a forty-foot stove was necessary to ripen one hundred pineapples. An average pineapple took three years to ripen and *c*. 1726 its total cost from the time of purchasing seeds was £80.

40 Hadfield, *Gardening*, 166, and C. H. Collins Baker and M. I. Baker, *The Life of James Brydges First Duke of Chandos* (1949), 103. See also George Sherburn, ' "Timon's Villa" and Cannons', *Huntington Library Bulletin*, VIII (1935), 143.

41 *Correspondence of Alexander Pope*, III, 233 and 453.

42 Ibid., IV, 360. See also IV, 405, 420; and Benjamin Boyce, *The Benevolent Man: a Life of Ralph Allen* (Cambridge, 1967), 114. On 25 March 1746, Pope wrote to Swift about the new fruits in his garden: 'I have good Melons and Pine-apples of my own growth. I am as much a better Gardiner, as I'm a worse Poet, than when you saw me: But gardening is near a-kin to Philosophy, for Tully says *Agricultura proxima sapientiae*' (*Correspondence*, IV, 6). Without documentation Hadfield, *History of Gardening*, 187, states that 'a year later [1742] Allen was advised not to take Scott's advice'. But Pope could not have been the unmentioned person since he fully approved of Scott's method. For Pope's activities as a gardener and ideas about gardening during Smollett's mature years, see Edward Malins, *English Landscaping and Literature 1660–1840* (1966), 26–51.

43 *Correspondence*, IV, 429.

44 J. L. Robertson (ed.), *The Poetical Works of James Thomson* (Oxford, 1908), 'Summer', ll. 253–4.

45 *The Art of Preserving Health* (1796), including a *Critical Essay* by Dr John Aikin, II. 334–40. Aikin commented on foods like pineapples as an example of a 'too luxurious diet' (p. 14). The medical aspects of pineapples were also discussed in scientific publications. For example, see William Bastard, 'On the Cultivation of Pine-Apples', *Philosophical Transactions*, LXVII (1777), 649–52, in which the author describes his

hothouse in Devonshire and the effects of the fruit on the body. Armstrong, a Scotsman who practised medicine in London, probably did not taste pineapples in Scotland. Dr John Hope, the Regius Professor of Botany at Edinburgh University and a populariser of Linnaeus in Scotland, allegedly grew in 1762 the first pineapples in Scotland, although I can discover no certain means of verifying this allegation.

46   *The English Physitian* (1684), 189–90. Pineapples were not mentioned in the edition of 1651, presumably because they were then unknown in England. Smollett referred to Culpepper's medical handbooks in several novels, and Joseph Addison listed the *Directory for Midwives* among essential books in an eighteenth-century 'Lady's Library'. See Donald Bond (ed.), *The Spectator* (Oxford, 1965), I, 155.

47   *Pharmacopaeia Universalis* (1742), 118. John Quincy made the same point in his *Complete Englrish Dispensatory* (rev. ed., 1742), 194.

48   *A Treatise of All Sorts of Foods, Both Animal and Vegetable,* trans. by D. Hay, M.D. (1745), 350 and 75–6. The signatures of several distinguished physicians of the Royal College of Physicians appear on the frontispage as approving the medical aspects of the book: among them are Edward Brown, Walter Charleton, and John Woodward, whom the Scriblerians satirised. For other comments by dietitians about the medicinal aspect of pineapples, see A. Cocchi, *The Pythagorean Diet, of Vegetables Only, Conducive to the Preservation of Health . . .* , trans. from Italian (1745), 74–6 and Sir Jack Drummond and Anne Wilbraham, *The Englishman's Food* (1939), pp. 228–9. Numerous comments about the danger of pineapples for pregnant women may also be found in Ephraim Chambers's *Cyclopaedia: Or, An Universal Dictionary of Arts and Sciences* (1728), article entitled 'Ananas', and in George Cheyne's *An Essay on Regimen* (2nd ed., 1740), pp. 76–7. Chap. xix of *Roderick Random,* in which the hero meets the French apothecary Lavement, makes it clear that Smollett was thoroughly familiar with the medical effects of different diets.

49   *Free Thoughts on Quacks* (1749), 164–5.

50   Alexander Campbell Fraser (ed.), *An Essay Concerning Human Understanding* (Oxford, 1894), II, 37–8. Although Locke was an expert botanist and did a great deal of plant research in the Oxford Botanical Garden in 1650–60, there is no evidence that he himself ever grew pineapples. See Kenneth Dewhurst, *John Locke (1632–1704) Physician and Philosopher: a Medical Biography* (1963), 8–9.

51   L. A. Selby-Bigge (ed.), *A Treatise of Human Nature* (2nd ed. rev., Oxford, 1928), 5.

52   I have borrowed this phrase from the excellent article of D. W. Jefferson, '*Tristram Shandy* and the Tradition of Learned Wit, *Essays in Criticism*, I (1951), 225–48.

53   Some of these have recently been studied and attributed to Smollett by G. S. Rousseau, 'Matt Bramble and the Sulphur Controversy in the XVIIIth Century: Medical Background of *Humphry Clinker*', *Journal of the History of Ideas*, XXVIII (1967), 577–90.

PART TWO

scientific

8

The discourses of literature and science (1)

As a disciple in the 1960s of the late Marjorie Hope Nicolson, I became familiar with her version of literature and science construed as a single field of study. She called it 'science and imagination', which for her meant the ways in which science had liberated the human imagination, often inspiring it to create great art. She harboured no sense whatever that bad science was ignoble or stultifying, or that good science was just another body of knowledge, as ideological and relativistic as any other.

But after the theoretical revolution of the 1970s and my own reading in theory, I could see that this discourse could not survive for very long with such a set of monolithic arrows – from science to literature and vice versa. So I decided to enter the field myself, since even then (in the 1970s) I knew I would be spending a large part of my own career grappling with this complex problem. The theoretical aspects of the reciprocity – whether posed in the tropes of connections, bridges, or arrows of direction – were not the only dimensions which attracted me. Much more urgent to one who taught in a large research university was the need to acknowledge and reverse the patent decline and transformation of the humanities from their fallen place in a post-disciplinary, micro-electronic information age. I realised that literature and science could play an important role in reviving these fading academic disciplines.

This essay was my first attempt to reconfigure the field by taking stock of its recent past. I was encouraged by the fact that the leading international journal in the history and philosophy of science, Isis, had commissioned it. If the historians and philosophers wanted to hear about this intersection, could our humanists be far behind?

> If the time should ever come when what is now called science, thus familiarized to men, shall be ready to put on, as it were, a form of flesh and blood, the Poet will lend his divine spirit to aid the transfiguration, and will welcome the Being thus produced, as a dear and genuine inmate of the household of man. (William Wordsworth, Preface to the second edition of the *Lyrical Ballads*, 1800)

Literature and science, or science and literature – the two are virtually synonymous – is a field that developed in America in the 1940s. This should not imply a lack of scholarship previous to this time (in 1882 Matthew Arnold published an essay entitled 'Science and Literature' that was so famous that the Americans asked him to deliver it all over the country) but rather dates the start of formal interdisciplinary programs dedicated to research in the field. It developed as the result of wide interest in intellectual history during the 1920s and 1930s, and especially as an outgrowth of interest generated by the circle centred on the History of Ideas Club at the Johns Hopkins University and their organ, a *Journal of the History of Ideas*. But the history of science also played a role: as this field developed between the two wars a certain small number of its members were alive to, if not openly interested in, the relations. By the 1950s the Modern Language Association of America had instituted a division on literature and science as a result of the efforts of Marjorie Hope Nicolson and several of her colleagues in the North-east, and by the early 1960s annual seminars on the subject were being held at convocations of the American Historical Association. Even *Isis*, 'devoted to the history of science and its cultural influences', shared in the bustle: until 1970 its editorial board saw fit to have a scholar, D. C. Allen, represent science and literature, and a brief article on the subject occasionally appeared.

What literature and science is in practice, however, is less easily grasped. To different types of scholars during the last three decades it has meant different things, and only a novice would believe that scholars today, if they could convene to discuss the question, would agree about any of the fundamental issues regarding its theory and practice. My charge here is to describe and document the major developments in the field since c. 1950, that is, during the last three decades. In doing so I naturally see the evolution from the vantage of a professional literary historian, but this angle of vision ought not to be too myopic, especially since there are no professional historians or philosophers of science who have made it their primary field of research.[1]

THE TRADITIONALISTS-PHILOLOGISTS

These scholars have conceived of the field as one in which scientific references in literature are documented. They annotate scientific references and allusions in Chaucer, Shakespeare, and Milton, for example, as well as study the frequency with which Shakespeare mentions words such as astronomy, alchemy, and astrology. Their ranks are composed almost without exception of literary historians (i.e. philologically trained students who took doctorates in language and literature); they very rarely have an interest in science *qua* science. Most traditionalists begin their exploration of literature and science as the result of research on a single literary author which uncovers a gap in scholarship in this area.[2] The traditionalist then recognises that he must become familiar with science in the historical age and social milieu of the subject, and eventually takes out time to do so;

but this is research done as a sacrifice – to illuminate the literary text, not to shed light on the science studied.

Before World War II, when the history of many sciences was still unwritten, literary historians pursuing this avenue of research found few comprehensive works on which to rely. They had to resort to any histories encompassing science in a particular age, and if these were inferior, that was their bad luck. Or else they had to read widely in the primary scientific documents of the time and formulate their own syntheses, which were often inchoate and naive.[3] Yet their projected audience remained the community of *literary* scholars, not historians of science or general historians who might be dissatisfied with the quality of generalisation and the degree of analysis found in their treatment. The available evidence, mostly verbal rather than printed, is that historians of science have had little interest in the work, especially the hard philological work, of the traditionalists.

THE THEORISTS

Composed mainly of intellectual historians with degrees in history and/or literature, the theorists filled a gap which they found in the work of the philologists and traditionalists.[4] The theorists questioned the value of the philological approach, not in so far as they were opposed to philological studies (e.g., of Donne's uses of alchemy or Pope's uses of light and colour), but rather because they believed the approach to be constricting. It could provide admittedly useful information about an author's use of specific terms, but it revealed little about his relation to science at large and did not distinguish among the various kinds of serious and popular science.

Since the theorist's argument with the philologist is somewhat abstruse, an example may serve to clarify the issue. Consider Pope's use of the term 'Great Chain of Being', a phrase that occurs in his poetry many times in this and other variant forms. The term, as it is now widely understood as a result of Arthur O. Lovejoy's researches,[5] has clear scientific implications. Yet all the traditionalist-philologist who was annotating Pope would provide or was competent to provide were verbal parallels, literary echoes, allusions, and a conventional definition of the term. He would accept whatever respectable definition he could find, without seriously questioning the type of respectability; and this definition he would consider 'standard'. Moreover, he would rarely dig beneath the surface of meanings to analyse them; and not being scientifically trained, he felt no guilt for not doing so. The theorist's caveat, then, related to the limits of the philological method in the process of integrating knowledge about literature and science. Lovejoy, considered in this context, would be classified as a theorist: he traced the concept of the Great Chain from its Greek origins to the early twentieth century, showing not only how it evolved but giving reasons why. He also considered all the logical arguments against his hypothesis and showed why these were untenable. This last task is the surest clue to Lovejoy's classification as a theorist rather than philologist.

By the early 1950s it was already somewhat disreputable to be a philologist in America. The New Criticism, in another context altogether, had dealt traditional philological criticism a blow from which it could not recover.[6] Fewer and fewer books written by traditionalists were getting published, and the new crop of graduate students entering the field was interested in theory more than in application. In the field of literature and science they (both the elders and the young) produced a spate of works that were different from those of their predecessors before the war.[7] Crudely speaking, the difference related to the degree of analysis required before any idea could be said to have been adequately treated by its author. Also, those now writing were more intrigued by the relation of ideas than their predecessors had been; and the study of science and literature was, if it was anything, a set of complex relationships. The diachronic approach, moreover, was quickly being abandoned for the synchronic, as students began to explore several authors and many centuries rather than limit themselves. The theorists, finally, engaged in the activities of historians of science, read their journals, and went to their meetings.

THE CASE OF MARJORIE HOPE NICOLSON

Marjorie Nicolson (1894–1981) is properly classified as neither philologist nor theorist, since her work embraces the best qualities of both; besides, her career has been so anomalous in relation to the evolution of literature and science since the end of the war that it must figure into a survey such as this one. Most important of all is the fact that fortune blessed her: in the 1940s. This should not imply a lack of scholarship previousverbal parallels, literary echoes, allusions, and a conventionalthe 1920s she accidentally stumbled upon Swift's sources for the scientific portions of *Gulliver's Travels* (especially part III, 'A Voyage to Laputa'), which set her off as a firebrand, blazing in her new career to conquer an uncharted field.[8] Secondly, she had been a student of Lovejoy's in her most formative years and learned from him that the philological approach was never adequate.[9] The most salient aspect, though, and the one most significant for our purposes is that she was the first scholar to make the field respectable: the first to prove by her appointment to prestigious university posts and endowed lectureships that science and literature was a valid field for research, publication, and professional advancement. It was not her traditional philological work on Milton that prompted the members of the

Modern Language Association in 1962 to elect her its president but her books on literature and science. This is also where she had received the most kudos,[10] a development serving to show less acclaimed scholars in the field had arrived.

Professor Nicolson trained dozens of scholars in the field for five decades: to mention just a few names, Ralph Cohen (Virginia), Ernest Tuveson (Berkeley), the late Rosalie Colie (Brown), Aram Vartanian (New York University), Peter Gay (Yale), Joseph Mazzeo (Columbia), French Fogle (Claremont), myself (UCLA) – all of reputable universities, and these scholars have trained others in the field. In almost every essay she wrote, in every book, Marjorie Nicolson made it abundantly plain that she was an intellectual historian first and foremost, not merely a literary historian but an intellectual historian. In her prefaces she professed not to be capable of overlooking the tremendous influence of science on literature. Accordingly, the best part of her writing career was devoted to studies of these influences: of 'imaginary voyages to the moon' on Renaissance and post-Renaissance writers; of the 'New Science' of the seventeenth century on the great writers of the age; of the *Philosophical Transactions of the Royal Society* on the Scriblerians; of Descartes on the Restoration; of Newton on the English Georgians and Romantics; of the cosmologists on English poetry from Pope to Coleridge and Tennyson; of eighteenth-century geologists on the Romantics. The student of Miss Nicolson's career would be right to assume this: if her already long academic life could have been extended by two or three decades, she would have brought her investigations up to the present time, showing the influence of scientists such as Freud, Heisenberg, and Einstein, for example, on modern authors.[11]

THE 'INFLUENCE' PROBLEM

One of the persistent criticisms of the work of Marjorie Nicolson and her students was that it had no sophisticated concept of *influence*. It treated of influence – the argument went – in one direction only: from science to literature, never the other way round. Something must be said about influence as a concept within the context of this field, as well as about the direction it takes, to grasp what happened in the 1960s. For here lie some of the reasons for the scepticism about literature and science that crept into discussions then.

Influence has always been a perfidious concept, particularly in the sense of the distance between the agent of influence and the influenced.[12] Most of us have little difficulty agreeing that character *A* may have exerted terrific 'influence' over character *B* in daily life, and we quietly assent to most points about personal influence in the biographies we read; but historical influence is a more complicated matter, especially when the two territories science and literature are seemingly alien and procedurally distant. The charge brought against some of Nicolson's work is that she assumed influence in cases in which it may not have existed and – perhaps more dire – that she failed to focus sharply the 'scientific' object. It was said she could not distinguish, for example in the case of Newton, between the real Isaac Newton and a fictional one imagined by his contemporaries, between Newtonian science as it actually exists in his printed works and as it was loosely perceived by the layman in the street, between 'hard Newtonianism', Newton's science with all of its technical and mathematical features, and 'soft Newtonianism', a watered-down version stripped of all its real complexity and reduced to easily grasped quips about colour, light, rainbows, and prisms.[13] These and other reservations were made primarily by historians of science; in this case they knew their Newton scholarship and could not be fooled. The literary scholars – philologists and theorists – could not see what difference the distinctions made; the important thing, they contended, was that *some kind* of Newtonianism had shaped (their term for influenced) the literature they studied. An impasse was reached as the result of two radically different focuses.

There is no reason to disbelieve on logical or epistemological grounds that literature and science affect each other reciprocally. That is, that each influences the other in just about the same degree, although conceivably in different ways. It is also probably valid to assume, although it would be practically impossible to prove, that science shapes literature to the same degree that imaginative literature shapes science.[14] Obviously the whole problem is largely one of semantics, yet semantics put aside, only the former has been studied in any depth. Literary scholars are understandably far more concerned about literature than about science, and most applied scientists as well as historians of science have not seriously considered the possibility that literature has shaped or can shape scientific developments.[15] The latter is an unexplored territory, probably the one in greatest need of cultivation right now and also the one requiring learning so vast that it is hard to imagine it in a single scholar.[16] In general, those interested in this latter type of reciprocity are neither literary types nor historians of science but social scientists, notably sociologists of science such as Robert Merton or Arnold Thackray.

THE GREAT ANOMALY: SCIENCE FICTION

Science fiction occupies a curious place in the recent evolution of the field. As a form of imaginative literature it has rarely been given serious consideration by literary critics,[17] nor has it engaged the critical energy of scientists who doubtlessly read and enjoy it. Nevertheless, it is the most natural literary form to be scrutinised by the serious student of the field, the one form in which questions about genuine influence dissolve or are easily minimised. The problem is ultimately political and sociological: American writers of science fiction do not normally hold prestigious academic posts, as do other types of authors (novelists, essayists, biographers, playwrights, poets), and consequently do not have the opportunity to inspire students zealous for academic recognition to write about science fiction. During the last decade, for instance, the number of college and university courses in science fiction has greatly proliferated,[18] but this has not yet produced sufficiently important *secondary* works instrumental in causing influential critics to change their minds. Nor do the few critics in the previous cases of reciprocal influence, this is an area in which genuine progress could be made by an intensified effort.

But a serious complicating factor exists in the relation between the *bona fide* scientific professional community and authors of science fiction, who rarely are members of that community. On many occasions the latter group has literally indicted the former with plundering its ideas; nevertheless, the scientists continue to deny the charge, and what is more, persist in depreciating the works of writers of science fiction. The cross-accusations almost always end in an impasse. It is even difficult to reach a tentative opinion about the matter; for anyone versed in the history of quackery in Western civilisation as well as anyone who personally witnessed, as I did, the debates about the theories of Immanuel Velikovsky held in Princeton in the 1960s, knows that the matter is less simple than it appears.[19] In fact the claims of each group have never been publicly debated, and there is practically no secondary literature on the subject.

ENTER STRUCTURALISM

No two literary critics will agree on a definition of the methodology now known as structuralism – the disagreement is actually inherent in the methodology of the method, paradoxical as this seems.[20] Nevertheless it is probably accurate to contend that it has recently affected studies in our field profoundly by maintaining that the views of our traditional theorists, mentioned above, are either insufficient or simple-minded, and by insisting that the field is intrinsically a study of philosophical assumptions and must therefore include contemporary philosophy in a major way.[21] None of this is easily grasped or capable of satisfactory demonstration. Perhaps no other writer has typified the structuralist intrusion better than Michel Foucault, all of whose books inherently deal with literature and science. Before his works began to appear in the 1960s scholars could be classified as practicing in the field in one of the areas listed above. Foucault's books deflected most serious students then (even Marjorie Nicolson, then in her seventies, read him) and had the further effect of transforming the old categories, in a sense rendering them obsolete. The question for someone writing about science and literature changed from 'what type of critic are you?' to 'how much self-consciousness do you have about your methodology?' It mattered little to Foucault and other structuralists what the particular methodology being used was; what counted far more was one's self-reflectiveness about his method, especially the variety of inner dialectic revealing the types of anxiety experienced by the scholar.[22] Paradigmatically speaking, the more self-reflective the better; the more the scholar could stamp his scholarship with his particular brand of self-reflectiveness, the greater the structuralist's respect for him.

All this produced the effect of alienating certain traditional scholars (some vestigial members of the Lovejoy-Nicolson school), especially those who professed to be more interested in their object (authors and scientific developments) than in their subjective selves, their dialectic about that object that casts the object into ancillary roles. Furthermore, an impression was given that structuralists were finally turning literary criticism into a science. This claim was not new, but many literary critics, believing themselves the champions of an ancient humanism now threatened by a science (more accurately, a pseudo-science, as they came to view structuralism) resisted the temptation to make a science out of literary criticism.[23] It now appeared to some that structuralism in the 1970s would finally end the debate about the validity of literature and science as a field. Others, discouraged, abandoned the ship before it could sink. As proof, the membership of the Literature and Science Division of the MLA took a sharp plunge from which it has still not recovered. By 1975 there were so few members of the division that the Executive Council dissolved it; it now exists on a probationary basis, but its future is very much in doubt. As further evidence, the annual bibliography of literature and science published in the quarterly *Clio* has been growing smaller at a time when the total number of annually published works is increasing.

CRITICAL PLURALISM

The term critical pluralism, commonplace today among literary scholars, denotes the multitude of approaches now employed. Whereas the universe of American literary scholars before 1950 could be divided into two hemispheres, critics and historians,[24] today there are many other types. To say that today there are traditional literary historians, neo-literary historians, vestigial New Critics, descriptive and textual bibliographers, biographers, journalists, structuralists (of all manner and types), hermeneuticists, psychological critics, neo-Freudians, neo-Jungians, formalists, just begins to describe the rampant proliferation. Each type has one or more journals

– and it is even difficult to categorise the journals accurately, since they now change their orientation without warning.[25] Besides, the mere designation of a taxonomy sidesteps the main point. The labels or tags by which one describes a literary critic now matter less, as has been suggested, than the critic's degree and type of self-reflectiveness. Yet the *effect* of this wave of pluralism is to deflect serious students (the few who remain) from literature and science, a non-valid field for all these approaches except the few traditional ones listed.

A structuralist or phenomenologist, for example, would never claim that his 'field' is literature and science. For him the concept of field is vapid, cannot be designated by signifiers (literature, science) that signify nothing themselves. Valid subjects include the study of origins, middles, ends, or, turning elsewhere, sets of relationships such as the application of structuralist linguistic concepts of syntax to a given poetics, that is, to a body of aesthetic mythology about literature.[26] But literature and science, no matter how narrowly conceived or how specific the case (e.g., Hutton's geological impact on Shelley), can never be a substantial region for exploration. Ultimately – structuralists and other contemporary philosophical types argue – the only serious questions asked by students of literature and science are questions of influence, and influence is a concept thrown out of court by the structuralist even before he begins: ejected because it is allegedly reductionistic, tedious, unrevealing about a deep substratum of belief, and of course because it happens to be unfashionable today. The structuralist or phenomenologist carrying on an imaginary dialogue with Marjorie Nicolson will fail to understand that her point about 'Newton demanding the muse' has any significance. What difference, he or she will inquire, did Newton or his theories make to Pope's couplet style, poetic syntax, or ironic mode; and precisely how, psychologically speaking, did Newton enter (if he did enter) into Pope's consciousness? How aware was Pope, moreover, as a cerebrating creature, as well as a consciousness possessing an unconscious, of the change Newton was effecting on his consciousness? Miss Nicolson answered none of these questions, nor did she try; she noticed the appearance of Newtonian terminology in Pope's imagery and was content to leave it at that.

A brief survey such as this one can have no adequate conclusion: it is after all by definition open-ended. Nevertheless I doubt that any meaningful discussion of the state of the field since 1950 written by historians of literature who takes great interest in the history of science would dare to omit mention of these seven factors, although they could extend the list to many more.

NOTES

1. Peter Medawar, the British Nobel laureate in physiology, and the late Jacob Bronowski, mathematician and biologist, are exceptions, but neither is or was a professional historian or philosopher of science. Medawar's sense of the study of science and literature is succinctly delineated in 'Science and Literature', *Encounter*, XXXII (1969), 15–23.

2. See, e.g. for Chaucer: Florence M. Grimm, *Astronomical Lore in Chaucer* (Lincoln, 1919); Walter C. Curry, *Chaucer and the Mediaeval Sciences* (2nd ed. enlarged, New York, 1960); Chauncey Wood, *Chaucer and the Country of the Stars* (Princeton, 1970); H. E. Ussery, *Chaucer's Physician: Medicine and Literature in Fourteenth Century England* (New Orleans, 1971). For Shakespeare: Clark Cumberland, *Shakespeare and Science* (Birmingham, 1929); I. I. Edgar, *Medical Practice and the Physician in . . . Shakespeare's Dramas* (Detroit, 1934); Robert R. Simpson, *Shakespeare and Medicine* (Edinburgh, 1959); Alan Dent, *The World of Shakespeare's Plants* (Reading, 1971). For Milton: Kester Svendsen, *Milton and Science* (Cambridge, 1956); Lawrence Babb, *The Moral Cosmos of Paradise Lost* (East Lansing, 1970).

3. A very early example is Alfred North Whitehead's *Science and the Modern World* (Cambridge, 1925); a relatively late one is James V. Baker, *The Sacred River: Coleridge's Theory of the Imagination* (Baton Rouge, 1957).

4. As in the case of the traditionalists, there had been work done by the theorists before 1950. An example is Sir D'Arcy Thompson's perceptive theoretical essay on *Science and the Classics* (1940).

5. See *The Great Chain of Being* (Cambridge, 1936). Also of interest here is William F. Bynum, 'The Great Chain of Being after Forty Years', *History of Science*, XII (1975) 1–28.

6. Influential works included the essays of F. R. Leavis and his Cambridge colleagues that appeared in *Scrutiny* (1932–49); also F. R. Leavis, *Revaluation* (1936) and *The Great Tradition* (1948); I. A. Richards, *Principles of Literary Criticism* (1925), *Science and Poetry* (1926), and *Practical Criticism* (1929).

7. Some representative works include Douglas Bush, *Science and English Poetry* (New York, 1950); M. H. Abrams, *The Mirror and the Lamp* (New York, 1953); E. H. Gombrich, *Art and Illusion* (1959); Aldous Huxley, *Literature and Science* (1963); Martin Green, *Science and the Shabby Curate of Poetry* (1964); Wylie Sypher, *Literature and Technology: the Alien Vision* (New York, 1968). I do not mention the by now celebrated controversy between C. P. Snow and F. R. Leavis over the 'two cultures' inspired by Lord Snow's 1959 Rede Lecture at Cambridge University; see C. P. Snow, *The Two Cultures and the Scientific Revolution* (Cambridge, 1960). An example of the new theoretical writing at its best is A. Dwight Culler, 'The Darwinian Revolution and Literary Form', in G. Levine and W. Madden (eds.), *The Art of Victorian Prose* (Oxford, 1968), 224–46.

8. In 1963 she wrote to me: 'My career would have been altogether different if I had been born forty years earlier or later.'

9. See M. H. Nicolson, 'A. O. Lovejoy as Teacher', *Journal of the History of Ideas*, IX (1948), 428–38.

10. The point is also made by Rosalie Colie in '*O quam te memorem*, Marjorie Hope Nicolson!', *The American Scholar*, XXIV (1965), 463–70.

11. Miss Nicolson was too much of a traditionalist to approve of the flashy form of Edwin Schlossberg's *Einstein and Beckett: a Record of an Imaginary Discussion with Albert Einstein and Samuel Beckett* (New York, 1973), but she would probably admire the author's courage and endorse the idea of such a book.

12. See David Hackett Fischer, *Historians' Fallacies* (1971), 'Fallacies of Causation',

13 Hard and soft Freudianism offers a more chronologically modern example: for the problems that ensue when Freud-the-object is not sharply focused see Adolf Grünbaum's *The Foundations of Psychoanalysis: a Philosophical Critique* (Berkeley and Los Angeles, 1984).

14 All that literary critics mean by 'imaginative literature' is distinguished poetry, drama, or prose fiction as opposed to secondary writing, no matter how distinguished.

15 This has been true at least since the middle of the 18th century. Joseph Warton, then a leading literary figure, wrote on many occasions about the reciprocity of literature and science but never imagined that literature could shape science. E.g., he notes in *Reflections on Didactic Poetry* (1753), I, 395, that 'we [English] have some elegant . . . specimens of this sort [poems with scientific content] in the *Musae Anglicanae*; such are the poems on the barometer, on the circulation of the blood, and on Dr. Hales' vegetable statics'.

16 The field science and literature would lend itself particularly well to team research, but this avenue has not been formally explored to my knowledge in America. Now see G. S. Rousseau, 'Discourses of the Nerve', *Literature and Science as Modes of Expression* F. Amrine (ed.) (Dordrecht, 1989), 29–60.

17 There are notable exceptions, but only a few are critics of any stature: see Kingsley Amis, *New Maps of Hell: a Survey of Science Fiction* (New York, 1960): Bernard Bergonzi, *The Early H. G. Wells: a Study of the Scientific Romances* (Manchester and Toronto, 1961): Robert M. Philmus, *Into the Unknown: the Evolution of Science Fiction* (Berkeley, 1970); E. Barmeyer, *Science Fiction* (Munich, 1972); Mark Rose, *Science Fiction . . . Critical Essays* (Englewood Cliffs, 1976).

18 Based on a survey of American and British college catalogues I made in 1975–7. I am grateful to Dr Roy Porter of the Wellcome Institute, London for his assistance in this chore.

19 See Immanuel Velikovsky, *Worlds in Collision* (1950), *Earth in Upheaval* (1955), and *Oedipus and Akhnaton: Myth and History* (1960). More recently, Curt Siodmak, author of the science fiction *Donovan's Brain* (1960), has charged that neurologists have stolen some of his ideas.

20 For discussion of the paradox see Jonathan Culler, *Structuralist Poetics* (1975), 152–69.

21 Such is the thesis of Denis Donoghue's important book *The Sovereign Ghost: Studies in Imagination* (Berkeley, 1977). See also Richard Kuhns, *Structures of Experience: Essays on the Affinity between Philosophy and Literature* (New York, 1970).

22 The proof about the difficulty of discovering agreement regarding the methodology of structuralism is evident in the case of Foucault himself: although most of his serious readers classify him as a structuralist, he continues to maintain that he is not a structuralist. See, most recently, Michel Foucault, *Language, Counter-memory, Practice*, ed. D. F. Bouchard (Ithaca, 1977), 15–18.

23 See Harry Levin, *Why Literary Criticism Cannot be a Science* (Cambridge, 1968).

24 The critics dealt with the textual aesthetics of particular passages; the historians chronicled literary evolution and accounted for literary movements.

25 An example is *Diacritics*: it began c. 1970 as a periodical devoted to reviewing books by structuralists and transformed itself in 1975 into a journal of 'new approaches, especially phenomenological, to art history'.

26 A 'poetics' has nothing to do with poetry; a given poetics deals with the system of criticism developed by a literary thinker and always implies theoretical rather than applicative writing. E.g. Aristotle's *Poetics* deals with Sophoclean tragedy. Two important recent examples are Tzvetan Todorov, *The Poetics of Prose* (Ithaca, 1977), and Mark Spilka (ed.), *Towards a Poetics of Fiction* (Bloomington, 1977).

9

The discourses of literature and science (2)

The Isis *essay (see Chapter 8) was widely attacked by structuralists and postmodernists who claimed that in configuring the field of literature and science I had still given too much weight to influence studies and arrows of direction; to building bridges and establishing connections between these types of knowledge rather than construing them in the light of the ideologies they both embraced. Many of my opponents would have been happier if I had reduced science to its discursive practices alone, without taking stock of its mathematics, instrumentation, and non-discursive practices or privileging cultural and historical mentalities in any way.*

I continued to resist the temptation. The idea of reduction was compelling, not least for its egalitarian view that all knowledge was discursive and lexically dependent without regard to external forms of referentiality. But this version, while attractive, seemed to ignore the difficult questions about literature and science as independent realms and then avoided the epistemological snares of their differences. Furthermore, it was altogether inadequate to the heuristic challenge presented within the academy by literature and science. In both this essay and the one in Isis, *I wanted a literature and science that would act as a diagnostic of the state of the humanities as much as anything else.*

It also seemed important to me in the mid-1980s, when this essay was written, to acknowledge the many *discourses of literature and science, not merely the most recent ones or the ones most endorsed at that time. The humanities were then still in decline. They were not only losing students but needed to put their houses in order. Their internal logic and methodology had become storehouses of vehicles for chaos and confusion. The only way, moreover, the emerging discourses of literature and science could have any practical effect on higher education was to recognise the need to transform the humanities in this post-disciplinary age.*

I tried to suggest this approach by surveying the rise of the discourse(s) of literature and science in the age of classicism from its (i.e., the discourses') pre-Kantian origins and post-Kantian rebirth; through its vicissitudes under Romanticism and as a neo-Romantic critique; extending into naturalism, positivism, quantum physics, and thermodynamics; arriving finally at our own Snowian and post-Snowian age of theory and social constructionism.

The essay introduced a volume dealing entirely with these discourses of literature and science.

As the Modern Language Association of America contemplates the publication of a centennial bibliography, it may be appropriate to reflect on the most elementary aspects of the subject: what *is* literature and science? If anyone asks why we should inquire into the status of something that already possesses an historiography a century old, as this bibliography will abundantly demonstrate, the reply must be that since 'literature' and 'science' are individually such fluid categories – constantly altering their borders and redefining their epistemological profiles – basic questions about their relationship can never be outdated. Indeed by definition, the relation continues to alter and must continue to do so.

A discourse combining literature and science begins in the seventeenth century in the midst of debates about universal language schemes. Before then 'science had of course appeared in 'literature', and all sorts of literary techniques, especially rhetorical devices, had been used by 'scientific' writers since the time of the Ancient Greeks. But it was not until the language projectors of the seventeenth century called attention to this particular blend that the discourse got started in earnest. Even Bacon, no devotee of universal languages, remains a figure prior to the discourse, although he figures in its prehistory for the remarkable manner in which he took all knowledge, certainly the sciences and the arts, for his personal province.

The aim of the projectors was the discovery of a single, 'universal language' capable of correlating all the diverse realms of nature.[1] This language would naturally not be an ordinary spoken language such as English, French, or German, or the then dead or moribund languages of Greek and Latin, but an entirely new and obviously man-made language composed of artificial symbols, ideograms, and other signs. The new language – itself only one of many discourses – would permit men and women of different nationalities to understand each other to a degree hitherto unknown, and would put an end to the long reign of cultural isolation that insulated Europe for centuries.[2] More importantly for the progress of civilisation, it would deflect men from interminable theological squabbles; at long last one common language would pave the way towards understanding the real, external world they inhabited. There was no sense within this programme that a universal language – assuming it *could* be constructed – would deflect men from God or induce atheism. On the contrary, the major reason for expending the effort to construct such a difficult artificial language was eventual discovery of the one and only true God: He who had created the cosmos from chaos deserved better attention than the religious squabbles perpetrated by even the best minds. In this restricted sense, these language schemes were theological: explicitly constructed to understand God better. There was no awareness then, as there is now, of a rift between science and theology, and, of course, in the last hundred years, between science and literature. For a theologian to be considered competent in that period, he had to be equally knowledgeable of science; the reverse was true for scientists, as the lives of all the Cambridge Platonists, John Ray, Boyle and Newton and many others make patent (Dillenberger).[3]

By the end of the century virtually all attempts had failed. Yet the long search was not unproductive. While searching, scientists and projectors (a general term for schemers of all types) had made some crucial, even extraordinary discoveries: not merely in particular sciences such as physics, chemistry and mathematics but in a domain we today would call the philosophy of science. They realised that no matter how far they still may have been from their goal, Nature herself was finite. Even the secular history of Nature throughout time was finite. That is, Nature was not an infinite body of knowledge that man could never hope to understand: a dismal future prospect for man to contemplate. It was rather codified in a vast but nevertheless finite set of laws and relationships that would gradually be revealed to man if he persisted. In this sense, the cumulative effect of the search for a universal language rendered an unprecedented optimism about the knowability of external nature.

But there was also a monumental hurdle that might prove insurmountable: namely, the theoretical *language*, or languages, in which such laws would necessarily be encoded. If just one universal language had been successfully constructed the prognostic for discovery and knowability might have been better since external reality could be artificially encoded in a *single* language universally understood. But repeated failures to construct a single language augured trouble: like Sisyphus with his rock, mankind seemed doomed to competing theoretical languages among which it was impossible to adjudicate which 'language' was better.[4] Spectators who distrusted this unprecedented optimism attributed the failure to postlapsarian *hubris*, arguing that as a consequence of the Fall man was saddled with diverse, competing languages – a veritable Tower of Babel toppling down under the weight of glossolalia.[5] Even those who were unconvinced of such necessitarianism had realised by the end of the seventeenth century that no single 'language' or 'discourse' would be universally privileged. The likelihood of competing, theoretical languages in the future seemed virtually assured; and while there was then endless controversy about the facets of cosmic knowledge that could be known with any degree of certainty, by the time of Locke and Berkeley there was little hope that an artificial, theoretical language could ever be successfully constructed. Too many impediments existed.

It may seem that this crucial chapter in intellectual history is indirectly related to the discourse of literature and science. Such is not the case; for only a clear sense of the relation of ordinary spoken languages and the much sought-for artificial one provides a clue to the precise manner in which Enlightenment thinkers became self-conscious of the theoretical languages they were already invoking (Struever). Furthermore, there had been much speculation before the eighteenth century about the grammatical rules of spoken languages in relation to those of 'universal' ones. Both types were compared and contrasted, as in schemes now well known to intellectual historians: comparison and extrapolation naturally provoked further analogies about so-called 'grammatical laws' of external reality. Did such laws exist according to the analogy of macrocosm-microcosm? If not, could it be that external reality differed in some essential way(s) from the second-hand representations of nature man could know? How had the grammars of ordinary spoken languages arisen in the first place? What was a 'grammar' if one language required it and another did not? These types of similar questions caused philosophic thinkers to be more scrupulous in speculative knowledge than they had been before. The end of the seventeenth and beginning of the eighteenth centuries reached a peak in this scrupulosity.

It is impossible to set a date to the collective sense of failure in the search for a universal language, but we do not stray if we imagine it as coming in the period of Locke's maturity at the end of the seventeenth century: Locke himself was a brilliant, original mind who would have succeeded in the quest for any type of original knowledge to which he applied himself. A trained and practising physician, he eventually discovered himself to be more interested in philosophy. More crucially for literature and science, he was uninterested in universal language schemes and never set out to discover one.[6] Instead, he embarked on a long intellectual quest – summarised in the *Essay Concerning Human Understanding* (1690) – to study, objectively and empirically, the workings of the mind of man. Here, too, the analogy of macro-microcosm was germane. Locke was aware that his exploration of the mind was as 'scientific' as the sidereal quests of all those – Kepler, Copernicus, Galileo, and, more recently, Newton – who had attempted to discover 'cosmic nature's laws'. In this sense Locke imagined himself to be but a type of Newton: and Newton realised on more occasions than one that he was a 'Locke of the planets'. The search for laws, or 'grammars', of external and internal nature were identical pursuits. Even more crucial for literature and science, each thinker realised that his discoveries would necessarily be expressed in a theoretical language. These languages would differ: Newton's principally in mathematical signs, in a new mathematical language of the calculus (fluxions), appropriate to the cosmic laws he was attempting to describe; Locke's in plain, unadorned English prose, stylistically simple enough not to distract the reader from complex ideas about the very notion of 'ideas'. Neither thinker was reticent to pronounce on other fields: poetry, politics, religion – for subjects or areas of knowledge were not then artificially separated out as they are today.⁷ Yet Locke and Newton – and they are but representatives of a type that was not uncommon at the time – were aware that the theoretical language each used differed from that of poets, statesmen, or theologians. Locke's *Essay* and Newton's *Principia* were no 'truer' than a great poem or major political pronouncement by Plato, Machiavelli, or Hobbes. What differed in all these were the languages, and necessarily the grammars, in which they were expressed. No one language was inherently better or truer than another. To privilege one would be no different from saying that a lion was better or truer than a tiger, or – in other realms – that a poem was better or truer than a mathematical theorem or proof. Quality and truth could not be measured in this way.⁸

It was in this intellectual milieu that the separation of the arts and sciences, and the accompanying discourse of literature and science, arose. What precisely was this discourse? The dramatic story of the separating of these two realms: their history of unity and the gradual rift that broke them apart.⁹ Chronologically speaking, the period between 1670 and 1820 is permeated with remarkable comments about a new disparity between the arts and the sciences. All these commentators noted perception of a new proliferation of theoretical languages. If many theoretical languages existed before the rift, their number seemed even larger afterwards.

This new awareness – certainly new in the period of Locke and Newton – entailed another grim prediction for the future which was debated in the eighteenth-century Enlightenment. Could it be that the persistent failure to discover a single universal language was not owing to any lack of human genius or imagination but to some defect in Nature? Perhaps to the absence of laws of nature altogether? That is, an inability to agree may not have derived from any historical or moral defect in man (i.e. the Fall, the corruption of language in the aftermath of the Fall, the necessity of competing theoretical languages after the Fall), but from a cosmic shell that ran on chance and whim rather than according to unchanging laws that were regular and predictable? Poets and radical philosophers had raised this possibility before 1700; now, in the eighteenth century, empiricists and rationalists of all sorts (not merely philosophers and metaphysicians, but scientists and theologians as well) also entertained the notion. Viewed in this sense, Blake the poet-artist, and Priestley the preacher-chemist, are intellectually much more similar than dissimilar; and it may be more profitable to compare them than to restrict Blake to the tradition of Spenser and Milton, and Priestley to that of Boyle and Lavoisier (Abrams, 331–9). Such comparison necessitates the discourse of literature and science in a contemporary phase. Nineteenth-century distinctions between Blake and Priestley as 'poet' and 'chemist' are less valid than either man would have recognised in his own time; in 1770 or 1800 the 'separation' was still new. The point to be gathered is not the disparagement of necessary shorthand labels (poet, chemist), but the urgent significance of an historical approach to two relatively recent categories: literature and science.

Meticulous study of the development of both concepts – literature and science – would be useful but would not provide, in themselves, an adequate context for understanding how the discourse of literature and science developed into the contemporary subject we know. The progress of knowledge is perhaps more crucial, especially the nineteenth-century sense that competing theoretical languages had been providentially privileged; so, too, philosophical positions about the regularity of the laws of nature. By the late eighteenth century a sense began to develop that theoretical language would change every time a revolution in delineating the laws of nature occurred.[10] This was not so much the perception (already mentioned) that man was saddled with 'theoretical glossolalia' as a sense that new information requires new description.

The perception has continued into our time, particularly among those who are now predicting that the micro-electronic revolution is creating a new language with its own vocabulary, syntax and grammar. Looking backwards then, the eighteenth century is the crucial historical period for literature and science. It was then that writers of all types first began to align themselves with one or another type of theoretical discourse and equally to conceive of themselves as social types. Such perception included economic and social dimensions, and although it is perfidious to conjecture whether linguistic awareness creates a need for social types or vice versa, it is nevertheless reasonable to observe that in the eighteenth century all types of thinkers then began to classify themselves into professional groups who had not done so before. Physicians and lawyers, statesmen and merchants, doctors and churchmen had existed for centuries; but philosophers, creative writers and critics (in our terminology) had not.[11] The development is crucial for the past, present, and future of Literature and Science. Without a firm grasp of these developing categories I doubt that anyone today would ever understand what literature and science is.

Yet it would be false to contend that the evolution or flow of science before 1800 had been linguistically determined in large part. In the seventeenth century it was believed, as John Ray said, that naturalists merely ought to observe Nature and describe their observations: not in numbers or universal languages but in plain English prose. Such is the spirit of the first history of the Royal Society by Bishop Sprat. Natural philosophy – the study of the natural world – accumulated an enormous mass of writing in the 150 years after Sprat published his *History*, much of it written in barbaric English prose that was sometimes illiterate and ungrammatical. The authors of these works did not always think of themselves pre-eminently as 'scientists' or 'literary writers'; they viewed their enterprise as related to truth and the advancement of knowledge.[12] As such, they were not different from those compilers of dictionaries who viewed their tasks as accumulation of the best available information. Samuel Johnson, the literary lion of his age, considered himself 'a scientist' of the English language; so too did his contemporaries: as scientific as Newton on stars or Locke on the mind of man. 'Scientist' for all these types denoted profound knowledge of the thing they were expounding: not merely knowledge of its present nature and status but also of its first causes (or principles) and origins.[13] All these men would have been baffled by notions that Newton was a 'scientist' while Locke was not, or that Johnson was less 'scientific' than either when matters of language were in question. Knowledge, true and accurate, verifiable and predictable, was the deciding factor, not the specific area where the knowledge happened to reside.

Debate arose – and here is a further point for literature and science – when precise scientific methods were called into question. The history of scientific method is not a subject that can be captured in the space of a few words but some of its modern roots are found in the rational-empirical debates of the Enlightenment. When Newton claimed '*hypotheses non fingo*' – 'I do not make hypotheses'[14] – he referred in part to debates over rational and empirical, and especially deductive and inductive, approaches. Much was at stake: the resulting 'science' derives from the approach, almost to the degree of cause and effect. The debates had gone so far that poets and their critics – Pope and Warburton as well as Wordsworth and Coleridge – were wondering whether a poem was a deductive or inductive artefact, and whether it contained only a *type* of truth. But as the eighteenth century wore out, empiricism clearly gained the ascendant hand, although it would be nonsense to posit that the rational approach to knowledge died out, especially inasmuch as all true knowledge is necessarily a blend of the two.

The cumulative effect was incorporation of empiricism into recognised, valid scientific method; nonempirical approaches, such as much, but not all, poetry, painting, and musical composition were deemed 'arts of a lesser kind' – rational pursuits completed in the privacy of one's own study without the need of observation and verification. Eventually a sense developed (here it must be remembered that we are narrating developments that occurred gradually over centuries rather than in single years or decades) – a sense that 'the sciences' were empirically oriented while the arts, or humanities, were not; and that because they were based on continued observation and verification – and in this sense public, masculine activities – the sciences were epistemologically more valid than their artistic counterparts. What began with Descartes in the *Discourse on Method* as inquiry over scientific procedure ended by the time of Lyell and Darwin – the mid nineteenth century – as a whole set of value judgments surrounding every aspect of scientific activity. This progress could not help but influence the fate of imaginative literature. Yet it is difficult, if possible, to authenticate these differences by proof or describe them accurately now. Our contemporary view represents the hardening of the differences to such a degree, that few today (except some serious historians of science) are willing to address the matter.

Two other broad cultural developments occurred after 1800 that radically affected the discourse of literature and science: Romanticism and positivism. If the former inverted the relation of God and man and permitted arch Romanticists to profess that 'whatever God really is, he is me', as Byron's heroes often do, the latter – positivism – maintained at the close of the century that all valid knowledge was scientific. Whether physical, aesthetic, or moral knowledge, it had to be scientific: that is, observable, demonstrable by proof, predictable, and capable of verification. Literature did not come out well when interrogated under the grid of positivism. Moreover in this progress from Romanticism to positivism, literature and science, viewed individually and together, experienced further traumatic reversals. The Romantics believed that the epistemological bases of literature and science were the same; this is why Wordsworth was not troubled by the rise of the scientist as a new professional type in the famous passage in the Preface to the *Lyrical Ballads*.[15] Yet the sciences – the Romantics claimed – omitted from their discussions the moral realm; consequently, science radically differed from, and was inferior to, the arts.

No *rapprochement* was possible so long as science refused to take cognisance of its own significance. Frustrated by its persistent refusals, the Romantics began to compose a new critique of science that showed its moral limitations and inherent methodological defects. In this new discourse added to the existing one, the Romantics generated a myth about the privileged status of literature (based on its taste, sensibility and refinement) and the unprivileged role of science (its mere utilitarianism and functionalism) under whose influence we today still labour.[16] If pioneer scientists and popular spokesmen for science like Lyell, Darwin, Faraday and, of course, T. H. Huxley had answered the attacks of Romantic liberal thinkers from Blake to Arnold, Emerson to Carlyle, the flow of nineteenth-century thought in this area would have been different. But they rarely did, leaving a door open for early positivists, from the arch-positivist Comte and his followers in France to Mill, Frederic Harrison, John Morley and dozens of other British positivist sympathisers who sought to heal the wounds of separation by privileging scientific knowledge as the *only* authentic type. The impact of positivism, with its endorsement of 'systems' and all things 'scientific,' was enormous on the discourse of literature and science.[17] Positivism sought to ground all human knowledge (not merely scientific) into separate sciences: the sciences of man as well as those of organic and inorganic nature (anthropology, psychology and sociology, as well as physics, chemistry and mathematics). If the discourse of literature and science appeared to have no guaranteed future as a developing theoretical language early in the nineteenth century, by the end, all philosophic writing, and certainly all allegedly valid knowledge, was necessarily of this variety. When Macaulay Posnett claimed in 1886 that the history of literature was 'a history of all the sciences', and the purpose of comparative literature was 'to establish a science of literature'[17], he uttered a fundamental positivist position. He also anchored criticism in a scientific approach: a place where it has remained for a century now, with little evidence of change.[18]

In our own century two other developments have proved consequential: first, the establishment of national literatures – English, French, German, etc. – as university subjects and then the development and recent flourishing of the history of science. Before approximately 1900 (the exact date varies from locale to locale) university students studied rhetoric, logic, and philosophy but not literary interpretation or literary criticism. Departments of individual modern languages are twentieth-century developments; the academic study of literary criticism, as distinct from primary literature, is even more recent.[19] The consequence of this relatively recent institutionalisation is greater than ever interest in the developing discourse of literature and science. No wonder that the subject seems to have taken off immediately after World War II.

History of science has exerted another type of pressure. It is even younger than literary criticism as a university subject. But it has claimed so many adherents who have performed so much good work that no serious student of literature and science (or of general cultural history for that matter) can dare afford to overlook it.[20] Whereas students early in our century could write without any first-hand knowledge of the actual history of science, no one today would get far if he did. Historians of science have realised, moreover, that they must become more sensitive to language and its history – especially the history of its imaginative literature – if they wish to chart their terrains adequately. Literary evidence may be problematic on many counts but no one can afford to overlook it entirely. The result of both developments is a new threshold of explanation for literature and science (and for the history of its discourse) unthinkable even fifty years ago when A. O. Lovejoy and Marjorie Nicolson were pioneering the field in America.

Since the middle of our century the perception of two cultures' –

crudely, the arts and the sciences – has also affected the discourse. All sorts of students who ought to know better project these so-called 'two cultures' back on previous civilisations without worrying if the dualism ever existed. It is practically schizophrenic to think that the Renaissance was divided in this way (Leonardo, Michelangelo, Vesalius, Galileo *either* artists or scientists?), yet even rigorous scholars continue to assume the dichotomy has existed throughout Western culture. Indeed the appeal of 'two cultures' has recently intensified owing to the magisterial advance of technology in our time and the sense of a new breed of 'technicians' who differ from 'humanists'. The perception has some validity, and was intensified by the now famous Snow-Leavis controversy over 'two cultures' (1959),[21] but historically speaking the rift – to the degree it is a genuine separation – is older.

The history of 'two cultures' before 1959 and the discourse of literature and science converge at many points; they are not identical but overlap in so many places – especially in the century from 1859 to 1959, from Darwin's *Origins* to Snow's *Two Cultures* – that it is futile to think of one without invoking the other. A century ago Thomas Huxley, the vigorous Victorian proponent of science and technology, claimed that 'there is one feature of the present state of the civilised world which separates it more wisely from the Renaissance, than the Renaissance once separated from the Middle Ages'. The passage is worth quoting:

> This distinctive character of our own times lies in the vast and constantly increasing part which is played by natural knowledge. Not only is our daily life shaped by it, not only the prosperity of millions of men depend upon it, but our whole theory of life has long been influenced, consciously or unconsciously, by the general conception of the universe which has been forced upon us by physical science. (132–3)[22]

A century later – today – we are still perplexed by many of these challenges. Literature and science, broadly construed, devotes much of its attention to these issues, not merely because it is an historically based subject but also in view of its interest in the linkings and separations that occur over long periods of time.

Today, the idea of 'two cultures' – whichever specific ones they may be – continues to be invoked and demonstrates that the concept is far from dead. Imaginative writers, as well as theorists, proceed on the assumption that critical discourse should also be divided into two, broad categories designated literature and science. Many literary critics now study the language of 'scientific' writings, past and present, as if these works inherently differed from non-scientific, yet the differences are assumed rather than made explicit and one wonders if these so-called differences have not been magnified and exaggerated. Conferences and symposia try hard to engage the ears of 'the others' (i.e. the 'other culture') by exploring their works and enlisting their scholars. Even some of our best literary critics who are not particularly interested in the discourse of literature and science have seriously explored the literary attributes of scientific writing. They do so to understand the workings of language, and with a recognition that scientific writing has been overlooked for too long. Professional writers also capitalise on the reading public's fascination for famous scientists of the past; for the private lives of figures such as Darwin, Einstein, Freud and Jung; as well as the ideal of so-called 'pure science' pursued by Galileo, Boyle, Newton and others. This activity, ironically, has little cumulative effect; has done little to obliterate the 'two-cultures' category. If it is valid to maintain that in the late seventeenth century there was widespread perception of two types of method, or discourse, in all writing – Aristotelian and non-Aristotelian – it is perhaps equally valid to contend that in our time hard and fast distinctions between 'literature' and 'science' endure. Nothing, after almost three decades, when Snow first made his pronouncement in the Rede Lectures at Cambridge University, appears capable of dissolving the notion. Both literary and scientific types seem to be comforted by the sense of an antagonist out there – if not an outright enemy, at least someone alien to their vision of reality. Few will take the time to explore these feelings and beliefs; perhaps too much is at stake.

A sense of the past and a view of history distinguishes these contemporary camps. Among the historians of science, an empiricist view of the past seems to be the only acceptable one. Likewise – in the other camp – for the heirs of New Criticism in the 1980s who also pledge allegiance to empiricism. For both groups, historical events are meaningful constructs whose causes and effects can be isolated and evaluated. If 'literary' works (novels, poems, plays, biographies, diaries, etc.) offer a more profound clue to past cultures than scientific writings, the past can also be understood *sans* literature, in terms of persons, places and events.

Others of our contemporaries – literary *and* scientific – deny it. Among both French literary critics and French historians of science another version of history has emerged that replaces empirical history with other varieties: with parallel developments in many types of discourse. Chief among these have been Bachelard, Foucault, Deleuze and Guattari, authors preoccupied with discourse itself and who cannot be classified merely as 'literary' or 'scientific'.[23] Their approach to literature and science differs radically from thinkers such as Barthes, Ricoeur and Derrida in France, Raymond Williams in England, the German hermeneuticists, and others elsewhere in Europe who are fascinated by the relation of discourse to history, but who have not been seriously interested in 'science' as a category possessing a significant past worthy of the best efforts of the human imagination. The former group (Bachelard *et al*), considered collectively, has diachronically studied the discourses of different subjects (medicine, economics, linguistics) to understand how subjects themselves are manufactured, i.e. how subjects come into being.[24] Foucault, in particular, and his mentor Bachelard, asked how such subjects as biology, economics, and linguistics are formed, and by so doing provided new maps of the past, especially for subjects like history of science and history of medicine.

Some of these French thinkers have also charted the 'manufacture of feelings', of complex new emotions, as subjects; thereby exposing themselves to the criticism of empiricists who maintain that their programme is impressionistic and relativistic: impressionistic as a consequence of the discourse they themselves (Bachelard *et al*) were generating; relativistic because the only type of 'truth' it (their programme) could discover was based on non-empirical and thereby questionable categories.[25] In *The Poetics of Space* Bachelard isolated the 'pre-scientific imagination', attempted to show that scientific thinking was so deeply ingrained in the primitive psyche that it is inaccurate to speak of a pre- and post-scientific age. Foucault showed in *Words and Things* 'subjects' or 'disciplines' (biology, economics, linguistics, etc.) arise, or are manufactured, at specific moments owing to identifiable discontinuities (*epistemes* is his term) with discourses of the past. In *Anti-Oedipus* Deleuze and Guattari have sought to document the 'histories' of such elusive emotional states as desire and longing, and have maintained that no empirical (i.e. scientific) methods could have been capable of assisting their inquiry even if they had been hardened empiricists. All these thinkers asked heuristic questions that could only be answered by heuristic methods. Their work has been sceptically viewed by many traditional-minded literary critics, yet embraced by others, perhaps most fervidly by those intrigued to explore the affective (i.e. emotive) side of human history.

Two thinkers in America. Kuhn and Feyerabend, have been particularly influential for the recently evolving discourse of literature and science. Like their French counterparts, they question the methodological bases of many historians, and replace straightforward diachronic history with more philosophical inquiry. If Feyerabend has forced questions about professions and professionalism (e.g. the relation of professional types to bodies of knowledge), Kuhn has gazed into 'the route of ordinary or normal science' and, more recently, into the linguistic dimensions of scientific writing.[26] Neither has pronounced on literature *qua* literature, or about the status of great imaginative literature from Chaucer to Conrad, Goethe to Mann, Dante to d'Annunzio, but such explications of great texts do not determine in themselves the future of literature and science, which is a more theoretical than practical activity. It is important to know which literary works contain scientific dimensions and how their authors incorporated this science. But the discourse of literature and science represents a wider domain than source hunting or topical allusion. More fundamentally, as I suggested in a survey of the state of the field a few years ago,[27] it is rooted in philosophical beliefs about the status of imaginative literature in relation to the whole of culture, and to notions about the liberating or enslaving effects of modern science.

Post World War II ferment of knowledge in these domains has generated a realist counterattack produced in the name of empiricism. All types of literary critics, especially traditionalists and historicists, have wondered whether literature and science is not too impressionistic a relation; more specifically, whether its degree of empiricism, cause and effect, source and influence, event and prediction, and the uses it makes of competing types of evidence (some of which are language-bound in extremely different degrees), do not invalidate it as an approach. And they have defined their scepticism against a varied background of critical pluralisms: all types of competing theorists trying to promote their particular brand of critical insight, whether hermeneutic or semiotic, deconstructionist or Marxist. Others who observe in the background, and who may not have participated in the debates of the 1970s regarding critical theory and the New Historicism (for example), wonder whether heads and tails have not been reversed here; convinced that the *contemporary* discourse of literature and science, like most other theoretical approaches now competing for attention today, has itself become highly scientised; so highly scientised, indeed, that pure literary theory, like pure mathematical theory, is itself a science – no different from particle physics or stochastic equations. It is something to ponder, particularly the emphatically theoretical bent of so many of our most famous professors of 'literature'. But another dimension exists and pertains to the demarcation between the many types of discourse and the rise of disciplinary histories. I mean, of course, the demarcation between a so-called scientific and non-scientific discourse, and (by disciplinary histories) the sense practitioners have that their disciplines or subjects (English, French, German, philosophy, history, etc.) could not be what they are today, if the unity or art (craft) of the discipline were removed. Upon examination these caveats dwell on matters of definition and delimitation: primarily whether 'literature' and 'science' can be separately designated as meaningful units.[28] The result is scepticism, even hostility. Yet when such doubt is viewed in the light of other types – for example, in the light of the claims of historians of science who profess to aspire to that knowledge which alone can be termed empirical – the problem of demarcating science and non-science, or history and non-history is just as large. Then, too, there is the literary critic's general hostility to things scientific to be coped with: not merely a fundamental anti-technological stance but virtual condemnation of scientists as creatures of a lower and more literal intellectual order. Critics may naturally deny the antipathy but it has been documented and no doubt exists. The curious thing is that, while critics are uncomfortable with 'science', they find a natural home in 'philosophy', as recent criticism shows. 'Literature and philosophy' thrives to a degree unknown today in 'literature and science'; yet the reasons are not at all apparent, especially since the philosophy of science is as *philosophical* an activity as can be imagined. Given all these obstacles, it is perhaps not surprising that literature and science has recently had to walk a delicate tightrope between the Scylla and Charybdis of criticism and scientism.[29] Much good work has been done in the last two decades – as the forthcoming MLA centenary bibliography of literature and science shows – at the price of constant self reflection and nagging doubt about the genuine identity of the discourse of literature and science.

If we inquire into the status of literature and science today, it is clear that its fate is tied to the vicissitudes and transformations of the contemporary critical mood. As has been suggested, the field gazes on *both* the realms of literary criticism and the history and philosophy of science: their pasts as shaping factors in their presents. Recent proliferation of literary methodologies – the by now much denounced pluralism – is causing confusion, it is true, but there is little reason to believe that empirical historical criticism is useless for literature and science. So long as source hunting and influence tracing do not become ends in themselves – limits beyond which literature and science ought not to proceed – traditional, as distinct from new, literary history can be a useful ally. The matter here is additiveness rather than exclusivity. Furthermore, historians and philosophers of science are increasingly coming over to the view that their proper study includes shifts in theoretical languages; the recent work of Mary Hesse and Roger Jones are just two cases in point (*Texts; Physics;* and Hoenigswald).[30] Granted, some historians of science continue to document problems and provide contexts without paying close attention to language; but their numbers continue to diminish. Finally, the prominence of theory as a respectable field for research has enhanced the scope of literature and science. Today more than ever, literary critics are interested in the problems of psychology and others of the social sciences. There is a sense that psychology and her sister social sciences can shed light on the most ploughed-over literature. This fascination results, primarily, from interest in the affective dimensions of the reading process and, more generally, from interest in all things affective.

Today, the discourse of literature and science hovers between radical empiricists who expect scientific explanations for all change in theoretical language, and imaginative rationalists like Foucault and his cohorts who have continued to claim that all empirical investigation is blind. Ironically, the dichotomies of the seventeenth century – some of which we consulted at the outset – endure. If we no longer systematically search for a 'universal' language capable of perfectly correlating external nature and artificial symbol, we still assign values to almost every type of theoretical language. Depending on our own vantage, we place a premium on this or that explanatory language without worrying too much about the consequences or alternatives. We call these languages subjects or disciplines (i.e. 'literature', 'psychology', 'economics', 'biology') and do not often realise the consequence of our shorthand labels. In one area, at least, we have progressed beyond any border imagined by seventeenth-century projectors. This is the country of metaphor.

It is clear that shifts in theoretical languages cause metaphoric displacement. Stated otherwise, metaphor is the key, more so than any other linguistic category.[31] Other features of language and style alter as well in these shifts; metaphor, however, takes the brunt. It matters not whether the particular discontinuous discourse centres on one type of content or another: what counts is the degree of literalness in relation to metaphoric density, as well as the nature of these metaphors. The discourse of contemporary philosophy differs from that of literary criticism, not so much in its rhetorical tropes and formal organisation but according to the uses to which it puts metaphor (i.e. analogy, simile, example, literal figure). If this is true, as I believe it is, then it would seem that a part of the future of literature and science will be determined by the fate of metaphor theory.

Today theorists of metaphor can be divided into two broad camps: dualists and unitarians – those maintaining that all language consists of combinations of the literal and the metaphoric versus those unitarians who argue that all ordinary language is necessarily metaphoric, i.e. that there exists no such thing as 'literal language'. Critics of metaphor are, in this sense, as divided as students of Literature and Science into camps of empiricism (Popper and the Popperians, Lawrence Stone, much other British empiricism) versus imaginative rationalism (Bachelard, Foucault, many French critics). Although the analogy is fatally flawed by virtue of nonparity (i.e. there is no necessary link between the two groups), the comparison may still be useful. If it is, then it may be fair to surmise that the future of literature and science must be a compromise: between unitarian and dualist views of *metaphor*, as well as rational and empirical views of *method*. Philosophers of language will determine the future of literature and science in this sense, no less than historians of science who have charted the principal maps of the field.

It is impossible to speculate with any degree of confidence on this future, but if these patterns continue, there is no doubt that literature and science will again shift its ground, and the transformations we have witnessed since the seventeenth century will be but a piece of a longer story begging for a narrative history.[32] The possibilities for change are probably great but none the less intriguing to ponder. If empirical literary history triumphs over new literary history and deconstructionism (whereby deconstruction is necessarily a rationalist activity based on *a priori* assumptions that annihilate empirical history), there is good reason to speculate that literature and science will have claimed more adherents than otherwise. This will be especially true if the history of science should also follow an empirical course in the near future. The position of deconstructionists and other anti-historicists is, in my view, fatal for literature and science; perhaps not in the short run among deconstructionists who tout the importance of metaphor, but surely in the long run when it will be evident that deconstructionism has disembowelled history of its past, and allowed it to exist only in tropes and lexia. Because deconstructionists endorse modern theory it may seem on the surface that they are the natural allies of literature and science. Such may not, alas, be the case. Given two evils – absence of theory, absence of history – the latter may prove more fatal. The discourse of literature and science can survive without theory; it will die without empirical history. Moreover, if deconstruction prevails, or some similar, derivative methodology that preoccupies itself with the present or the nearpresent only, the future of literature and science as a narrative account of pre-1800 discourses will be in danger. Critics and scholars will continue to talk about 'the science in poetry' or 'the language of scientific theories', but the development and maturation of the field as a worthwhile academic discipline will be less assured.

Even more consequential in my view is the future of metaphor theory. By now it is practically commonplace that all theories of metaphor depend upon a prior theory of meaning. On the degree to which meaning is construed to be stable or unstable, universal or changing, all subsequent theories of meaning will depend. Unitarians and dualists among theorists of metaphor disagree principally upon the literal meaning of a word in relation to these stable or unstable versions of meaning. Words such as spirit, sense, or energy may be literal or metaphoric – it is alleged – to the degree that their original meanings continue to endure. According to the unitarians, metaphors undergo life cycles (birth, maturation, senescence, death); networks of metaphoric relations exist among constellations of words in any given, ordinary, spoken language. These constellations, or patterns, can be isolated and traced; they are capable of being treated in terms of – again by analogy – their status in the life cycle.

The dualists deny the life cycle altogether and hold up, instead, universal metaphors. Yet any theory of metaphor that denies, as does that of the dualists, a life cycle of eternally evolving and dying metaphors jeopardises the status of literature and science by diminishing the basis on which the so-called independent realms of pure imaginative literature and undiluted science can be meaningfully connected. Via metaphor as its focus, the discourse of literature and science remains at least stable and manageable. Deprived of this unitarian privilege, its impediments – problems of demarcation, source and influence, arrows of reciprocity, intended meanings versus subconscious ones – pile up. Indeed, the impediments magnify themselves into grotesque intellectual nightmares.

Any serious students of literature and science can do meaningful work by diachronically tracing the evolution of even a few metaphors in literary and scientific texts; alternatively, they can analyse the life cycle of just one metaphor (plus the attendant constellation of satellite words) over blocks of time. But they cannot proceed if the operative language theory minimises the significance of metaphor in comparison to other tropes; or if they construe metaphor as something so stable and culturally noncontextual that it is denied a life cycle based on ever-changing social and intellectual conditions. The debates of linguists who concentrate on metaphor theory will therefore be most consequential for the future of literature and science. If the dualists should prevail, a new fear, not altogether different from that now posed by deconstructionists, will arise. This anxiety will have to be addressed when the ground of literature and science necessarily shifts again.

NOTES

1   Although there were dozens of these schemes in the seventeenth century, the best known in England was John Wilkins's *Essay Towards a Real Character and a Philosophical Language* (1668); see M. M. Slaughter, Knowlson, and the very perceptive essay by Borges showing the significance of these attempts for poetry and science.

2   For discussion of the intellectual dangers of cultural isolation see Harry Levin, 107–10.

3   Science and religion have appeared to be so disparate, even at war, in our time that we can barely imagine they had ever been otherwise. John William Draper, a distinguished nineteenth-century British research chemist resident in New York, decorated with professorships and elected to many scientific societies on both sides of the Atlantic, wrote a best-seller on the subject which argues that there is no conflict between religion and science. More recently, E. A. Burtt, the author of many studies on the metaphysics of science, has written that 'a religion and science brought together in one [person] is the deepest need of the modern world' (141).

4   Swift's masterpiece, *A Tale of a Tub*, assumes this background and invokes satiric language pointing to fierce competition from other sects and groups; the books of Christopher Hill dealing with England in the Interregnum and Restoration have described these numerous sects and their party interests.

5   Glossolalia, the existence of many tongues, flourished to an unprecedented degree in England during the Interregnum, especially during the decade of the 1650s, owing to the sudden proliferation then of religious and scientific sects; its progress during the seventeenth century is very much a part of the rise of the discourse of literature and science.

6   Locke wrote to some of the Fellows of the Royal Society that he was disgusted with literary language: dissatisfied with 'the cheat of words'. His doubt about universal language schemes is stated unequivocally in the *Essay* (3.ii.2): 'I am not so vain as to think that anyone can pretent to attempt the perfect reforming of the languages of the world, not so much as of his own country, without rendering himself ridiculous. To require that men should use their words constantly in the same sense, and for none but determined and uniform ideas, would be to think that all men have the same notions and should talk of nothing but what they have clear and distinct ideas of: which is not to be expected by any one who hath not vanity enough to imagine he can prevail with men to be very knowing or very silent.'

7   Locke pronounced on almost every subject and wrote about distinctions between 'wit and judgement' (i.e. imagination and reason) that are now known to every student of literary history; but it was the effect of Lockian psychology on aesthetics in the eighteenth century that changed the course of the future, especially the ways in which poets after Locke began to creep into their poems much more than they had before. Like Locke, Newton also cast a wide net especially pronouncing on history, alchemy, and apocalypsis, as well as on language and poetry when he bravely called poetry 'a kind of ingenious nonsense'; see Buchdahl; Manual (*Portrait, Religion* and Dobbs).

8   There is a long history, from the Renaissance forwards, of commentary on 'poetry as a type of truth' – on the poem as the embodiment of truth; for perceptive discussion of the subject see Rasmussen and Miner, although Miner perpetuates some false dichotomies, such as that 'poets and painters create' while 'scientists and critics discover'. Long ago George Henry Lewes, the friend and adviser of George Eliot, realised that imagination was equally the basis of art and science; see his *Principles of Success in Literature* (1898), chap. III.

9   Empirical scholarship has not produced work on this separation, except for the books of Marjorie Nicolson; even here the discussion is peripheral to the main one of her books. Lacking is detailed study of the way natural philosophy – study of the natural secular world – slowly appropriated 'science' for its own purposes; eventually making itself virtually synonymous with 'science'. Part of the difficulty is that a concept of 'science' as 'a connected body of demonstrated truths' did not emerge until the eighteenth century (see *OED*, 'science', various meanings). Throughout the eighteenth century, systematic thinkers on a wide variety of subjects not thought to be 'sciences' today continued to look for the 'scientific principles' of their subject. See, for example, Sir John Hawkins's *General History of the Science and Practice of Music* (1776), a five-volume attempt to place music on an equal footing with 'the other Newtonian sciences'.

10 This has been one of the *leitmotifs* of the work of Kuhn as well as the various responses to Kuhn's work in Lakatos.

11 The emergence of the 'professional writer' and 'creative writer' in our contemporary sense is discussed by Gross, Rogers, and Cannon in her discussion of 'professionalisation'.

12 George Cheyne, the popular eighteenth-century physician and writer on health, is a perfect example. According to which of his activities is he to be categorised or labelled? Although trained in medicine and a practising doctor all his life, he was also an ardent Newtonian scientist, perpetual mystic, adviser to Pope, Richardson and many other great English writers, prolific author of works about health and diet, and best-selling writer in his own right. Fielding the novelist took issue with his style and claimed it was ungrammatical gibberish.

13 The idea was developed by Plato in the *Gorgias* (the passages dealing with *techne*, 465.a.2–466.a.3), developed by Aristotle in discussions of *episteme* and *scientia* in various of his treatises, and carried into the Middle Ages by didactic-encyclopaedists; it emerged in the Renaissance as the earliest definition of 'science' (*scientia*). Coleridge,

The discourses of literature and science (2)

in the nineteenth century, insisted on strict Platonic requirements, demanding in public and affirming in print that 'scientist' must be reserved for a select few who understand the first cause of things, not attached 'to every Fellow, who has made a lucky Experiment'.

14   Newton's letter to Oldenburg (1672).

15   The language of the passage is worth citing: 'If the time should ever come when what is now called science, thus familiarised by men, shall be ready to put on, as it were, a form of flesh and blood, the Poet will lend his divine spirit to aid the transfiguration, and will welcome the Being thus produced, as a dear and genuine inmate of the household of man' (preface to the second edition of the *Lyrical Ballads*, 1800). But critics have not agreed about Wordsworth's attitude to the relation of science and poetry; one school maintains that ultimately Wordsworth, despite his protestations, was no friend to science and continued to cultivate a poetic ethic based on 'love not science'; another school finds these oppositions forced and maintains that, whatever Wordsworth's reservations may have been, he was finally a friend of both. Wordsworth's idea of their relation was not, of course, static and unchanged over time; see Smyser. A model study of the discourse of literature and science during the Romantic period is found in Trevor Levere. Equally a model for the earlier periods is Schatzberg.

16   See Marx 'Reflections' and 'American', both responses to the 'gnostic ideas of Roszak'.

17   Positivism, as a movement, begins to invade literary studies in the 1880s, at the moment when critics (Brunetière in France, John Addington Symonds in England) begin to call for 'a science of literature' modelled on the biological sciences.

18   Since the 1880s and the application of positivism, literary theory has also grown increasingly scientific, whether according to the laws of one science or another. There is even an historiography of the subject that now forms a main part of the last century's discourse of literature and science; see, for example, Dingle, the works of Jacob Bronowski and Sir Peter Medawar, and the many books by René Wellek.

19   In the Anglo-American world there have been two traditions: at Cambridge a school of New Criticism that arose in reaction to the dilettantism of Walter Raleigh and Edmund Gosse at Oxford, and a school of philological explication in America that altered when New Criticism was imported by I. A. Richards *et al.* in the 1930s.

20   Douglas Bush, then Gurney Professor of English Literature at Harvard University, realised this decades ago when noting the crucial significance of the history of science for cultural history: 'During the nineteenth century the great watershed between the medieval and modern era was commonly taken to be the Renaissance of the fifteenth and sixteenth centuries. In our time the watershed has been moved up to the seventeenth century 'Enlightenment.' In this large change of focus the history of science has played a large role'; see Bush, 'Science', 29; *English*; writing under the influence of A. N. Whitehead (*Science and the Modern World*), Bush would continue to stress the crucial importance of science as a shaping force in the development of modern English literature. Yet Bush never distinguished the levels of history of science: as the record of daily experimentation, speculative theoretical assumptions, and the public's image of the previous two.

21   I.e., Snow's Rede Lecture of 1959 – *The Two Cultures and the Scientific Revolution* – Leavis's various replies after 1960, and the thousand pages on the subject in the decade after 1960. For discussion of the pre-1959 history of the controversy, see Rousseau, 'Peril' and Johnson. For an appraisal of its more recent interpretation, see G. S. Rousseau, ''Till we have built Jerusalem': the Berkeley Symposium and the Future of Literature and Science', (chap. 10 below), and Lance Schacterle, 'What Really Distinguishes the "Two Cultures"?', both in G. S. Rousseau (ed.), *Science and the Imagination: Proceedings of the 1985 Berkeley Symposium on Science and the Imagination*, a special issue of *Annals of Scholarship*, IV (1986), 1–22 and 83–94.

22 See Huxley. Huxley's lecture replied to Matthew Arnold's attack denouncing the inclusion of science in university curricula; the curious parallels between the two-cultures controversy of 1859 and 1959 have not been drawn in any detail, but for discussion of Huxley and Arnold see Paradis, and the collection edited by Paradis and Postlewait. For two very different approaches to the pre-1959 history of the controversy, each a brilliant defence of its own position, see Berlin, lamenting the split (80–110), and Himmelfarb.

23 Even a casual glance at Bachelard's *L'Air et les songes: essai sur l'imagination du mouvement* (1962), *La Formation de l'esprit scientifique* (1938), *La Philosophie du non* (1960), and *La Psychanalyse du feu* (1949), makes the point; as do Foucault's books and Deleuze and Guattari's *Anti-Oedipus: Capitalism and Schizophrenia* (Minneapolis, 1983), with a preface by Michel Foucault.

24 The question has been avoided in American universities as a consequence of the arrangements among academic departments reluctant to poach in the groves of non-departmental colleagues. Demarcation extends to printed scholarship at every level and is of enormous significance to literature and science; for demarcation in science, see Lauden (119–28) and Burkhardt. For demarcation in discourse that calls itself literary as dinstinct from medical, see Rousseau, 'Medicine'.

25 For example, Stone's fierce attack on Foucault and his 'Exchange' where Stone claims to have discovered five serious flaws with 'Foucault's useless model' and concludes, even after the exchange, that Foucault 'has not directed himself to the central issue of my criticism, namely his pessimistic evaluation of Enlightenment thought, and the institutions and professions that grew out of it'. See also the negative Anglo-American reviews of *Anti-Oedipus* (n. 27). I have attempted to explain the ambivalent reception of Foucault in the Anglo-Saxon world ('Death').

26 See Kuhn and Feyerabend. Kuhn's recent interest in the different languages of theoretical discourse represents a move towards the subject matter of literature and science.

27 See a companion of pieces by Rousseau, and, more recently, a theoretical essay that discusses the implications of this interdisciplinary approach for disciplinary approaches to the subject ('Towards').

28 The demarcation problem between science and non-science represents the most urgent problem for literature and science. The degree to which scientists concede that art and science overlap – cannot be neatly distinguished – determines the future of our subject. This is why it is crucial for scientists to become involved in the dialogue. No matter how tempting it is to do so, the demarcation problem ought not to be considered apart from literary history and the history of science; fortunately, it cannot be excavated from the pre-1959 history of the two-cultures controversy. In my view few subjects are more significant for literature and science than the future of demarcation.

29 Scientific criticism has elicited a backlash from critics who, like Susan Sontag, are 'against interpretation' – who claim that as a consequence of its scientific pretensions criticism no longer is in touch with the primary literature it professes to interpret. During the period 1940–70 another attack came from Leavisites who believed that 'theoretical science' – of whatever type – is necessarily inferior to great primary literature because it shuns morality and dwarfs the creative imagination: see Leavis (135–60). Some say that both attacks have now been answered and finally put to sleep, but the battle is far from won.

30 See Hesse, Jones and Hoenigswald. My discussion of metaphor owes much to Professor Hesse's paper which I was able to consult in manuscript, and to discussions with her in spring 1984 about the unitarian-dualist debate.

31 Many toilers in the vineyards of literature and science did the work of historians of science before there were practitioners of this subject, and apologised for science before scientists designed to discuss these matters in public. More recently, certain

historians and philosophers of science, especially in France and England, have been doing the work of literary critics; one thinks of Bachelard and Foucault in France, and of Mary Hesse and Roy Porter in England. But the trend is turning, as historians of science and literary critics now begin to come *together*, especially in the newly formed international Society for Literature and Science (SLS), to discuss these very interesting overlaps and to ponder how scientific discourse is imbued with literary devices and how literature, especially but not merely literary criticism, is permeated with scientific procedures and approaches. And the day cannot be far off when it will be evident to administrators and educators that historians of science need to study literary interpretation formally, in colleges and universities, and that literary critics interested in this subject need some advanced training in the history of science.

32 See Reiss, whose narrative account represents one controversial version.

WORKS CITED

Abrams, Meyer, *Natural Supernaturalism* (New York, 1971).

Berlin, Isaiah, 'The Divorce Between the Sciences and the Humanities', *Against The Current* (New York, 1980).

Borges, Jorge Luis, 'The Analytical Language of John Wilkins', *Other Inquisitions, 1931–1952*, trans. Ruth Simms (Austin, 1968).

Buchdahl, Gerd, *The Image of Newton and Locke in the Age of Reason* (New York, 1961).

Burkhardt, Frederick, *The Cleavage in our Culture* (Boston, 1952).

Burtt, E. A., 'The Conflict of Science and Religion – and Their Reconciliation', *Religion in an Age of Science* (New York, 1929).

Bush, Douglas, 'Science and Literature', in H. H. Rhys (ed.), *Seventeenth Century Science and the Arts* (Princeton, 1961).

— *Science and English Poetry* (New York, 1950).

Cannon, Susan Faye, *Science in Culture* (Folkestone, 1978).

Dillenberger, John, *Protestant Thought and Natural Science* (1961).

Dingle, Herbert, *Science and Literary Criticism* (1949).

Dobbs, B. J. T., *The Foundations of Newton's Alchemy* Cambridge (1975).

Draper, John William, *A History of the Conflict Between Religion and Science* (1872).

Feyerabend, Paul, *Against Method* (Atlantic Highlands 1969).

Gross, John, *The Rise and Fall of the Man of Letters* (1969).

Hesse, Mary, 'Texts Without Types, Laws Without Lumps', *New Literary History*, XVII (1985), 31–48.

Himmelfarb, Gertrude, 'In Defense of the Two Cultures', *American Scholar*, L (1981), 451–63.

Hoenigswald, Henry M. and Linda Weiner, *Biological Metaphor and Cladistic Classification: an Interdisciplinary Approach* (Berkeley, 1986).

Huxley, Thomas H., *Science and Culture* (1880; rpt. 1886).

Johnson, Martin, *Art and Scientific Thought: Historical Studies towards a Modern Revision of Their Antagonism* (1944).

Jones, Roger Stanley, *Physics as Metaphor* (New York, 1983).

Knowlson, James, *Universal Language Schemes in England and France 1600–1800* (Toronto, 1975).

Kuhn, Thomas S., *The Essential Tension* (Chicago, 1977).

— 'On the Relations of Science and Art', *Comparative Studies in Society and History*, XI (1960), 403–12.
— *The Structure of Scientific Revolutions* (2nd rev. ed., Chicago, 1970).
Lakatos, Imre, *Criticism and the Growth of Knowledge*, 2 vols. (Cambridge, 1970).
Lauden, Laurence, 'The Demise of the Demarcation Problem', in Robert S. Cohen and L. Lauden (eds.) *Physics, Philosophy and Psychoanalysis* (Dordrecht, 1983).
Leavis, F. R., ' "Literarism" versus "scientism": the Misconception and the Menace', *Nor Shall My Sword: Discourses on Pluralism, Compassion and Hope* (1972).
Levere, Trevor, *Poetry Realized in Nature: Coleridge and Early Nineteenth-Century Science* (Cambridge, 1981).
Levin, Harry, *The Myth of the Golden Age in the Renaissance* (Bloomington, 1969).
Manual, Frank, *A Portrait of Isaac Newton* (Cambridge, 1968).
— *The Religion of Isaac Newton* (Oxford, 1974).
Marx, Leo, 'American Literary Culture and the Fatalistic View of Technology', *Alternative Futures*, V (1980), 45–70.
— 'Reflections on the Neo-Romantic Critique of Science', *Daedalus* (1976), 61–74.
Miner, Earl, 'That Literature is a Kind of Knowledge', *Critical Inquiry* II (1976), 487–518.
Muller, Herbert, *Science and Criticism* (New Haven, 1943).
Paradis, James, *Thomas Huxley*, (Cambridge, 1982).
— and T. Postlewait (eds.), *Victorian Science and Victorian Values: Literary Perspectives* (New York, 1981).
Posnett, H. M., *Comparative Literature* (1886).
Rasmussen, Dennise, *Poetry and Truth* (The Hague, 1974).
Reiss, Timothy, *The Discourse of Modernism* (Ithaca, 1982).
Rogers, Pat, *Grub Street: Studies in a Subculture* (1972).
Rousseau, G. S., 'The Death of Foucault', *Literary Review* (Sept. 1984), 13–15.
—'Literature and Medicine: the State of the Field', *Isis*, LXXII (1981), 406–24 (chap. 7 above).
— 'Literature and Medicine: Towards a Simultaneity of Theory and Practice', *Literature and Medicine* V (1986), 152–82 (chap. 2 above).
— 'Literature and Science: the State of the Field', *Isis*, LXIX (Dec. 1978), 583–91.
—'Medicine and Literature: Notes on their Overlaps and Reciprocities', *Gesnerus* XLIII (1986), 33–46.
— 'The Peril of Princes: Those Two Other Cultures', *Denver Quarterly*, VII (1972), 20–45.
Schatzberg, Walter, *Scientific Themes in the Popular Literature and the Poetry of the German Enlightenment, 1720–1760* (Berne, 1973).
Shuttleworth, Sally, *George Eliot and Nineteenth-Century Science: the Make-Believe of a Beginning* (Cambridge, 1984).
Slaughter, M. M., *Universal Language and Scientific Taxonomy in the Seventeenth Century* (Cambridge, 1982).
Smyser, J. W., 'Wordsworth's Dream of Poetry and Science', *PMLA*, LXXI (1956), 269–75.
Sprat, Thomas, *The History of the Royal Society* (1667).
Stone, Lawrence, 'Madness', *The New York Review of Books*, XXIX, 16 Dec. 1982, 28–36.

— 'Exchange with Michel Foucault', *The New York Review of Books*, XXIX, 3 March 1983, 42–4.

Struever, Nancy, *The Language of History in the Renaissance* (Princeton, 1970).

10

The discourses of literature and science (3)

In August 1985 a national Society for Literature and Science (SLS) was officially launched in Berkeley, California at an international congress for the history and philosophy of science, where a symposium on science and the imagination was featured. The purpose of SLS was to facilitate dialogue about the two realms: explicitly, to build sturdier bridges and evaluate the complex connections. It also undertook to bring together students and critics, scientists and technicians, who would not normally talk to each other. The last thing SLS wanted or contemplated, when redefining their connections, was the reinscription of these individual fields and their literary representations as discrete entities (i.e., literature, science) or a revitalisation of the disciplinary divisions they designated. I was proud to have been one of the founding members. This essay served as the introduction to the published volume of the SLS symposium which appeared in Annals of Scholarship: Studies in the Humanities and Social Sciences *in New York in 1986.*

My tone, as the reader will soon grasp, is both enthusiastic and imperialistic. We founding fathers and mothers were then, so to speak, in an empire-building mood. Nevertheless, I could see that literature and science was headed for rough waters in our pluralistic, post-disciplinary age, when the academy had been greatly theorised, and when every post-discipline, such as the academic one implied here, purported to be on the verge of making grand discoveries. On the other hand, I thought the risk worth taking. Now I find myself more cautious about its transformative powers, but I still labour under the belief (illusion?) that literature and science can play an important role in making the university a more egalitarian workplace. The essay's title, 'Till we have built Jerusalem', suggests the kind of institutional reform I hope will eventually occur, although it may not be implemented in my lifetime.

1985 actually represents something of an *annus mirabilis* for literature and science. Thirty years ago the idea of literature and science practised as a *separate* field or discipline, cultivated in and of itself, and waiting to be tilled by scholars on many continents, unburdened by Diltheyan caveats about the incommensurability of the two types of discourse accumulated by each field – Dilthey's *Geisteswissenschaften* and *Naturwissenschaften*[1] – would have impressed most academics as an impossible dream. Or, if not literally impossible, then at least sufficiently remote from mundane likelihood that it may as well have remained in the domain of the impossible dream.

Furthermore, thirty years ago C. P. Snow (not yet Lord Snow), was pondering, as we now know from the various biographies of him that have been written, his now-famous Rede Lecture about 'the Two Cultures', which he would deliver four years later, in 1959, in a college in Cambridge, with the most influential living critic of English Literature – F. R. Leavis – sitting in the audience. Three years after that, in 1962, intellectual warfare between the two men and their camps of followers and devotees broke out when Leavis published a scathing riposte to Snow's Rede Lecture,[2] resulting in a monumental paper war whose extraordinary significance we have not even begun to appraise with historical objectivity, and from whose fallout and aftermath a name has been given to the intellectual temper of the period following 1960: 'the two-cultures decade' or 'the two-cultures controversy.'[3]

In the years intervening between 1955 and 1985 literature and science – for our purposes now, considered *individually* – have been continuously redefined: the former through the ever more sensitive lenses of diverse types of critics who have reinterpreted literary greatness and revalued the achievement of its authors; the latter, science, through a sociological and anthropological grid that now makes it practically impossible to entertain the old, narrow (i.e. internalist) position that views all science *merely* as a branch of pure mathematics, as a system of quantifications free of historical contingency and social pressure; or, to return to the Diltheyan debate about incommensurability, as a discourse that is pure and untarnished by subjectivity because it is a faithful mirror, even a perfect representation, of nature. Or, to alter the metaphor from the mathematical and the Diltheyan (or the Popperian) to the poetic: after thirty years of revaluation and redefinition, science is no longer viewed as an objective discourse that mirrors Nature in the sense of poets who sang that Newton alone had looked on beauty bare, on a kind of everlasting, unchangeable truth; on a type of absolute truth, frozen into aesthetic perfection in the sense of Keats's 'Cold Pastoral' in the 'Ode on a Grecian Urn';[4] in brief, into a science unbound by the laws of time and place and the vagaries of human nature and societal variation.

The record of these thirty intervening years since Snow delivered his catalytic Rede lecture, and Leavis his inflammatory rejoinder under the guise of the Richmond Lecture, forms much of the recent history of the newly developing subject literature and science, as do the prolific debates for or against separation. This is not the appropriate place to rehearse the history of these debates, or the record of all those modern debaters who have taken sides with Bacon and Newton on absolute truth and the making of hypotheses, or more recently, those who had sided with or against

Popper, Thomas S. Kuhn, Richard Rorty, and, to a lesser extent, with or against Paul Ricoeur, Paul Feyerabend, and Stephen Toulmin. The annals of this thirty-year war – it has been a type of ongoing debate rather than a war – form a library in themselves, with dramatis personae including Aldous Huxley, Jacob Bronowski, Arthur Koestler, Sir Peter Medawar, and many others. And it needs to be said not merely that it has been a sexist debate to a large extent – no female except Marjorie Hope Nicolson ever entered it, and even she only did so to a limited extent – with the males arguing out a whole repertoire of other matters under its umbrella; but also the fact of the debate itself is very much a part of the intrinsic recent history of the new subject of literature and science.

Yet it is inconceivable that the *history of science*, which in 1955 was still struggling to define itself and integrate its accumulated wisdom into the academic programmes of American and European higher education,[5] should have then been in any position to invite an even younger newcomer, literature and science – then not even a fledgling and certainly not an organised subject with structures of its own – to launch a society under its own auspices. And yet in 1985 the new Society for Literature and Science, which now consists of more than three hundred members, was invited to launch itself at an international congress, a mark of the triumph not only of the new subject but also of the academic entrenchment and development in the last thirty years of the history of science.

By rehearsing this background I do not suggest that these subjects form the matter of the essays in *Annals of Scholarship*. In point of fact they do not. There is no discussion here of the monumental problem of separation – the Diltheyan question – or of the tradition of Snow and Leavis, although Lance Schacterle has treated an aspect of 'the two-cultures controversy'. My reason for introducing these essays historically, and setting them into a context that glances at the recent fortunes of the subject literature and science, is rather to explain how these essays came into being in the first place, and then to suggest reasons for the frame of mind their individual authors revealed in 1985 and 1986. That is, briefly in the space allotted I intend to paint a canvas of literature and science sufficiently broad to allow the essays to speak for themselves without their authors' having included a context; and most importantly while speaking for themselves, to explain why their authors are not optimistically blind to the seemingly intractable continental divide between literature and science, even though they manage to thrive on their particular side.[6] For the tide is turning against the monolithic divide itself. A quarter of a century ago, in the heat of the Leavis – Snow debates, one would have been given the best odds that the sciences and humanities could never happily be brought together. There was no way for Leavis as literary critic to appreciate what the scientist in the laboratory or the technologist in the space shuttle was about. Now, almost thirty years later, things are different; and it is not unusual to find all manner of literary critics not only appreciative of the techniques and discourses of the scientist, but also profoundly appreciative of the scientist's enterprise, and enthusiastic and curious about every aspect of his modes of thought and expression.[7] If the reversal has demonstrated more interest *from* literature *to* science than in an opposite direction, there have nevertheless been a substantial number of scientists, or scientific mentalities, who are seriously interested in making literary connections with their own work and even writing about problems in literary criticism.[8]

A crucial barrier enforcing the bifurcation has been levelled. But the opposition has not been silenced – far from it, even if it is no longer possible to isolate the demarcation as if it was there from time immemorial and thereby pretend that because it had always been this way, it will necessarily remain so. About thirty years ago Hyman Levy, an English, Socialist, professor of mathematics in the University of London, and his collaborator, Helen Spalding, a teacher of English literature, wrote a book called *Literature for an Age of Science*.[9] It was a noble, if somewhat Marxist, attempt to persuade readers in both camps that imaginative literature was an undervalued asset to modern society and could be used to shape cultural goals for the better.

> While there has in fact been a scientific revolution in which people have become aware of the powers of science, and scientists themselves are attempting to use these powers with fuller consciousness of their effect on society, there has been no comparable revolution in literature, and consequently no comparable advance in understanding by writers as a body of the power they wield, and its effects for good or ill on society.[10]

But the authors assumed the divide to be cut in marble and never considered, perhaps because they did not view themselves as theorists, the possibility of breaking it down. Today a Levy or a Spalding would probably know better; and even if their contemporary counterparts had wanted to proselytise in favour of an undervalued body of imaginative literature, they could not possibly frame their argument in this way without becoming the targets of ridicule.

For one reason, the historians of literature and historians of science have been too busy to permit such false historiography from surviving,[11] for if anything the record today is just the reverse: not until the middle of the eighteenth century was the divide firmly established, and in previous periods, even as recently as the seventeenth and early eighteenth centuries, it was assumed that no divide existed at all.[12] With this commonality in mind, writers as diverse as Richard Burton, Sir Thomas Browne (who made little distinction between his identity as a physician and writer),[13] Butler, Swift, Akenside (whose medical dissertation on epigenesis is still considered a classic of the epoch, while his *Pleasures of Imagination* continues to be studied for its formidable role in the development of English romanticism),[14] Johnson (called by some the first modern psychiatrist for his probing examination of madness and insanity in *Rasselas*),[15] Erasmus Darwin (also a physician as well as a prolific poet who was immensely influential in *both* medicine and literature),[16] and Coleridge (who has been called a significant biologist as well as the greatest critic writing in the language)[17] – to mention only a few specimens within the English tradition – have pronounced confidently in both domains without worrying about any intractable division.

But that day is gone – wiped away, though not for ever, by romantic ideologies professing that the languages of the one side could no longer be comprehended by the other.[18] And despite brief encounters during the last century, when thinkers like the positivists tried once more to develop 'sciences' of virtually every field, the divide has appeared as if it were more or less firmly entrenched. Michel Foucault broke down the barriers once again in the same chronological period that the followers of Snow and Leavis were debating the two cultures, each (English and French), it seems, without awareness of what the other camp was arguing. Foucault was himself uninterested in Snow and Leavis; he probably had barely heard of either man, let alone read him. And the last thing on his mind was intervention into a debate that was uniquely English in origin and development. But Foucault's theory of multiple discourses, in various of his books,[19] overlapping in their languages and embedded in the metaphors of other disciplines, is a uniquely potent French contribution to an ongoing English debate about 'two cultures'. And it has finally served – by the 1980s – to make the divide seem even less intractable than it would appear if one had merely followed Snow and Leavis from the 1960s.

The effect of such interventions in the name of breaking down barriers has been an impression – perhaps erroneous but nevertheless felt – of a turn of the tide. Therefore, in serious theoretical debates today one must substantiate, at the least, any ingrained resistance to the building of bridges, for those who would not only *point* to the division as given fact continue to lose ground.[20] Having occupied positions of professional primacy for so long, those who merely continue to insist on the inevitability of the demarcation have not been driven out: almost everywhere they are still powerful and authoritative. But they no longer represent the exclusive mandarins of the 1960s/1970s. Furthermore, the rise of interdisciplinary studies in the academy during the last two decades, and the ripple effect of this interdisciplinarity in professional organisations and non-academic segments of the society, further strengthen the impression that the tide is turning against those who simply reiterate, rather than substantiate anew, the old Diltheyan position about a permanent divide. We may be approaching a return to a monism of the humanities and the sciences as opposed to the belief in a fundamental dualism of veritable contrary states which peaked when Leavis attacked Snow in 1962. The divide is no more permanent today than it was in the late eighteenth century, when the tide was going in the opposite direction: from dozens of bridge builders, as we have seen above, to a new epoch of specialisation and particularly developing in the nineteenth century.

The authors of the essays in *Annals of Scholarship* (all deeply committed to the developing academic discipline literature and science) are sensitive to the issue of demarcation, but have written with an intention to build bridges between the two camps, especially between the two types of discourse each has given rise to. But these essays, considered collectively, possess no inherent unity or theme. The authors come from diverse backgrounds and institutions, ranging from academic universities to institutes of science and technology, and they work in different national literatures and chronological periods. Their original instructions before speaking in Berkeley were to address the two domains in any way they wished, either theoretically or practically, and to be sufficiently eclectic as to make a specific point, or set of points, in a brief period of time. The dice were not loaded: no participant was prodded to speak out against demarcation and for bridge building, if his/her personal inclination lay in another direction, and a predictable inclination to remove the barrier between the domains was not a precondition for an invitation to participate in the Congress.

Yet it is curious, though perhaps in the end it is random, that none of the participants addressed himself directly to the history of the demarcation. As even the casual reader will notice while reading, there are numerous glances at the Leviathan – the barrier – and glances conceptualised in multiple ways, but no one appears to have been sufficiently confident to take on the demarcation problem itself. Given this obvious lacuna, which no single essay could fill even if it aspired to (since it has formed the central concern of *Geisteswissenschaften* and *Naturwissenschaften*, at least since Kant's time), though it is again perhaps random, that virtually all the essays deal with application rather than theory.

There is a merit and virtue in this fact. The authors of *Annals of Scholarship*, collectively having assumed that bridges needed to be built, went about their business without calling attention to their methods (a state of affairs arguing that more rather than less theory is needed in the developing academic discipline literature and science). Some of the contributors, both in Berkeley and now in their expanded versions, may have concluded that a theoretical discussion was premature; that not enough specific legwork on specific authors and specific works has been done to validate any general theoretical discussion about the demarcation problem. Others may have believed that the task should not be assigned to literary critics, which all the contributors in this volume are, but it better performed by philosophers and historians of science.[21] Yet however virtuous this procedure may be — and I certainly have no intention to criticise it — the student of literature and science inevitably recognises that there is more urgent business at hand. This involves the most fundamental aspects of the developing subject and asks basic questions rather than specific ones about particular authors and their works, particular discourses and their relations to other specific discourses. These are questions of a general rather than specific nature: what is literature and science? For whom and to whom is it addressed? Which groups are the natural audiences of the subject? Where should its professional and academic loci be placed? How can the demarcations between the humanities and the sciences be better defined, and what professional factors need to be considered for those who elect to build bridges rather than fortify the barriers of the demarcationist position?[22]

Yet even without this theoretical dimension of the problem, the essayists in *Annals of Scholarship* demonstrate trends and suggest patterns worthy of notice. Blake, as is clear from Peterfreund's opening essay dealing with the gnostic resonances of Newtonianism, is a crucial figure in the debate. This is not merely because Blake's recurrent 'energy' remains an indispensable concept to both the humanities and the sciences — lying on both sides, as it were, of the demarcation; but also because Blake's discourse, in his poetry and visual representations, and in more recent critical discourses generated in response to his works, has cast itself so fundamentally within the terms of energy.[23] It is therefore not surprising that four of the essayists should find themselves referring to Blake and energy in one form or another, and not without good reason, given Blake's concept of force, spiritual power, and cosmic explosions. Twenty years ago Jacob Bronowski, already mentioned as a recent, important figure in the discussion of literature and science, singled out Blake as central to his own interests.[24] It is becoming increasingly clear that Blake looms over the discourse of literature of science as do few other English authors of any historical period. It is therefore predictable that Nelson Hilton, a leading commentator on Blake, should here have selected Blake as the lens through which to discuss the perception of science, and that Mark L. Greenberg should have chosen Blake around whom to erect the framework of an argument about the relation of artistic form to the technological development of a period of society, in this case in relation to one of Blake's incontestable masterpieces: *The Marriage of Heaven and Hell*.

But even these reasons do not go far enough to explain the preeminence of Blake as the author of choice (though not an imposed choice) in this volume. Weightier reasons lie beneath the facade he has developed as a difficult artist – poet who was a monolithic rebel against the Enlightenment's representative Urizenic figures of Newton and Locke – an archfiend defying an ever-creeping, ever-sweeping, Industrial Revolution, whose technological effect would necessarily result in the death of the imagination as the natural consequence of human creativity gone amuck while 'dark, Satanic mills' overrun the land. As Leavis himself has written in the introduction to *Nor Shall My Sword*, his meditations written in response to Snow and the two-cultures phenomenon: 'Blake is committed to knowing where knowing is impossible'.[25]

This overreaching, quasi-Marlovian, attitude is above all what distinguishes Blake from the other great artist – poets of his time, even from Milton, and ultimately places him – Leavis contends – in the camp of Voltaire and Diderot, even of Newton and Locke, whose representations (or emanations) he so detested. I think our contributors would concur that it also places Blake in the camp of modern, positivistic, technologico-scientists, who are also 'committed to knowing where knowing is impossible'. The fact that Blake does not know where to stop in his insatiable quest for intellectual knowledge, as Leavis has again suggested in this same essay, is the aspect of Blake that renders him essential to Literature and Science. It is *au fond* a Christian aspect: Blake's relation to the Fall and the possibility of knowledge in a postlapsarian world in which the scientist – in the sense in what we have come to use the word 'scientist' since William Whewell first employed it in the early nineteenth century with all our modern connotations – has usurped the Godhead's prerogative as his own, and conferred upon his own type of knowledge, and without any limits or warrant, the remarkably arrogant name of absolute truth. Leavis's point, like T. S. Eliot's in his own essay on Blake,[26] is of course that Blake ultimately fails; that he is not the supreme proto-Romantic artist – poet because on balance he cannot envision Eternal Man after the apocalypse: 'The Eternal Man and Jerusalem can't even by Blake be *imagined*; there can be no presentation of them in terms of "minute particulars."'[27] Yet this failure and loss (Los too) does not here prevent Leavis, any more than it has prevented our academic Blakeans today from acknowledging Blake as central, vital, even essential to the two-cultures controversy, or the *one* culture, as many of us would prefer to call it, and as Leavis did indeed call it.[28]

This primacy of Blake, rendering him of supreme importance to the ongoing two-cultures debate and setting him apart from his proximate poetic lineage, which extends from Milton and Dryden to Wordsworth and Coleridge, should not occlude the only slightly less impressive ways in which he played a principal role in the emerging discourse of literature and science. First, with his axioms about infinity alone residing 'in Definite and Determinate Identity', he figures into (or ought to figure into) contemporary debates about commensurability and incommensurability;[29] and whatever else these debates may be, they are principally topics of the first importance to historians and philosophers of science. It is true that for Blake 'Definite and Determinate Identity' were matters primarily related to the definition of identity as distinct from selfhood; to the constellation of imaginative – intellectual properties as distinct from the egotistical (dare one say Wordsworthian?) categories of personality. But Blake's words, as well the profile he has come to represent, have often been interpreted metaphorically, and we would profit a good deal in the contemporary debates about commensurability and incommensurability, especially in discourse, if we received Blake's highly original criteria for the intellectual attainment required of the artist – poet.

Second, and no less crucial although of a different nature, is Blake's supremacy in the sexual – more accurately the *politico*-sexual – dimensions of the recent discourse of literature and science. For we have collectively begun to gain an awareness, however procrustean it may be, of the extraordinary phallocratic basis of all critical, and theoretical, discourse; not merely the one pertinent to the evolving discourse of literature and science. As Eva Keuls has recently written: until recently 'the story of phallic rule at the root of Western civilization has been suppressed'.[30] And others (the name Evelyn Fox Keller comes to mind) have been vigilant in demonstrating how such 'phallic rule' has played a significant role in the development of modern science.[31] But his his own way and however ultimately inadequately, Blake has already taken a stand on these issues that remains integral to the developing critical discourse of literature and science. No wonder then that in his essay below Hilton writes of Blake's 'sweet Science' in distinction from the 'phallocratic and militaristic "hard science" ' of Newton and the Newtonians. As everyone knows, in Blake's Jerusalem the sexes have ceased to exist: Man is Woman, Woman Man; and a type of androgynous utopia prevails. Far more crucial than the discovery of Blake's sources for this androgynous vision in a Christian heritage that embedded androgyny in an occult alchemical framework, thereby casting it out of orthodoxy as a pariah of the most offensive type,[32] is Blake's suggestion (again metaphorical) of an androgynous literature, and androgynous science, in Jerusalem. Today, we can go one step farther: an androgynous critical discourse that is neither opprobiously designated as 'feminist' or 'masculine' or 'sexist'. That is (more specifically), a developing discourse of literature and science incapable of being branded as either having derived from a sexist basis, or having been generated by a member of a particular sex. Such androgynously generated theory can only add to the health of the developing subject.

Finally there remains, still in relation to Blake, the crucial matter of certainty in knowledge and the predictability of events. Leavis, as we have seen, has expressed Blake's position with the simplicity of a song sung in the state of innocence: 'Blake is committed to knowing where knowing is impossible'; and in Hilton's argument that 'the crux of the matter for Blake is perception', he quotes from Richard L. Gregory, the contemporary physicist: 'If complete prediction, as supposed in principle possible on the Newtonian account, could be realised, the observer would disappear.'[33] But two strains have coalesced into one here, and even apart from these two there are others that need identification if we are to recognise how central, indeed how indispensable, Blake is for literature and science. Observation, logic, reason – all within the gaze of the observer: these remain the three cornerstones of any rational science, whether in Plato's or Kant's definition of a science, and they necessarily determine the basis of every scientific theory of predictability in just the sense that Gregory has commented on complete prediction in physics. Yet they are also central concerns in Blake's intellectual thought; indeed, it is impossible to discuss Blake as an intellectual thinker without constant recourse to these three cornerstones – the veritable Graces of his nonsystematic system of thought. But we still hover in a state of near darkness with regard to Blake's position on these three crucial concerns, and if we remain less than certain how Blake himself came to observe the natural things of this world he later 'imagined' in one or another of his various states, we know even less – pitifully less – about his processes of logic. If we can agree – echoing Professor Hilton – that perception remains at the heart of Blake's psychology, especially the degree to which an objective gaze is possible and the amount of attention the observer can humanly bring to his natural object, it nevertheless remains unclear how Blake proceeded from physical science to cognitive science (in our sense, from physics to psychology), or how he came to believe that absolute certainty of knowledge was impossible for Fallen Man as well as Eternal Man. That in his own idiosyncratic way Blake shattered the legendary Cartesian dualism of mind and body, and shattered it more successfully than many of his empirical predecessors in the eighteenth century did, there can be no question; in so doing he cut a figure, whether he knew it or not, that was as significant (again it is tempting to say indispensable) as any empirical psychologist of the period. Still – and it is an interesting comment on the current state of scholarship as well as proof of the need for further development of literature and science – the relation of Blake's mature intellectual thought to these three categories (observation, logic, reason), so necessarily inherent in any scientific system of thought, remains to be conceptualised first, and then delineated.

In the current milieu, an all-embracing technology, which practically defies definition and accurate labelling since it seems to reach out to everything in contemporary civilised life, appears second only to energy (that is, to energy as a pervasive concept) and to Blake as a crucial figure in the continuing dialogue of literature and science. It is by now a commonplace, of course, to make careful distinctions between science and technology; to begin in the knowledge that the gap between science (*scientia*, theory, hypothesis, models) and technology (*techne*, craft, skill, state of the art) is as large as that between science and the humanities, or science and the arts. But historically the demarcation between science and technology has been blurred, and literary critics still have not been willing to ask the broad philosophical question, how has literature conceptualised technology, as *distinct* from science, and they have been unwilling to do so in rigorous, historical terms. Almost twenty years ago the New England literary critic and professor Wylie Sypher wrote a book (a thought-provoking, good book) about literature and technology;[34] but for the few reviewers who paid it any attention it may as well have been entitled *Literature and Science*, so nebulous was the distinction then in literary camps between science and technology. More recently, this state of affairs has improved, and a number of works have placed technology in sharp relief against its rather distant cousin science.[35] In *Annals of Scholarship* science and technology are never invoked interchangeably, whether in Mark Greenberg's discussion of Blake's printing techniques, Lance Schacterle's treatment of the theme of the material benefits of scientific progress as conceptualised by Snow, Donald R. Benson's comparison of the techniques of the painter with the physical theories of space, or George Slusser's researches into the images of 'soft' technology (in this case placing emphasis on its possibly disturbing consequences) in contemporary French science fiction and science writing. The authors seem to agree that technology has had at least as vigorous and liberating an influence on human creativity (again one thinks of Blake and Leavis) as did the more theoretical science. It may then be that, in time, literature and technology will come to designate a rather different discourse from the one I have been invoking as literature and science. Such predictability notwithstanding, it is perfectly clear, even now, that much more work needs to be done among the historians of science, as well as more among the general cultural historians and literary historians, to define the nature and extent of this gap between science and technology.

It is also not surprising to discover two of the essayists – Schacterle and John Woodcock (a critic of modern literature at Indiana University who has championed the significance for literature and science of Margaret Atwood) – gravitating to Snow and assessing his contribution to make their own contributions to the continuing dialogue of literature and science. More noteworthy than their different approaches is their implied sense that the two-culture controversy, in whatever shape or form it may appear, remains an intrinsic topic of literature and science. If Schacterle's approach, like Peterfreund's, is historical and even hermeneutic in the weight it places on history as the category leading to truth, Woodcock's is anything but historical. Schacterle maintains that the significance that was widely attributed to Snow's Rede Lecture subsequently bore little, if any, relation to any point Snow actually intended to make in 1959, and thereby diminishes Snow's biographical significance for literature and science. Woodcock, by contrast, seems to accept at face value the whole two-cultures debate, and invokes it as the backdrop and context in which to set Margaret Atwood's novel of 1979, *Life Before Man*. 'In any given work,' Woodcock heuristically writes, 'Snow might have asked, how much attention is being paid to the science and technology that are increasingly part of the human condition, and how fair, how rounded, is the portrayal?' And throughout this introduction to the essays I have been attempting to indicate, however cursorily, my agreement with this attitude, as well as my belief that there was a two-cultures controversy long before the 1950s or 1960s. For social and political phenomena (like capitalism, communism, and individualism) have existed long before they were given these names, and the two-cultures controversy – as I have suggested elsewhere[36] – has been around, in one form or another, for a long time. It was evident when (moving chronologically backwards) Huxley, Arnold, and others thrashed the subject out publicly in the 1880s;[37] it was present (if not explicitly labelled in this way) in the immediate aftermath of Darwin's *Origin of Species* (1859) when Tyndall, Whewell, Ruskin, Carlyle, and others debated the matter;[38] present at the turn of the nineteenth century when Coleridge and Wordsworth wrote the by now very famous, and often quoted, sentence about 'Poetry and the Man of Science' in the 1800 and 1802 *Preface to Lyrical Ballads*; present, even earlier, in the mid-eighteenth-century when Pope, writing apocalyptically at the end of *The Dunciad*, exercised his binary imagination to divide the universe of learning into two hemispheres of 'Arts and Sciences'; present more generally in the eighteenth century, when Diderot and Swift anticipated the whole future of technology by conceptualising mad scientists plying their schemes in antisocial seclusion; and probably also implicit, if not actually endowed with a label or specific name, or consciously and palpably felt, long before the Enlightenment.

The cumulative effect of these papers suggest the multifarious influence that science and technology have had on the human imagination: as influence, as source of themes and images, as topic for exploration (especially in science fiction and science fantasy), as a catalyst for anxiety, and – not least – as cultural antagonist demanding action and reaction. But if these authors agree on the need for drawing a general distinction between science and technology, and concur that both science and technology have liberated the human spirit rather than enslaved or shackled it, they nevertheless manifest no clear pattern in the methodology they invoke. Some of the papers continue existing directions and tendencies, while others introduce new topics for research or inaugurate new avenues of research within the developing discourse of literature and science.

Yet these approaches hardly exhaust the possible approaches to literature and science. As a consequence of the exigency of space in this issue, as well as the necessary limit in the number of participants who could be invited to Berkeley, whole nations and epochs do not appear here. Nothing has been said about literature and science, however anachronistic the label itself may be, *before* the period of the European Renaissance, and lands that lie to the east of France are barely ever mentioned. Perhaps more consequentially for the ways in which a volume such as this one eventually are contextualised, none of the authors has written (for example) in a purely theoretical topic such as a model for literature and science, or a theory of metaphor and analogy that would give rise to such a model. It may indeed be premature to attempt to generate a single model when so much legwork – toil in the vineyards of the kind suggested in the essays by Peterfreund and Hilton – remains to be done, and when there is, regrettably, no single reliable source that has retrieved the developing discourse of literature and science itself: the primary texts from Bacon onward.[39] For more or less continuously from the seventeenth century, an ongoing debate about the arts and the sciences, their individual status, their relations, their differences, their similarities, has evolved, forming a rich life of his own and gradually annexing to itself, like a complex organism in an imperialistic mood, a vast archive. Not until these annals are surveyed, and the reinterpreted as a single, developing discourse, can we hope to have a substantially new model for literature and science. When that day arrives it will entail a new literary history as well as a breakthrough in the province of critical theory. It exceeds my warrant here to conjecture from which contemporary critical camp such a new model may arise, but if the weighty essays by Peterfreund and Hilton count for anything, it may well be the hermeneutic critic (a type of latter-day Husserl or Cassirer?), grounded in history and all things historical but for ever vigilant to the tides of critical theory, who makes the breakthrough.

NOTES

1 For Dilthey's classic separation see Wilhelm Dilthey, *Selected Writings*, ed. and tr. H. P. Rickman (Cambridge, 1976). Stephen Toulmin, Paul B. Armstrong, and myself are three contemporary critics who have vigorously argued *against* the divide; see Toulmin, 'The Construal of Reality: Criticism in Modern and Postmodern Science', *Critical Inquiry*, IX (Sept. 1982), 106; Armstrong, 'Understanding and Truth in the Two Culture', *Hartford Studies in Literature* (1984), 70–89; G. S. Rousseau, 'The Peril of Princes: What ever happened to Those Two Other Cultures?', *The Denver Quarterly*, VII (1972), 20–45, 'Literature and Science: the State of the Field', *Isis* (1978), 583–91). Larry Laudan has argued admirably about the epistemological problems involved in the demarcation itself: see his 'The Demise of the Demarcation Problem', in R. S. Cohen and L. Laudan (eds.), *Physics, Philosophy and Psychoanalysis* (Dordrecht, 1983), 111–27. Among modern literary critics who continue to think and write as if the divide still exists are the vestigial New Critics, many traditional literary historians, and – most numerous among these groups, if not also most influential for the numbers of the young they instruct – most college professors of English, American, and foreign literatures. For a specimen of the latter variety, arguing *for* the divide, see Hans Eichner's misguided essay, which the Modern Language Association of America in its infinite collective wisdom saw fit to award its prize for the best article of the year published in *PMLA*, 'The Rise of Modern Science and the Genesis of Romanticism', *PMLA*, CXCVII (Jan. 1982), 8–30.

2 F. R. Leavis, *Two Cultures? ... The Significance of C. P. Snow, being the Richmond Lecture, 1962. With an Essay on Charles Snow's Rede Lecture by Michael Yudkin* (1962). Leavis's ideas on the two cultures were more fully developed in *Nor Shall my Sword: Discourses on Pluralism, Compassion and Social Hope* (1972), ultimately an even more devastating attack on Snow than his Richmond Lecture. For Leavis's ideology as it affected the debate, see Gerry Watson, *The Leavises, the 'Social', and the Left* (Swansea, 1977).

3 The evolution of his controversy is a crucial domain for all those interested in the development of literature and science as a subject. The most complete bibliography of the issue is found in Paul Boytinck's *C. P. Snow: a Reference Guide* (Boston, 1980). Aldous Huxley's *Literature and Science* (New York, 1963) was a response to Snow – Leavis, an attempt to find 'a middle road', and a rehearsal of the genealogy of the debate at least since the appearance of T. H. Huxley and Matthew Arnold. Important commentary is also found in D. K. Cornelius, *Cultures in Conflict: Perspectives on the Snow – Leavis Controversy* (Chicago, 1964); Martin B. Green, *Science and the Shabby Curate of Poetry: Essays about the Two Cultures* (Westport, 1978).

4 John Keats, 'Ode on a Grecian Urn', l. 49, in which the poet suggests the essence of an art-for-art's aesthetic in which truth is beauty, beauty truth, as in this example, where the equation would be truth as well as beauty, and that is all one needs to know.

5 The secondary literature is primarily pedagogical, even propaedeutic, but for a useful survey see: David Knight, *Sources for the History of Science* (Cambridge, 1975). Michel Serres's proclamation in *Hermes: Literature, Science, Philosophy* (Baltimore, 1982), demonstrates some of the antipathy to the history of science which the natural sciences have continued to demonstrate: 'I maintain that the history of science is not worth an hour's trouble if it does not become as effective as the sciences themselves' (39).

6 Contemporary pluralism has emphasised the degree to which the divide is intractable among the various critical camps, as well as the extent to which literary theory and criticism have become 'scientific' in our time. For example, the hermeneuticists, considered collectively, are eager to build bridges over the divide, as are students of the theory of models (Mary Hesse) and metaphor (whether unitarians or dualists in contemporary metaphor theory is irrelevant). But the vestigal New Critics and the

7   See, for example, H. R. Gavin (ed.), *Science and Literature* (Lewisburg, 1983), especially the two opening essays: J. Neubauer, 'Models for the History of Science and Literature', 17–37, and J. M. Curtis, 'Epistemological Historicism and the Arts and the Sciences', 38–62. A fine example of this application to an individual author is found in Ian F. A. Bell, *Critic as Scientist: the Modernist Poetics of Ezra Pound* (1981), who begins his discussion with a quotation from Pound's own *ABC of Reading* (1934): 'The proper method for studying poetry and good letters is the method of contemporary biologists'(1). An early collection that was years ahead of its time in bridging the gap is Wallace Ferguson (ed.), *The Renaissance* (New York, 1962). Other works include (chronologically): F. S. C. Northrop, *The Logic of the Sciences and the Humanities* (New York, 1947); Alan Ross (ed.), *Arts versus Science* (1967); W. T. Jones, *The Sciences and the Humanities: Conflict and Reconciliation* (Berkeley and Los Angeles, 1967); William Davenport, *The One Culture* (New York, 1970); J. Opper, *Science and the Arts: a Study in Relationships 1600–1900* (Rutherford, 1973); Harold Harris (ed.), *Astride the Two Cultures: Arthur Koestler at Seventy* (1975).

8   The obvious figures are Norman Thomas, S. J. Gould, Richard Feynmann, but see also Alan Lightman of Harvard and the Smithsonian, a physicist, whose recent collection of essays in *Time Travel and Papa Joe's Pipe: Essays on the Human Side of Science* (New York, 1986) is an attempt to break down the barrier.

9   Hyman Levy and Helen Spalding, *Literature for an Age of Science* (1952). Reviewers tried to discredit the book by calling its authors Marxists and Communists; see, for example, J. G. Weightman's negative review essay in *The Twentieth Century*, LIII (March 1953), 226–9, and his complaint that 'Professor Levy contributes not infrequently to Communist periodicals' (226).

10 *Ibid.*, vi.

11 See Roy Porter, 'The Historiography of Science', in *A Companion to the History of Science* (1986).

12 See E. S. Shaffer, 'Literature and Science: Towards a New Literary History', and G. S. Rouseau, 'What is Literature and Science?' both in G. S. Rousseau (ed.), *Literature and Science*, forthcoming as a special issue of *Hartford Studies in Literature*, XIX:1 (1987), 37–52, 1–24.

13 J. S. Finch, *Sir Thomas Browne: a Doctor's Life of Science & Faith* (New York, 1950); G. S. Rousseau, 'Literature and Medicine', *Literature and Medicine*, v (1986), 152–81 (chap. 2 above); Jonathan Post, *Sir Thomas Browne, Writer and Doctor* (Boston, 1987).

14 Akenside dazzled his contemporaries by his outstanding proficiency in both medicine and literature; his 1744 medical dissertation at Leiden, dealing with the preformationist controversy and epigenesis, caught the attention of Gaubius and the leading Dutch anatomists, and was immediately recognised in the medical world.

15 *Rasselas* was required reading for all medical students majoring in psychiatry at the Johns Hopkins Medical School until the 1930s, since it was considered a 'classic' of psychiatry, and it continues to evoke commentary in some of the nations's leading psychiatric clinics.

16 See Desmond King-Hele, *Erasmus Darwin and the Romantic Poets* (Cambridge, 1985).

17 No less eminent a scholar than Joseph Needham, the great expert on Chinese science at Cambridge University, has written perceptively on 'S. T. Coleridge as a Philosophical Biologist', *Science Progress*, XX (1926), 692–702.

18 For evidence that it has been wiped away, see Wylie Sypher, *Literature and Technology: the Alien Vision* (New York, 1968); G. S. Rousseau, 'Science and the Discovery of the Imagination', *Eighteenth-Century Studies*, 3 (1969), 108–35: and Richard Macksey

and Eugenio Donato (eds.), *The Languages of Criticism and the Sciences of Man: the Structuralist Controversy* (Baltimore, 1970).

19 Especially in Foucault, *The Order of Things*, tr. anon. (New York, 1970).

20 The point is further substantiated by three annual meetings of the theory session within the Division on Literature and Science (Modern Language Association of America, 1984–6), chaired by G. S. Rousseau and entitled, 'Is a Theory of Literature and Science Possible?'. Those who have assumed a great barrier without arguing their reasons have not gotten very far.

21 This caveat needs to be seriously considered, even though it falls victim to the very problem of demarcation it seeks to address; and yet, some of those who have written most perceptively about the *theoretical* aspects of the barrier have not worried about their home discipline. See an important essay by the late distinguished art historian Erwin Panofsky, 'Artist, Scientist, Genius: Notes on the "Renaissance-Dämmerung"', in *The Renaissance* (see n. 7), 123–81.

22 These fundamental philosophical questions are addressed by four scholars in *Hartford Studies* (see n. 12): in order of appearance, G. S. Rousseau on what Literature and Science is; Stuart Peterfreund on the recent history of Literature and Science; E. S. Shaffer on literature and science as a new literary history; and John Neubauer on the future of literature and science.

23 Morton D. Paley, *Energy and the Imagination: the Development of Blake's Thought* (Oxford, 1970).

24 J. Bronowski, *William Blake and the Age of Revolution* (1972); see especially the discussion of space and goemetry on 139ff. And yet the authors of a book such as the recent anthology edited by Nelson Hilton and Thomas Vogler, *Unnam'd Forms: Blake and Textuality* (Berkeley and Los Angeles, 1986), which professes to be on the cutting edge of Blake scholarship, mention energy but never deal with Blake and science.

25 F. R. Leavis, *Nor Shall My Sword*, 19.

26 T. S. Eliot, *Selected Essays* (New York, 1960), 275–80.

27 *Ibid.*, 18.

28 See F. R. Leavis, 'Luddites? or There is Only One Culture', in *Nor Shall My Sword*, chap. 3, 75–100.

29 For Kuhn on commensurability and incommensurability, see T. S. Kuhn, 'Comment on the Relations of Science and Art', *Comparative Studies in Society and History*, XI (1969), 403–12.

30 Eva C. Keuls, *The Reign of the Phallus: Sexual Politics in Ancient Athens* (New York, 1985), 1.

31 Evelyn Fox Keller, *Reflections on Gender and Science* (New Haven, 1985).

32 No specific source could do justice to this vast province of scholarship, but the works of the late Frances Yates did as much as anything to demonstrate to what degree this hermetic tradition, while intrinsic in orthodox Christianity, nevertheless represented the unorthodox.

33 See Richard L. Gregory, *Mind in Science: A History of Explanations in Psychology and Physics* (Cambridge, 1981), 251.

34 Wylie Sypher, *Literature and Technology: the Alien Vision* (New York, 1968).

35 See Wilfred Eastwood, *Science and Literature: the Literary Relations of Science and Technology* (1957), who does not distinguish between science and technology; T. de Laurentis *et al.* (eds.), *The Technological Imagination: Theories and Fictions* (Madison, 1984); Joan Digby (ed.), *Permutations: Readings in Science and Literature* (New York, 1985); John Holloway, 'Poetry for the Technologist', in *The Colours of Clarity* (1964).

36 See G. S. Rousseau, 'Are there Really Men of Both Cultures?', *The Dalhousie Review*, LII (1971), 1–25.

37 R. H. Super, 'The Humanist at Bay: the Arnold – Huxley Debate', in U. C. Knoeflmacher and G. B. Tennyson (eds.), *Nature and the Victorian Imagination* (Berkeley and Los Angeles, 1977), 230–9.

38 For several of these controversies see the essays by R. A. Donovan and P. L. Sawyer in J. Paradis and T. Postlewait (eds.), *Victorian Science and Victorian Values: Literary Perspectives* (New York, 1981).

39 By primary texts I do not mean anthologies of secondary criticism such as the collection (excellent as it is) complied by Theodore Jennings and entitled *Science and Literature: New Lenses for Criticism* (New York, 1970). There have been numerous brief anthologies of the primary works but nothing complete and authoritative; see, for example, Patrick Brostowin, *Science and Literature: a Reader* (Boston, 1964).

11

Repenser Bachelard

Foucault's death in June 1985 came as a blow both to me and the Clark Series on mind and body, which I directed. No one else could have contextualised the age-old dualism and the discourses it secretly engendered with the same insight and flair. I subsequently wrote several articles about his death and its implication for post-disciplinary enquiry.

Yet not long afterwards, it became apparent to me that Gaston Bachelard, Foucault's teacher and mentor, who had died a generation ago but was still being read and discussed in France, had ceased to be studied or written about in North America. The English translations of his works in print, and Foucault's unpredictable achievement, had not revitalised Bachelard's own reputation in the way I thought he deserved.

I published this essay in an appreciative mood in 1988 in Annals of Scholarship: Studies of the Humanities and Social Sciences. *My intention was to show that in some ways Bachelard was as important as Foucault, despite the different disciplinary contexts in which the two figures had worked.*

> Ours is a culture of premature ejaculation (Jean Baudrillard, *Forget Foucault*, 24)

Born in 1884, dead in 1962, a contemporary of Eliot and Pound, Gaston Bachelard wrote nearly thirty books, most of which deal with the philosophy of science, more specifically with the philosophy of physics, and at least seven of which deal with literary theory and literary criticism construed in the broad sense. Of these, the best known are *The Psychoanalysis of Fire*, *The Poetics of Space*, and *The Poetics of Reverie*. All three were translated into English and issued in inexpensive paperbacks in the 1960s by the Beacon Press, the first two with eloquent if regrettably brief introductions by Etienne Gilson and Northrop Frye. More recently – in 1985 – the Beacon Press issued a half-century anniversary translation of *The New Scientific Spirit*, originally published in 1934, as Hitler was consolidating his Nazi party and purging from it traitors of every type. Further works of Bachelard have since appeared in English translation, and plans are afoot to issue others in an ongoing Bachelard Translation

Series.[1] It would be false to assume that bachelard had two phases, or that he migrated from one field (the philosophy of science) to another (literary theory), or vice versa, for the two kinds of books, in my classification, were written simultaneously: *The Psychoanalysis of Fire* originally appeared in 1938, four years after *The New Scientific Spirit*, and throughout the 1940s and 1950 Bachelard continued to address questions in both subjects and embodied his meditations in books. It is true that in this last active decade or original writing, the 1950s, his interest in the philosophy of science waned and he persuaded himself that studied on the imagination, the imaginary, *l'imaginaire*, that is to say the intersection of human imagination and the natural world, were more pressing, and this is the period when he composed *The Poetics of Reverie*, which first appeared in French in 1960. But it remains false to assume that there was a clear pattern or linear progression in his long intellectual career from the one to the other, as false as to surmise that any single, unifying metaphysical pattern can be imposed on his oeuvre, or any boundaries neatly distinguishing an early and a late Bechelard the demarcations of which would be – crudely speaking – *science* and then *literature*, or a first phase of philosophy of science followed by a second phase of literary criticism. His imagination never perceived these domains as separate. All his intellectual life Bachelard devoted himself to *theory*, the precise domain of the theory being much less significant than its inherent methodologies and logics.

This much said, it is equally crucial to note that throughout his intellectual life Bachelard desired to create a science of human imagination, in much the same way that Kant and many neo-Kantians have yearned for one, especially a science of the *pre*scientific imagination as it universally responded to the four elements of fire, water, earth, air – this fervid enterprise despite his perfectly solid knowledge as a chemical physicist that these four elements are not the elements of Nature. Bachelard's monographs on Lautréamont, Shelley, Nietszche, Huysmans, and others were attempts to confront the resonating human imagination as it responds to, and then creates and recreates these element as images, symbols, archetypes. In doing so Bachelard had no interest in literature as text, as *explication de texte*, or as a text possessing a prescribed genetic organisation or predetermined structure, much less as a genre, a pattern of rhetorics, or fabric of narratives. he was not a 'literary critic' in any traditional sense that we use the term, and it impedes understanding rather than enhances it to reconstruct him three decades after his death in 1962 as if he had been some neglected literary critic. What needs to be said about Bachelard does not lie in the domain of traditional literary criticism at all.

If it is a misguided activity – as I suggest here – to attempt to locate any binary roles of (crudely speaking) *literature* and *science* in Bachelard's imagination and writing career, it is nevertheless entirely appropriate in my view to inquire into his continuing agendas and programmes, never naively assuming that they were ideologically neutral or ideologically free, despite his being among the most politically neutral of twentieth-century thinkers, a biographical fact that may have diminished his reputation in the short run. Bachelard's conscious agendas were based on the view that words are prior to thought, and that there can be no meaningful meditation outside linguistic categories, a programme he executed, and even exalted, to the degree of believing – in his memorable phrase – that 'words dream'. He believed, moreover, that material elements possess archetypal aspects of the material universe, and that what needed to be constructed more than anything was 'an archaeology of the soul'; as crucially, he thought that imagination precedes and must be considered – philosophically speaking – prior to perception. As a consequence of these three fundamental it is appropriate to *repenser* Bachelard as the last and perhaps the greatest of modern vitalists responding to Bergson, and also as one of the last of the Romantic prose-poets in the tradition of Blake and Shelley, Jung and Freud: forever inquiring into the workings of the archetypal creative imagination in relation to those privileged four elements and their four humours, as well as teasing out the metaphors and analogies that attach to one or another type of imagination.

But these were merely Bachelard's fundaments. Beyond these and the three programmes which they elicited, Bachelard was fiercely opposed to both idealism (whether of a Kantian or Hegelian variety) and to positivism, and virtually all his works demonstrate to what degree he saw both idealism and positivism as coeval menaces. When puncturing the one he often found himself opposed to the other, as in his sustained critiques of Descartes and his fierce polemices against the contemporary philosophical Establishment in France. Long before Richard Rorty proclaimed a type of 'death of philosophy' in our own time and depicted the hubris of the professional establishment of philosophy, Bachelard was dismantling its structures by chipping away at its ideological and positivistic pillars, demonstrating that scientists and poets (in the loose sense of 'poet' as maker), not philosophers or theorists, cause the genuine 'ruptures' (Bachelard's continuing term) or mutations in the evolution of culture. On this axiom about *scientists and poets* versus philosophers and theorists, all his work stands or falls. This agenda in itself – assuming now that Bachelard had never proclaimed anything else or written anything more – would have elevated him to the first importance for the developing domain that we have called *faute de mieux* literature and science; for exalting laboratory scientists and practising poets in these heroic terms as well as purporting, contrary to popular mythology, that in this potentially iconoclastic activity *poetries and sciences* share the commonest ground. What Rorty wrote in the 1970s for an Anglo-American readership, Bachelard conjured an half century ago when fewer readers attended to these matters or pursued these topics.

The implications of Bachelard's agenda for literary theory are equally germane. First and perhaps foremost in this category, he was an imperialist of discourse, so to speak, long before the flourishing of what we today – in the aftermath of modernism and postmodernist discourse – call discourse theory or textual theory. Bachelard believed that *all* written and spoken language was metaphoric and analogical and that for this reason discourse should be invoked under the rubric of a single entity. In this taxonomy and phylogeny encompassing the different kinds of language and the degree to which languages embody specialised vocabularies and metaphors, no one could have been more opposed than Bachelard to the New Criticism of the Agrarians or of I. A. Richards (in principle, that is, as Bachelard was presumably unaware of their programmes). Bachelard's configurations, especially his discourse embracing the domains of mythology, cosmology, folklore, anthropology, poetry – the images and symbols of which resonated with equal vitality for Bachelard – endeared him most, among literary theorists, to Northrop Frye. Himself no traditional New Critic, Frye, who knew rather little about science and who wrote on the most obsolete of Romantic assumptions about the gulf between literature and science, could never understand how a physical laboratory scientist like Bachelard could possibly want to show that the four elements must *always* be the elements of the most primitive imaginative experience when Bachelard knew perfectly well that those four elements – fire, water, earth, air – are not the chemical elements that constitute Nature. There was much confusion here in Frye's conception of Bachelard. But after the Beacon Press published Bachelard's books in English translation in the 1960s, there were others who were confused. Some noticed Bachelard's affinities with René Wellek and Austin Warren on the function of images and imagery, and tried to tame Bachelard's thought (i.e. purge it of its deep scientific base) by construing him as another French *literary* critic; still others read Frye's four introductory paragraphs to *The Psychoanalysis of Fire* and claimed to have detected more than endorsement by Frye, but a veritable similarity between Bachelard's concept of archetypes and Frye's own archetypes in the *Anatomy of Criticism* (1957). These were rough proximities – their differences were greater: Bachelard was no taxonomer or critic of *literature* but an altogether different kind of mentality, or conceptualiser, for whom we have as yet no adequate label. Some day, when literature and science has become a more advanced field than the primitive exercise it now remains, we will know what kind of naming can be appropriate for the likes of a Bachelard.

Bachelard was as opposed – on theoretical grounds – to Freudian psychoanalysis as he was to excessive idealism and positivism. For this antipsychoanalytical stance Sartre praised him exuberantly, even if for the wrong reasons, and Georges Poulet, the phenomenological poet-theorist of interior time and space, congratulated him in 'Gaston Bachelard et la conscience de soi'.[2] As Bachelard grew older he identified increasingly with Jung and Jungian archetypes rather than with Freud's idealistic science of sex. The Bachelard who made 'states of reverie' the crystal touchstone with which to interpret human imagination and its stratified languages had little in common with a positivistic psychoanalysis that claimed 'scientifically' to purge the imagination of those very primitive and prescientific essences which Bachelard wanted to identify and retain. Bachelard's implicit cultural heroes and heroines were not those persons who dreamed of cures for so-called individual derangement but those imaginaries – the fictional Ophelias, Rousseaus, Novalises, Swinburnes – whose states of reverie (prescientific, primitive, symbolic, organic, vitalistic, inherently childlike) embodied the quintessential alchemy of human imagination. These were the real and fictive figures of the past to be consecrated and sung. On them, culture's best creative voices, Bachelard bestowed his taxonomies: his Prometheus Complex, Empedocles Complex, Novalis Complex, Hoffman Complex, always with an eye on the Greeks and the Germans.

But not even a tentative estimate of Bachelard, like this one, can be formed until his authentic biography and historical reception is consulted. This is a complicated labyrinth that historians will someday chart in detail, for eventually there will be not one but many biographies of Bachelard. If it is valid to affirm that Bachelard would be even more obscure than he is now if the Beacon Press had not begun translating his works into English, it is also true that his continuing reception in France is as important.

When the fifty-six-year-old Bachelard was called to Paris in 1940 to fill the Chair in the History and Philosophy of Science at the Sorbonne, he was still an obscure name even in his native country; certainly a rather apolitical figure who – so far as is now known – played no role in the Resistance or, alternatively, pro-Nazi collusion in France. It was the *appropriation*, or importation, of Bachelard's thought into Marxist materialism that made Bachelard a cult figure in France, long before his student Michel Foucault was recognised. François Dagognet and François Pire both wrote monographs about Bachelard's philosophical thought in the mid-1960s, but these (now) obscure dissertations did little for Bachelard's international reputation.[3] In actuality, Bachelard was blessed, as it were, not by these monographs but by his students, who included – among others – Georges Canguilhem (the philosopher of medicine), Jean Hyppolite (the historian of philosophy), Foucault (who worshipped in his temple and who continued to *repenser* Bachelard as the most formidable influence on his early phase), Jacques Lacan, and most importantly, the Marxist Louis Althusser. And it is Althusser whom historians and biographers will eventually demonstrate to have been the seminal figure in forging Bachelard's career both in France and in the English-speaking world. Althusser, without ever having heard of Thomas Kuhn, and conceptualising materialist history before Foucault's epistemes were committed to print, invoked Bachelards's cultural 'ruptures' and 'mutations,' his 'epistemological breaks' and discontinuities, not arbitrarily but importing them, so to speak, for his own historical materialism and Leninist agenda. The pages of *Reading Capital* (1968) and *For Marx* (1965) are filled with Bachelardian historical ruptures and mutations; and Althusser's books were the ones on the shelves of every literate Class-of-'68er on both sides of the Channel as Vietnam dragged on and as the Boulevard St Michel erupted in flames in May 1968.

Among the Class of '68 was Dominique Lecourt, who had been Bachelard's student before his death in 1962, and who afterwards presented himself to the seminar of the then very young Jacques Derrida. Lecourt was the stereotype of the Parisian university radical of the Class of '68. He had lived through the events that set Paris aflame in 1968, endured its *ancien régime*, so to speak, and was arrested more than once for political activity conducted against the state. Demonstrating by day and by night reading Althusser, Lecourt also appropriated Bachelard's methodology in the doctoral dissertation that he was writing under Derrida's direction. His subject was Marxism and Epistemology, his aim to demonstrate that Bachelard's epistemology leads neither to positivism nor idealism (indeed, that this was the very agenda to which Bachelard was *opposed*) but rather to historical materialism, and that Bachelard, like Althusser, must be placed in the 'materialist camp'.[4] Lecourt thereby appropriated a methodology and converted it into a political agenda: Bachelard as rigorous philosophical Marxist.

Lecourt acknowledged Derrida's guidance in the dissertation (142), especially in the chapter dealing with Bachelard's poetics (a difficult one dealing with what we would call literature and science), and explained why among Bachelard's many books written over three decades *The New Scientific Spirit* remained – and would continue to remain – the most important (155). Yet it was not Derrida, but the geographically distant American historian of science Thomas Kuhn who proved to be the revelation for Lecourt – as it turned out a *negative* revelation – for reasons that will become evident in a moment. Lecourt completed his dissertation on Bachelard a few months after the spring 1968 student riots. Sometime between then and 1972–3 he discovered a French translation of Kuhn's *Structure of Scientific Revolutions*, (1962; 2nd ed., enlarged, Chicago, 1970). Lecourt's analysis of Kuhn's book demonstrated that the American *capital-*

ist philosopher was (unlike Bachelard) a naive ideologue masquerading, for all his paradigms and 'routes to normal science', as a world-class epistemologist. Persuaded that an extended contrast between Bachelard and Kuhn was now essential to the argument of his Paris dissertation, Lecourt prepared a long new introduction to the 1975 English translation of the thesis eventually published by New Left Books as *Marxism and Epistemology* which concludes on this emphatic contrast: 'I repeat . . . the one [Bachelard] is, timidly and confusedly but indisputably, ranged in the materialist camp, the other [Kuhn] is inscribed in the orb of idealist philosophies' (19). Bachelard emerges as the good guy and Kuhn the villain, Bachelard's oeuvre constituting the epistemological proof that Althusser's prophetic Marxism was philosophically sound; and it appears to have mattered little to Lecourt that Bachelard himself had been so apolitical, to such degree that there is not a single mention anywhere in Bachelard's writings of his political inclinations. But Lecourt was a Class-of-'68er whose idealist agenda was not determined by anything that can crudely be designated as facts. Kuhn-the-capitalist had served his purpose as a foil to Bachelard all too well.

The English edition of *Marxism and Epistemology* sold out within two years. By 1977 even the reprint was depleted; by 1978 or 1980 it was a book for which one would have paid dearly to have a copy. The point for Althusser, Lecourt, and their Marxist brethren of the mid-1970s was not the validity of Kuhn's sociology of science or philosophy of science, was not even whether Kuhn's 'old idealist question about the objectivity of scientific knowledge' amounted to anything important (19), but rather the attitude that Kuhn's capitalist idealism could never lead to anything but more capitalism and the political status quo. This is the message that Lecourt wanted to export across the English Channel via New Left Books, even if it meant holding up the British edition of his book.

What then does this brief sketch amount to, adorned – as it must be – with considerations of Bachelard's students and his Anglo-American reception? And why should the interpretative community composed of scholars who work in literature and science take note of these chronologies and personalities, or as I have so rhetorically framed my position, why rethink, '*repenser* Bachelard'? An alternative approach would have posed the rhetorical question 'who reads Bachelard now?' To provide but one example, Richard Rorty need not refer to Bachelard in the index of *Philosophy and the Mirror of Nature* because (1) he may be unaware that he has replicated, however differently, Bachelard's polemic against professional philosophy, or (2) Rorty's work is so nationally tuned and so finely calibrated to cater to the needs of an Anglo-American establishment in philosophy that he may not even have read a word of Bachelard.[5] Furthermore, whatever Bachelard's deserved niche, he is clearly a problematic figure who speaks to us much less now than he did in 1970–5 when his influence in both France and the English-speaking world peaked, and when the prestigious French Colloque de Cérisy, for instance, selected Bachelard as their figure to be studied for 1970.[6] When I. Bernard Cohen, addressing the community of historians of science, comments in *Revolution in Science* that 'the French school has not as yet had a significant influence in the English-speaking world on the mode of considering science's history', he is thinking primarily of Bachelard.[7] Cohen's point remains unassailable: it is Kuhn, Popper, Feyerabend, and a handful of others who have influenced the community of Anglo-American historians of science, not Bachelard. Nor is Cohen's identification and recognition surprising when even a Foucault continues to be viewed suspiciously in that community, no matter how reverentially consulted upon occasion. Yet – and this continues to be my main point in suggesting that we *repenser* Bachelard – it is Bachelard who had anticipated virtually *all* the sources of contention debated by this Anglo-American interpretative community, and anticipated them long before Kuhn wrote. Bachelard's agenda mandated that we explore the degree to which scientific theories are ideologically laden, and probe how and when 'revolutions' occur in science; Bachelard demonstrated that science is not value-neutral, as so many of its practitioners believe; and he emphasised the extraordinary degree to which scientific research is a collaborative community effort based on commonly held but unspoken assumptions that are anything but ideologically neutral or value-free. These concerns continue to be the sources of the deepest tension and confusion in contemporary Anglo-American science, and Bachelard anticipated them a generation before Kuhn did. The revisionist corrective is not intended to diminish the impact of Kuhn's insights, but rather to elevate those of one of his predecessors. Moreover, the same Bachelard who anticipated Kuhn in these ways and whose students compared him to Kuhn constructed a poetics of reverie and intuited that ultimately *the poet and the scientist*, not the critic or the philosopher, are the genuine 'sleepwalkers' – in Koestler's sense – who change things. This was rare wisdom in an early twentieth century still plunged in darkness when the demarcation of science and nonscience was in question. For Bachelard had grasped the plight of the imagination in the twilight of that interregnum culture between the two great European wars, and had done so in a milieu before Kant's critiques became influential and before Husserl's 'Crisis of the European Sciences', drafted in the fever of crisis (1934–7) as a last-ditch effort to define phenomenology, had made a dent in France.

The implications for literature and science are, if anything, of a grander nature. Viewed on balance the largely apolitical Bachelard seems to me to be a more considerable figure than – for example – the late Sir Peter Medawar, who was both a Nobel Laureate in Physiology and a frequent commentator on the overlaps of literature and science, or the very much living Jean Starobinski, who was a practising physician and practising psychiatrist before he became a literary theorist, and who, like Bachelard and Poulet, has also been interested in reverie and dream states. Neither of these men could ever claim to possess Bachelard's scientific or philosophical sophistication. Even so, the fact remains that Bachelard is not much read any more, even in interpretative communities where one would think him terrifically important, and the reasons why beg to be understood. The Lecourt who struggled to exalt Bachelard for Marxism may be forgiven for indulging in sheer naïveté when framing the question whether the literary or the scientific Bachelard is the 'real' one: 'Which Bachelard? Are there two? Two in one: prodigious duplication or disturbing duplicity. Should we celebrate the achievement of a rare all-rounder? Or rather be disturbed by such an acute contradiction: could one really, without loss, divide one's interests to such an extent?' (142). It is the old question of the all-around Renaissance Man, or at least the paradox of the 'two cultures' and all other binary oppositions. Lecourt juxtaposes duality against unity, claiming that in his study of Bachelard he will 'attempt to discern what in Bachelard's epistemological work "called for" the construction of his poetics' (143). But Lecourt cannot deliver what he promises – alas – because his own agenda is epistemologically so biased. He understands the aporia in Bachelard's epistemology of the imaginary, yet confesses to having done nothing more than fill it with yet another one. Despite the defect I would rather have 'Lecourt's Bachelard', however partisan, than nothing; would rather find readers of Bachelard today than English editions that have gone out of print.

But when we have had in American theory since 1970 a Kuhn and a Rorty – the one arguing for 'routes to normal science' and 'paradigms', the other affirming that the profession of philosophy is fraudulent, even effectively dead – then I suppose that it is easy to forget that Bachelard raised these issued and posited at least partial solutions decades before, even if he had no original theory of language to compete with Saussure; easy to forget the value of Bachelard's versions of prioritising. If one is going to ask rhetorical questions, then why not inquire: when the tradition of C. S. Pierce and the new pragmatists in America is as strong as it is today, who needs a Bachelard? The implicit response rings true: one cannot argue against the existence of this all-American pragmatism without opening oneself to charges of excess and obsolescence. But contexts other than philosophy remain in which to assess Bachelard's enduring legacy. When the new discipline of literature and science is emerging at the pace it currently is, and when literature and science has made the strides it has in the last few years, where are our *literature and science readers* of Bachelard? After all, Bachelard, not Foucault, was the professional scientist; Bachelard, not Foucault, the laboratory researcher who knew what he was doing in scientific research and who revealed to scientists what they ought to have learned long ago; and Bachelard the figure – admittedly among other Europeans in the early twentieth century – who ought to be speaking today to the practitioners of the new literature and science.

Jean Baudrillard advises us to *oublier Foucault* – forget Foucault.[8] I say *repenser Bachelard* – rethink Bachelard, and I do so fully aware that the suggestion about this deceased foreigner comes at a time when everyone's politics and ideology count for so much.

NOTES

1   Nine titles by Bachelard have been translated into English in all, several of these out of print; the Bachelard Translation Series is an undertaking of Spring Publications of Dallas. See also Colette Gaudin's fine translation of a selection of Bachelard's writings entitled *On Poetic Imagination and Reverie* (Indianapolis and New York, 1971; reissued by Spring Publications in 1987) – I owe much to her Introduction to the original edition. Mary Tiles's *Bachelard: Science and Objectivity* (Cambridge, 1984) is the most extended treatment in English of Bachelard's epistemology and philosophy of science to date. Astute as Tiles's study is, it does not concern itself with literature or literary theory and has no bearing on my argument here. 'Bachelard', Tiles notes on the first page of her preface, 'also wrote extensively on literature and on poetic expression, and is better known in the English-speaking world for these works. Although I have indicated points at which these apparently divergent concerns make contact, I have not explored this dimension of Bachelard's philosophy, although it would be necessary to do so for any exegetical account.' It would indeed. More relevant to my discussion is Mary Ann Caws, *Surrealism and the Literary Imagination: a Study of Breton and Bachelard* (The Hague, 1966), and Gaudin's Preface to the new edition of *On Poetic Imagination and Reverie*.

It is crucial that I emphasise my interest in Bachelard's *Anglo-American* reception as distinct from his European one (especially in Switzerland), and the neglect of his work among *English-speaking critics*. His French reception differs from the Anglo-American one, emphasising once again what an ocean of otherness exists between the interpretative communities of these places. In France serious study of Bachelard continues, books about his work appear with predictable regularity, as in Jean Lescure's two recent books: *Un Été avec Bachelard* (Paris, 1983) and *Bachelard aujourd'hui* (Paris, 1986).

I must also clarify at the outset that my purpose is not a general discussion of Bachelard's theories of science and literature but focuses specifically on the nature of his Anglo-American reception. The former of these activities cannot be undertaken in the space of a short essay, but were I investigating them at length I would stress, of course, that Michel Foucault and Louis Althusser (discussed below) were not Bachelard's only disciples; that Georges Canguilhem and Michel Serres also were and the ways in which they were. Canguilhem remained a devoted follower, writing prefaces to his posthumous *Études* (Paris, 1970) and *L'Engagement rationaliste* (Paris, 1972) and carrying on Bachelard's unique tradition of medical theory (as in *On the Normal and the Pathological* (Dordrecht, 1978). Serres's work – see especially *Hermes: Literature, Science, Philosophy* (Baltimore, 1982) – is unthinkable without Bachelard; yet, curiously, there are almost no references to him in Serres's writings and this would have to be explained, as would Bachelard's relation to various Continental schools of theory and criticism. I would adumbrate, for example, his relation to

Sartre's and Merleau-Ponty's theory of the imagination, and especially to the Genevan school of thematic and phenomenological criticism (especially Georges Poulet, Jean Rousset, Jean Starobinski). And I would explore, naturally, his relation to Foucault, and explain how the career of the latter greatly eclipsed his own in the Anglo-American interpretative community.

Bachelard's bibliography, like Foucault's, poses hurdles inasmuch as he constantly revised and republished works under different titles, even rewriting previously published books. Some of these knotty puzzles, have been addressed by J. Rummens in 'Gaston Bachelard: une bibliographie', *Revue internationale de philosophie*, LXVI (1963), 492–504. But the fact remains that excepting introductions to Gaudin's anthology, Tiles's book, Patrick A. Heelan's and Arthur Goldhammer's very brief prefaces to the English translation of *The New Scientific Spirit* (Boston, 1984), and Stephen W. Gaukroger's 'Bachelard and the Problem of Epistemological Analysis', *Studies in History and Philosophy of Science*, III (1976), 189–244 (which again says nothing about literature or literary theory), little has been written *in English* on Bachelard. For a few additional indications, see Gaudin's revised 'Selected Bibliography' in *On Poetic Imagination and Reverie*, lxi–lxvi. Among the bravest writings on Bachelard is an unpublished booklength manuscript by Annie Petit, a French philosopher of science, entitled 'Bachelard, ou l'art d'être grand-père', written in 1970–1. This work is both appreciative and critical of Bachelard's philosophy: approving of his leaps and insights into various fields, while remaining suspicious of his radical vitalism. Petit locates Bachelard within the evolution of recent positivistic thought and demonstrates how avuncular he was in both the Existentialist and Positivistic movements. It is very much to be hoped that this work will eventually be published. Very much needed, in addition to Petit, is a study of the 'grand-père' and the epigoni he spawned, especially in France: the famous disciples (Foucault, Serres, Canguilhem, Lecourt), as well as a group of less well-known figures. For the uses to which Bachelard's literary theory has been put by an art historian, see Barbara M. Stafford, *Voyage into Substance: Art, Science, Nature, and the Illustrated Travel Account, 1760–1840* (Cambridge, 1984).

2   Georges Poulet, 'Gaston Bachelard et la conscience de soi', *Revue de Metaphysique et de Morale* LXX:1 (1965), 1–26.

3   See F. Dagonet, *Gaston Bachelard, sa vie, son oeuvre, avec un resumé de sa philosophie* (Paris, 1965) and F. Pire, *De l'imagination poétique dans l'ouevre de Gaston Bachelard* (Paris, 1967). See Michel Mansuy, *Gaston Bachelard et les éléments* (Paris, 1967), 374–80, for Bachelard's influence on Parisian critics.

4   Dominique Lecourt, *Marxism and Epistemology: Bachelard, Canguilhem and Foucault*, trans. Ben Brewster (1975), originally published in Paris by J. Vrin in 1969; all parenthetical references will be to the translation.

5   Here I wrote from conjecture rather than verified fact: I have no evidence whether or not Rorty has read Bachelard, nor do I know why Bachelard nowhere figures in *Philosophy and the Mirror of Nature* (Princeton, 1980).

6   *Bachelard: Colloque de Cérisy* (Paris, 1974).

7   I. B. Cohen, *Revolution in Science* (Cambridge, 1986), 559.

8   Jean Baudrillard, *Forget Foucault* (New York, 1987); originally published as *Oublier Foucault* (Paris, 1977).

12

Science books and their readership in the High Enlightenment

The emerging discourse of literature and science in which I had been so immersed during the 1980s was often generated without attention to readership. Yet despite recent critical insight into Enlightenment writers and their readers, I discovered there was a striking paucity of commentary about British science books.

I tried to remedy the gap by combining theory and practice. But I soon found myself in a quagmire of conflicting facts and opinions about the readers of these books. No one seemed to have a grasp on readership at all, let alone scientific readers and the specific ways in which their reading habits differed from those of other readers. I tried to separate the strands of the argument but am no longer certain that I was successful. The essay's weakness is its lack of attention to genre in relation to the reader's expectation. I could not see then – around 1980–1, when I was composing the essay – to what degree this set of reading expectations was shaped by the reader's perusal of non-scientific literature, especially chapbooks and Biblical commentary, as well as short stories and the novel.

The essay appeared in Books and their Readers, *edited by Isabel Rivers (New York, 1982), 197–255.*

> If science helped to give birth to the printed book, it was clearly the printed book that sent science from its medieval habits straight into the boiling scientific revolution. . . . It was of course the rapid dissemination of knowledge to whole new classes that created the modern new attitudes to both science and religion . . . (Derek de Solla Price, 'The Book as a Scientific Instrument', *Science*, CLVIII (1967), 102–4[1]

> . . . many books filled with profound philosophical [scientific] reasonings are every day published in *England*; but correctness and elegance in Writing, and a just taste in Architecture, Painting and Sculpture, are there still in their infant state. (Madame du Bocage, 25 May 1750)

ENGLISH IDEOLOGY AND 'NATURAL PHILOSOPHY'

It is not surprising that natural philosophy – the study of the secular natural world – should have begun to make its largest strides in the

Restoration period. Natural philosophy already possessed a significant history by 1660 or 1680; this is clear merely from the repertoire of attitudes held towards Aristotle, Bacon, Descartes, and Malebranche.[2] And it was discovering itself increasingly successful in deflecting students of moral philosophy and theology, perhaps nowhere more evident than in the remarkable career of Boyle, who had been attracted to both moral and natural philosophy and who pronounced on both subjects, but who found himself gradually enticed by the latter at the expense of the former. In the Restoration period, natural philosophy enjoyed its own philosophy – what we today would crudely call 'the philosophy of science': a belief that increased study of the subject would eventually improve the lot of common man. As a consequence of this 'philosophy of natural philosophy', and also because historians of natural philosophy then were often experimenters themselves – as were many of the early Fellows of the Royal Society – there was little interest in natural philosophy as a platform for political propaganda but great curiosity about it as an activity holding the potential for changing every aspect of man's sense of the universe about him. This new curiosity was more widespread among Britons than it had ever been before, and various explanations have been put forward to account for it. Puritanism in particular has been shown to have played a seminal role in the dissemination of the new curiosity about natural philosophy,[3] and the blend of social and political life in the years between the Great Revolution and the Glorious also contributed something distinctly unique,[4] as did the relation of the sexes and the notion that science was a masculine activity. Recently Simon Schaffer has brilliantly argued that the whole topography of natural philosophy in the Restoration can be attributed to monolithic interest in Newtonian matter-theory,[5] and in some fundamental sense in which 'natural philosophy' in the Restoration and early eighteenth century was a 'programme' *opposed* to progress in science, and that despite a shift in terminology the two types of 'natural philosophy' must not be viewed as similar in any way. Moreover, one school of natural philosophy – especially those who proposed universal language schemes – was clearly opposed to all thinking that was poetic and imaginative, and perhaps this group needs to be linked with the last. Finally, French scholars – Bachelard, Pêcheux, Canguilhem, and especially Foucault – have argued that natural philosophy then was the *object* of an 'archaeology' aimed at unravelling significant world cosmologies, and not matter how hard one tries to combine this view with the previous three it is ultimately different.[6] These divergent approaches all have merit, and compel a modern student of 'eighteenth-century science and its books' to be vigilant when deciphering such a complex subject. Yet despite their differences, these three varieties of natural philosophy – as Newtonian matter theory, as hermetic pursuit, and as the object of an 'archaeology of knowledge' – share common ground. To the educated man in Restoration and eighteenth-century England, and perhaps to a few women, it was the most exciting of all subjects, ancient or modern, 'the genuine way to truth', as Malebranche had clearly decreed, and as Boyle repeated, the most certain means by which to understand the universe. But it was still 'God's' universe, and there was as yet no twentieth-century notion that what man does not understand – how the universe came into being and where it ends – he called 'God'. Perhaps deism flourished for just this reason; and its popularity may have increased and waned in response to tensions felt and anxieties experienced over the degree to which the whole universe was 'God's'. What therefore began on the return of Charles II as an ideology to which a certain number of 'Baconians' adhered,[7] was transformed by the turn of the century into a widespread agreement about the soundness of natural philosophy – in whatever way it was construed – as the only route to 'truth'.

Once the scientific *via sacra* was established it also became fashionable. To cultivate 'natural philosophy', no matter how loosely defined, was to bring status to oneself, such as that enjoyed by the men who were collectors.[8] Yet these men and women collected specimens – rocks, shells, seeds, flowers, even instruments – in a religious and ideological climate which seemed incapable after 1660 of separating God and the natural universe. There was no inherent contradiction between religion and science, as there would be in the nineteenth century and even more emphatically in our own. Robert Boyle's best-known book, *The Christian Virtuoso*, renders this point patent even in its title: *Shewing that by being Addicted to Experimental Philosophy, a Man is rather Assisted, than Indisposed, to be a Good Christian*, and the extreme notion of 'natural philosophy' as an 'addiction' is purposefully intended. Also, the rapid dissemination of deistic ideas in the Restoration period – and among all social classes – rendered it fashionable as well as 'pious' for men to explore the whole universe through which God had decided to reveal himself.[9] Furthermore, widely held if contradictory beliefs about the 'argument from design' – the notion that the more one understood the physical universe, the better one would then comprehend the wondrous planner who had brought all this universe into being – and the 'argument from the limitations of reason' – the idea that no matter how much man penetrated into the physical universe, his limited reason would never permit him to penetrate the deepest mysteries, not even with God's aid – ensured a common perception that God had aggressively invited men to try to unlock the secrets of the universe. Man consequently needed to cultivate humility and to implore the diety to assist him in two very distinct directions.[10] Exploration of God's creation was perhaps the best way to understand the moral as well as the physical universe; but no sooner had man begun to explore it, to experiment with its elements, than he began to align himself with other explorers, both amateur and professional. Man discovered the human element in the activities of 'natural philosophy', especially when he began to ponder and write about his experiences. The emphasis, therefore, of some recent students on the inferior quality of the 'new science' produced by the eighteenth-century *virtuosi* – those so-called amateurs who would be increasingly ridiculed after 1700[11] – is in a sense beside the mark. It is a 'Whig' approach to cultural history that also errs by minimising scientific activity as human experience – and we shall also see how it can cause modern scholars to misunderstand 'the rise of scientific books' after 1660. More consequential, surely, than the quality of science was the awakened interest of the *virtuosi* in a realm – the physical – in which they had been lethargic before.[12] For this reason the 'collections' of eighteenth-century men – of which books constitute a part – differ from those of their seventeenth-century predecessors. The 'Virtuoso-class', as Pope referred to himself,[13] now began to collect rocks and shells, telescopes and microscopes, books and pamphlets, instruments and clocks, and to gaze at the stars and planets, as well as to perform experiments of all types – to do everything to quench its wonder about this physical universe that was living proof of God's goodness. And fortunately, for the first time in economic history, a large segment of British society could afford to indulge this natural and religious curiosity. 'Literary men' were as sedulous in their avocations as 'professional scientists' because they are never independent of the cultural environment, and also in part because it was now so fashionable to be part of Pope's Virtuoso-class'.[14] The Baconianism which derived from the *Novum Organum*, and which had figured so prominently in the manifestos of some serious experimenters in the Restoration period,[15] was now enhanced by current religious, ideological, social – even fashionable – factors. Not to endorse natural philosophy in some form or shape among the fashionable select few was to be a misfit, out of tune with the whole spirit of the age.

The Royal Society, which was formed in 1660 and received its charter shortly thereafter, also played a significant role in the common man's thinking about 'natural philosophy'. But books, not the Royal Society, as de Solla Price has also noticed in the epigraph of this essay, transformed the common man's attitude to science and shaped the 'modern' beliefs we have opposed to the 'medieval'. Historians have studied the influence of the Royal Society in greater detail than there is space for here,[16] yet sometimes without noticing to what extent interest in this 'colledge of learned men' itself evoked an interest in *books* about natural philosophy. The reading of books, as Sprat and other early historians of the Society reflected, was in no way a primary purpose of the organisation; indeed, its motto, echoing Horace, *nullius addictus iurare in verba magistri* – 'bound to the word of no one' – mandated just the opposite: that the word counted for little and that language was too ambiguous to be trustworthy as evidence of anything.[17] Nevertheless, despite the anti-linguistic stand of the early Royal Society, its scientific and non-scientific members were addicted to books no less than Boyle's 'Christians' were addicted to 'experimental philosophy'. Both the professional scientists in the society, especially the physicians – of which there were many – and the literary men were book collectors, some collectors of note. Furthermore, the earliest Fellows themselves wrote papers and books which were recognised from the outset as of immense importance, and no contradiction between distrust of the word and application to it was seen. As early as 1679 Hobbes maintained that *The Philosophical Transactions* had veritably superseded academic textbooks, 'And as for Natural Philosophy, is it not remov'd from *Oxford* and *Cambridge* to *Gresham-College* in London, and to be learn'd out of their *Gazets*?[18] These *Philosophical Transactions*, especially the *Abridgements*, continued to be reprinted throughout the eighteenth century – even Swift used a cheap 'penny edition' – and to find avid readers. They were widely advertised in newspapers and magazines, and aimed at large audiences. More important, the authors of these 'scientific' papers themselves read books – which we would call 'science book'.[19] If it is true that the largest single group of Fellows consisted of physicians, then the point about buying and collecting science books gathers even greater strength, for physicians were the one group, even more than university professors and scholars, who could afford to buy science book that were usually not cheap.[20] As the cults of 'natural philosophy' ideologically grew stronger and more fashionable, and as more men and women[21] began to dabble in as well as grow serious about them, the need for such books increased. Whether the period from 1660 to 1700 represents an unparalleled epoch in the demand for and subsequent production of science books is a topic to be explored below; at this point it is necessary only to understand why there was such a sudden increase of books published about 'natural philosophy' after the Restoration, and what role organised bodies such as the Royal Society played in this development.[22]

The ideology of a given culture is always elusive and difficult to gauge. Karl Mannheim has written that 'it is only when we more or less consciously seek to discover the source of [a collective society's] untruthfulness in a social factor, that we are properly making an ideological interpretation'.[23] This may be requisite, but there is no doubt that by 1700 or 1720 the various roles of natural philosophy – even in its protean shapes – had altered from those demonstrated at the opening of the Restoration period, and that books lie at the centre of the change. The evidence to support the contention is abundant: the establishment of the Boyle Lectures is perhaps the weightiest, but it must not be forgotten that stipulated in the lecturer's agreement was a clause requiring him to submit a text, however much expanded over the original monthly sermon, which could be printed. The further fact that the most distinguished chemist of the age endowed lectures to establish a proper relationship between 'natural religion' and 'natural philosophy' speaks plainly for itself, as do the titles of the early lectures.[24] Margaret Jacob has shown that the Boyle Lectures 'were controlled by the latitudinarians and in particular by Thomas Tenison . . . and John Evelyn', and, furthermore, that 'in the early eighteenth century the lectureship served as a podium for latitudinarian thought and for Newtonianism'.[25] No reason exists to doubt the assertion; yet even so, the Boyle Lectures in their *printed* version, as distinct from the monthly sermon, were a powerful collective weapon for disseminating Newtonian thought. Without the printed version, it is unlikely that the 'modern new attitudes to science' of which de Solla Price speaks could have been adopted in such a relatively brief period of time; and the Boyle Lectures – as we shall see – are but one of several examples. With powerful minds such as Boyle and Newton assisting, however indirectly, in the wedding of science and religion; with the diligence and industry of such men as John Martyn, the Royal Society's first printer; with the new literacy of the common man making gargantuan strides during the early decades of the eighteenth century;[26] and with the international reputation the British were establishing for themselves – as Madame du Bocage comments in her travel book – as the purveyors of science and reason over taste and elegance,[27] it is not surprising that by 1750 or so natural-philosophy books should have been the most sought after of all printed books. It is true, a few members of the literary establishment continued to savage and even sabotage the practitioners of the 'new science' and all they represented, but in almost every case – such as that of the Scriblerians – these attacks were made against pedantry and corruptions of learning rather than in defiance of the new widespread cults of natural philosophy.[28] Addison's attitude is typical of these attacks; a spiritual son of Dryden and a proponent of the new blend of Protestantism and Newtonianism, he nevertheless fears the 'dulness of the Royal Society' and fails to see the point of 'collecting Nature's refuse':

> they [the Royal Society] seem to be in a confederacy against Men of polite Genius, noble Thought, and diffuse Learning; and chuse into their Assemblies such as have no Pretence to Wisdom, but Want of Wit; or to natural Knowledge, but ignorance of everything else. . . . When I meet with a young Fellow that is an humble Admirer of the Science, but more dull than the rest of the Company, I conclude him to be a Fellow of the Royal Society.[29]

Yet Addison never minded that Newton had been 'a Fellow' of this society. Perhaps 'wit' is after all the clue: by 'wit' Addison designates 'quick natural ability expressed in refined language'; his objection about 'Want of Wit' centres on a lack of linguistic facility – the very facility against which so many early Fellows rebelled. Addison could not understand why learned men who could be cultivating their 'wits' were instead wasting precious time collecting specimens: 'It is indeed wonderful to consider, that there should be a sort of learned Men, who are wholly employed in gathering together the Refuse of Nature, if I may call it so, and hoarding up in their Chests and Cabinets such Creatures as others industriously avoid the Sight of'.[30] But if we remember that these scientists also wrote *books* about their activities, and even about the specimens they had 'hoarded', then it is not so hard to comprehend Addison's drift. A Fellow the Royal Society is not dull because of the *source* of his wonder – Addison himself gazed in steady wonder at the stars night after night – but because of linguistic failure, because he cannot compose a good book. Pope and the Scriblerians made the very same charges in *The Memoirs of Martinus, Scriblerus*, just as Pope and Warburton were to charge much later in their joint venture, *The New Dunciad: Notes and Variorum* (1742). Yet not one of these men can justly be said to be 'anti-science'. All these attacks on science – by Swift, Addison, Pope and many others – issue from a segment of the literary milieu worried about the infringement of 'natural philosophy' on 'the territory of wit', yet not out of any opposition to the pursuit of science itself or to the hopes for progress science was capable of offering mankind.[31] To what degree the attitudes to science held by literati of the early eighteenth century – both Grub-Street hacks and great writers – were shaped by their political ideology, is a subject that has not yet been studied.[32] But their acid criticism of 'natural philosophy' almost always occurs within the context of books, printed books. Had the scientists of the early eighteenth century not paraded their learning in treatise upon treatise, the Popes and Swifts would have left them alone. But they would not stop writing – making books. Pope's pronouncement on all pedants in the *Dunciad* could be made to apply as readily to them: 'Forever writing, never to be read.'

EDUCATION IN AND POPULARISATION OF 'NATURAL PHILOSOPHY'

Such a state of affairs naturally changed the course of eighteenth-century education, but not at once. At the opening of the Restoration period, a student at any of the better schools – even old schools such as Eton and Winchester – would not be instructed in 'natural philosophy' subjects except for arithmetic and geometry, which had been taught in English schools at least since 1500.[33] At approximately the same time the utility of the ancient languages was being called in question by educational propagandists who were increasingly seeking to instruct children through 'nature' and 'the concrete'. But they were a small minority and barely noticeable. The anonymous author of *An Enquiry into the Melancholy Circumstances of Great Britain* (1740), who advocated overthrow of the 'dead language' in preference for more useful subjects like the principles of business and 'numbers', would not have found many followers in 1660, but as the seventeenth century wore out and the new century began, more and more educators were turning to this point of view.[34] Locke's philosophy also encouraged this point of view, as did the new needs of the expanding middle class; but actual reform and change proceeded slowly, as a survey of the whole period 1660 to 1800 demonstrates. As late as 1684 Reeve Williams clearly failed to be appointed master of the mathematics school in Christ's Hospital – then a leading school – because he was 'merely a mathematician' and not a classical scholar, on the premise that boys who studied maths ought to be thoroughly proficient in the grammar and vocabulary of the ancient tongues.[35] At most schools mathematics was not part of the normal curriculum, and there was consequently no need for mathematics textbooks. Dr Busby of Westminster, the London school attended by so many *virtuosi* in the Restoration period, taught maths throughout his tenure as headmaster, using Oughtred's *Clavis Mathematicae* and Isaac Barrow's *Euclid's Elements*,[36] but he was unusual. In most other schools students could learn mathematics only by paying a tutor extra sums, as even Joseph Banks, the naturalist, found himself compelled to do a century later.[37] When Thomas James, and headmaster of Rugby who published a school edition of *Euclid* in 1791, reminisced about Eton College in the 1760s,[38] he used this occasion, as he had previous ones, to urge the College to include arithmetic and geography in the main curriculum on the grounds that they are crucial; and he advised that 'these sciences' be taught 'on a Holyday, or Half Holyday',[39] so that other subjects would not lose time. But around 1700 there was little sense of the belief Bentham would have in the next century that 'the sciences' are ultimately more useful subjects than any others because 'they strengthen the mind, preclude superstition and lay the basis for future employment'.[40] The problem of change was administrative as well as philosophical. The early inclusion of Locke in the syllabus at Cambridge shows that the university at least tried to modify its curriculum, whereas Oxford apparently did not. But even Cambridge did not alter its programme of studies enough, and the dissenting academies continued to attract all types of students who were not in fact dissenters precisely because of the altered shape of their curricula. The grammar schools, unlike Cambridge, were usually restricted by their charters and statutes – as is evident in some of the above instances – which required them to teach the classics. A celebrated eighteenth-century law suit involving the statutes of the Leeds Grammar

School demonstrates how difficult it often was then to teach non-classical subjects such as 'science'. In view of this administrative state of affairs and educational philosophy, it is not surprising that few science textbooks for school use exist in our period, at least not until the late eighteenth century. Editions of Euclid were common, and they continued to increase as the eighteenth century wore on.[41] And texts in arithmetic, such as Oughtred and Barrow, and in algebra – especially John Wallis's *Treatise of Algebra* – followed as seconds in numbers of editions, but apart from these, what we would call 'science text books' for daily use in schools were not yet known.[42] The situation *vis à vis* science instruction in dissenting academies is somewhat different, certainly by the middle of the eighteenth century; but even in these schools students rarely purchased the books which their masters either read or on which they had based their lectures.[43] The facts then are these: certainly some books existed, but they were not used; masters and students alike complained of the unavailability of science books; and most schools did not teach subjects included under the heading 'natural philosophy' – not so old itself – so there was seemingly no need for the books. The situation was something of a Catch-22: science books for school use were not needed, and because they were not needed they did not exist in any great numbers.

Science instruction in colleges and universities was different, even if the universities had fallen into the decline about which much has been written. Christopher Wordsworth, the Edwardian Fellow of Porterhouse, captured the essence of academic life at the university when he described how Richard Watson became a professor of chemistry at Cambridge, and the point sheds light on science books at Cambridge and Oxford.[44] Watson had grown tired of mathematics and thought he could achieve fame – in the 1760s – by turning himself into a professor of chemistry, then the most fashionable of scientific subjects. 'He had never read a single word on the subject', Wordsworth comments using all sorts of reliable sources, 'nor seen a single experiment in it'. Therefore 'he sent for an operator from Paris, and buried himself . . . in his laboratory . . . and was soon able to interest a very full audience [at the university of Cambridge] of all ages and degrees'.[45] If the account appears hyperbolic, the substance of Watson's transformation probably is not; the episode supports the point that in a half century or so – from 1710 to about 1760 – science in Cambridge and Oxford had not altered. Gunther notes that in 1661, Newton came up to Trinity 'with nothing in mathematics',[46] and then cites an anonymous critic who accused the English universities of neglecting science except 'when some exceptional person has honoured the university'.[47] A century later, in 1759, Goldsmith made the same point, arguing that the English universities neglect learning, especially natural science, and exist merely to ensure 'the best chance of [one's] becoming great'.[48] Yet Goldsmith thought the real problem to be the schools rather than the colleges: 'might natural philosophy be made their pastime at school, by this means it would in college become their amusement'. Natural philosophy, then, was a problem after all, and had continued to be one for almost a hundred years. Hundreds of comments like those of Wordsworth and Goldsmith can be accumulated; there is no dearth of material. The issue about science and the science books students used in the universities is not whether the books are available, but rather the *extent* to which they were purchased and consulted and how they were read if they were read at all. No mystery exists about the titles and publishers of these science books: Christopher Wordsworth and others have carefully outlined the studies of typical students at Cambridge, the books they read and at what stage they read them: for example, 'Books in use at *Cambridge* about the year 1730, for *Arithmetic, Algebra, Geometry, Physics, Mechanics,* and *Hydrostatics*'; 'A list of Books used at Cambridge aboutthe year 1730 for *Optics* and *Astronomy*'; 'Daniel Waterlands [sic] Advice to a Young Cambridge Student With a method of study for the four first Years 1706–40'.[49] Certain books were used, but the issue, surely, is rather in what quantities these books were published, at what prices, and how far the economics of printing was a determinant of use. Wordsworth affirms without authority that by 1710 'the study of natural philosophy was extensively diffused in the university, and copies of the *Principia* were in such request that . . . one which was originally published at *ten shillings* was [now in 1710] considered cheap at TWO GUINEAS'.[50] Two trends seem clear from the available evidence: first, as the eighteenth century progressed the sciences grew increasingly important at the universities and a certain expansion of the use of science textbooks occurred, and secondly, the adoption of Newtonian principles in the first decade of the century spurred the composition and publication of university textbooks.[51] If the first point seems apparent, the second is somewhat deceptive, at least so far as cause and effect are implicated. That is, was the dissemination of Newtonianism in the universities the excuse for composition of new texts based on these Newtonian principles, or was the sudden need for textbooks the excuse for merely inserting or intruding Newtonian principles? If one contends that the spread of 'natural philosophy' as a taught subject in universities in the early eighteenth century is but the master's – Newton's – campaign to have his principles universally adopted, then the problem is evanescent. But if not, then a question looms about an increase in science books after 1700 directly aimed at university students.[52] After the mid-century, the growth of 'life sciences' – especially botany and chemistry – is rapid and concentrated in the universities and, even more so, in dissenting academies such as the famous one in Warrington.[53] Historians of science have charted the terrain of the life sciences after 1750 rather well, but the puzzling question relates to the early century and to the so-called 'early books' of science published between 1700 and 1710, when a certain flowering is observed. There may indeed be other explanations for this surge than Newton and Newtonianism, but if Newton's campaign was as concerted as modern scholarship would have us believe, then the sociologists may be correct after all about the social origins of scientific knowledge.[54]

The availability and use of science textbooks in military schools is even more nebulous than in the universities. Early eighteenth-century military academies did not teach scientific subjects and required few books. Sir Herbert Richmond, the naval historian, comments that 'in the Navy, there was little in the way of book-learning', and that 'a boy's education depended on a captain under whom he served'.[55] This trend is confirmed by many sources, although sometimes the boy copied out of printed books as distinct from his master's lecture notes. In 1750 Robert Sandham wrote to his mother from Woolwich, 'I have written [i.e. copied out] all Mr Muller's Artillery which is 40 octavo pages'.[56] But John Muller was then the headmaster at Woolwich, a teacher called by Boswell in the *Life of Johnson* 'the scholastick father of all the great engineers of this country', and the book referred to is his recently printed *Treatise Containing the Elementary Part of Fortification . . . for the Use of Royal Academy of Artillery at Woolwich* (1746). F. G. Guggisberg, a captain himself and later on a distinguished military historian, is more specific about the eighteenth century:

> The difficulty of teaching in these early days [the eighteenth century] was greater than at present, owing to the scarcity of printed books of instruction. The cadets themselves had none to guide them in military subjects, and could only learn by copying the masters' MSS, and drawings, making notes from their lectures, and carefully acquiring by memory the practical part of the sciences. In mathematics there was considerable improvement in this respect, as several treatises existed on the various branches.[57]

The 'scarcity of printed books' is curious, yet this is the strain constantly repeated among all types of educators of the period. Considering the subjects the young cadets needed to study – especially in mathematics and astronomy – there certainly were plenty of existing texts: editions of Euclid, Oughtred, all types of geometry textbooks, a variety of books about geography, cosmology and navigation. But the masters continued to complain that printed books were unobtainable, and that this is why the students must copy their assignments out of the master's notes. In view of the disparity between fact and complaint, it may well be that the masters saw how they could advance themselves: what better reason to tidy up their lecture notes and peddle them to booksellers in the name of need?[58] Even if this is what actually happened, such publications cannot have found many students in military schools, for the royal Naval Academy at Portsmouth Dockyard, for example, was started in 1729 with only a handful of students, the Royal Academy at Woolwich in 1741 with space for only 20 cadets, the Royal Military College at Sandhurst in 1799 with not many more.[59] Students in all military academies combined could not have totalled many more than 300 by 1800 – a figure that must give pause to the student of eighteenth-century book production. Furthermore, although instruction in mathematics and navigation apparently was required, this was of an elementary level only, and the Newtonian texts being adopted in colleges and universities were never used. Finally, more difficult scientific subjects such as hydrostatics, chemistry and physics were never introduced into the formal education of an English cadet until the mid-Victorian period,[60] so there can have been no need whatever for these books.

England's coffee-houses, her 'penny universities' rather than her military academies, colleges, or universities, were the authentic source of the popularisation of 'natural philosophy' in the eighteenth century, as well as the origin of many of its books. So much evidence exists to support the affirmation that it is impossible to compress it into a small space.[61] The epoch marks the heyday of the coffee-house and neighbourhood academy as institutions of learning, and the primary reason may be that the new widespread literacy *c.* 1700 increased the demand – a veritable hunger – for knowledge among adults. Such is the contention of one historian of education,[62] and others have joined in the chorus. J. T. Desaguliers, the indefatigable populariser of Newton and an erudite mathematician in his own right, has explained why there was a sudden demand in the early eighteenth century for published lecture notes. In the preface to his *Physico-Mechanical Lectures*, which he had delivered from 1712 to 1717 in a 'private academy' in Little Tower Street and later published, he wrote: 'the following papers being only minutes of my lectures for the use of such gentlemen as have been my auditors, were printed at their desire; to save the trouble of writing them over for every person. Therefore, I beg all such readers as have not seen my course of experiments, to pardon my want of method ... and desire them not to expect a full account of all the experiments made in the course.'[63]

The formula executed was straightforward: the printed book was a student's manual, and as such had to be brief, clear, simple, accessible to all readers, uncluttered by irrelevancies and remarks on methodology, and reasonably priced. Desaguliers's success in both lecturing and then selling his 'lecture notes' as a printed manual – a 'science book' – was followed by others. Examples include Whiston, whose brilliant lectures dazzled Pope and whose similar albeit expensive illustrated manual, *The Copernicus* (1715), Pope purchased; Thomas Watts and Benjamin Worster who also lectured at the Tower Street Academy and who printed manuals; James Stirling at the Bedford Coffee House and many others.[64] These itinerant popularisers worked hand-in-hand with the printers and booksellers, often delivering the lectures on their (the printer's or bookseller's) premises, as in the case of some of these just mentioned.[65] Profit from this type of publication – perhaps the most significant form of popularisation of natural philosophy – was obtained by all three entrepreneurs, lecturer or teacher, printer, and bookseller, and all three exploited a ready market. By the 1730s public lecturing was an established business, a profession with a modest guaranteed income for the teacher, as can be gleaned from several 'projects' and 'manuals' such as the *Proposals for a Course of Chemical Experiments with a View to Practical Philosophy* offered in 1731 by Shaw and Hauksbee,[66] who soon afterwards went on tour in the provinces with considerable success. By mid-century a large number of scientific societies and book clubs had come into existence, all of whose members demanded printed science books and could now afford one or two. The published lectures of 'the Professors of Gresham College' continued well into the eighteenth century and satisfied a small group of readers who had been lucky enough to obtain advanced training and who could understand technical language; the more typical adult reader who had not been fortunate enough to attend an academy or university, and whose total exposure to science derived from the public lectures of popularisers like Desaguliers and Whiston, wanted shorter and easier books than a lengthy treatise by a Gresham professor. Yet no books were so popular as the manuals used for courses of public lectures which the reader had already attended or which he was then in the process of hearing. Not even a repertoire of 'how-to-do-it' books for businessmen – works such as John Ayres's *Arithmetick: a Treatise Designed for the Use . . . of Tradesmen* (1710) – could compete with the coffee-house manuals. If only as much were known about the science books never intended for public lectures – books meant to be read in the solitude of one's own chamber – the scholar could speak with much greater authority on the subject of eighteenth-century science books.

The extraordinary aspect of the eighteenth-century coffee-house – especially in the period before 1720 – is the clear way in which its proprietors and all those involved in its commercial life adopted *natural philosophy* as the chosen subject. Perhaps this reflects nothing more than popular taste, yet it is astonishing to learn that the general public preferred 'science' to all other subjects. Two recent social historians have attributed much of the scientific progress and industrial advance of the era to the men and women who convened in the 'penny universities': 'scientific advance came *mainly* from the activities of groups and societies meeting at first in the London coffee houses, and from public lecture courses given by scientists who were often of some note in their day'.[67] This was no doubt true, but if a significant number of the science books of the era were the published versions of these courses, in 'penny universities' and 'natural philosophy book clubs', then it is essential to understand the genesis of these books. The same social historians have provided a club: 'these clubs', they note, 'were patronized largely by medical men, dissenting ministers and manufacturers . . .' They may have been 'patronized largely' by such men, but members of other professions frequented them as well. Moreover, all those who frequented the coffee-houses and joined clubs for *social* in addition to scientific reasons – writers, artists and politicians – are omitted from this consideration, especially greater writers such as Pope on whom the influence of Whiston was enormous. The crucial issue then is not numbers but types, and there is little doubt that for every Pope or Steele who became an habitué at Buttons to hear Whiston or Desaguliers lecture, dozens of other social types – men and women from many different walks of life – frequented the coffee houses to 'be seen' and to quench their curiosity, to familiarise themselves with the 'new science', and in many cases to apply this knowledge to commerce and industry.[68] The reasons for attendance are so different that it is impossible to condense them to two or three, and for just as many different purposes readers bought these 'coffee-house books'. From the instructor's point of view – whether speaking in a coffee-house, private home, or small academy such as that in Tower Street – the task before him was equally manifold: first and foremost to fill the room and *sell* his published lectures (this is why the booksellers were so attentive to itinerant lecturers as a new professional type); then to cater to the most influential group in the audience, those genuinely capable of arranging for another set of lecture-demonstrations; and lastly to arouse more interest in natural philosophy so that all members of the audience would return next time.[69] As the eighteenth century wore on and the coffee-house audiences proliferated, various groups splintered: those interested only in the application of the new science formed clubs like the Spitalfields Mathematical Society (1717) where they could discuss and compare their 'inventions'; others concerned with one branch only – mathematics – established organisations such as the Northampton Mathematical Society (1721) and the Manchester Mathematical Society (1718); the Spalding Gentleman's Society (1717) was composed primarily of literary 'gentlemen' who reckoned themselves 'projectors' and virtuosi in their spare time. The most important development for science books is that in every case formation led to some type of publication, and this is why any account of the spread of science books that omits discussion of these penny universities is inchoate and can never satisfactorily explain why these books were suddenly in such great demand.[70] A new type of book had come into being for a brand new readership.[71] Much of the financial profit was divided between itinerant lecturer and bookseller, who were identical in certain instances, but there had been nothing like this new consortium – lecturer, bookseller, habitué or club member – in the previous century.

Viewed in the single dimension of 'content' this new book changed over the decades.[72] If the early (1700–20) species of the genre reflect an audience new and unaccustomed to natural philosophy, later specimens at mid-century do not. Mechanics, hydrostatics, pneumatics and optics continued to be popular throughout the century, but astronomy and mathematics waned as 'natural history' – botany, biology, and zoology – overtook and gradually supplanted it. Furthermore, all these books grew lengthier as the century wore on and as their authors assumed a readership far more adept in natural science than earlier readers had been. A comparison of these popular books with the Boyle Lectures, which continued to be published throughout roughly the same period (from 1691 to 1735), clinches the point: whereas the Boyle Lectures are large expensive tomes (the average volume cost over £2) filled with religious and metaphysical disquisitions and speculations, these popular lecture books were short, practical, accessible, adjusted to the literal-minded reader. Their style is usually chipped and plain, composed of short sentences and with figurative language played down as much as possible. The vocabulary employed is easy, and when technical or difficult words are used they are instantly defined. Experiments demonstrated in the coffee-house are rehearsed and simplified, and mechanical instruments – such as Whiston's *The Copernicus*, a perpetual calendar of the skies permitting the user to predict precisely when future eclipses would occur – are diagrammed and fully explained. Engraved plates and other charts and diagrams contain clear legends and simple explications. The reader understands from the outset that the book in his hand is not a work of art or philosophy – nothing shaped and meditated upon – but a manual, a how-to-do-it book for those who have been attending the private demonstrations and listening to the lecturer.

The dissemination of natural philosophy – more accurately of 'cults of natural philosophy' of which popular science books constituted only one aspect – also intensified a need for dictionaries and almanacs of all types, inexpensive as well as dear, handy as well as decorative and handsome. Historically speaking, it would be false to contend that these widespread cults in themselves created a market for such books for such dictionaries had also been written, although in fewer numbers, in the Restoration period: Richard Read's *Secrets of Art and Nature . . . being the Summe and Substance of Natural Philosophy* (1661o); Richard Blome's *The Gentlemans [sic] Recreation* (1686); Richard Neve's *Apopiroscopy: . . . a Faithful History of Observations and Experiments* (1702); John Harris's instantly popular *Lexicon Technicum* (1704),[73] and several others. But the dissemination of natural philosophy after 1710 or 1720 caused an unprecedented consumer-interest in popular science books, a consumption that continued into the early nineteenth century and that stemmed from the typical Englishman's new commercialisation of leisure, in Professor Plumb's phrase, and that could be witnessed after 1800 in such popular 'science books' as John Paris's *Philosophy in Sport Made Science in Earnest: being an Attempt to Illustrate the First Principles of Natural Philosophy* . . . (1827). As several cultural historians have noted, the eighteenth century was an epoch of general dictionaries;'[74] but it has not always been clear to what extent this propensity to compile dictionaries was actually channelled into the 'arts and sciences', that is, to dictionaries of scientific subjects as distinct from religious, historical, or other general types of dictionaries. From beginning to end, every decade of the eighteenth century brought to light an increase in the number of these compendious works – compiled by John Harris, Ephraim Chambers, Robert James – for reasons having as much to do with profit for the bookseller as with the public's craving for scientific knowledge. For example, both Ephraim Chambers and his publisher, David Midwinter, reaped huge profits from the *Cyclopaedia* – best described as a popular 'science dictionary' – which first appeared in 1728 and went through a dozen editions and revisions by 1779. William Strahan realised what a best seller this dictionary was, and bought the rights to reprint it in 1778, eventually adding to his fortune by doing so.[75] Robert James's *medicinal Dictionary* (1743), to which Johnson contributed, reveals a similar history based on immediate success and subsequent reissue yielding vast amounts of money to the printer and bookseller, as do with the various scientific dictionaries of Richard Brookes from which Goldsmith plundered so much, as we shall see, in his *History of the Earth and Animated Nature*. There is not enough space here to list even a fraction of the mid-century, let alone the late eighteenth-century, dictionaries; such a list, taken together with an analysis of the royalties gained by author and bookseller, would soon render evident to what extent the new literate public was buying these books.[76] By contrast the encyclopaedias of this period are perhaps not so prolific as the dictionaries, but they contain more scientific information than any previous compendia. As natural philosophy grew increasingly technical and specialised, dictionaries and encyclopaedias, like other popular works, grew more specialised too: no longer general compendia of 'arts and sciences', or even more narrowly of 'natural philosophy', dictionaries after 1720 or thereabout were more specifically dictionaries of mathematics, dictionaries of astronomy, dictionaries of geography, and so forth.[77] Medical dictionaries incorporating hard words – words unknown to the layman, such as those found in a handbook of diseases or in the pharmacopaeia – had appeared from the Restoration onward, so this continuation into the eighteenth century was not a new development, perhaps because medicine served a different social role, but the fragmentation of science dictionaries into specialised works marked the beginning of a publishing trend that continues to the present day. The role of the British press in this dissemination of science through popular dictionaries has not been fully explored, but scholars – even historians of science – are beginning to realise that the surge in science c. 1700–50 coincides with another development in publishing, namely 'the scientific press in transition'.[78]

If popular knowledge about natural philosophy was disseminated in dictionaries and encyclopaedias, it was also circulated in periodicals and magazines for which the eighteenth century has already become famous, and nowhere more so than in women's magazines such as the long-lived *Ladies' Diary* which commenced publication in 1704. It is true that a certain amount of 'women's science' had existed in the Restoration period. Marjorie Nicolson has demonstrated what a terrific impact 'Mad Madge' – the racy Duchess of Newcastle – had on the Fellows of the Royal Society, and has also demonstrated what an inspiration Lady Conway was to scientific thinkers.[79] But natural philosophy was nevertheless still a 'male endeavour' in the Restoration period and 'Madge' – called by G. D. Meyer 'the first English woman who wrote about science' – was the exception rather than the rule. Most serious scientists as well as *virtuosi* were men and conceived of scientific activity as 'male activity', perhaps as a consequence of the physiological reasons Malebranche offered in his *Search after Truth* (1694). Yet this situation began to change when John Tipper, an enterprising schoolmaster in Coventry who had grown tired merely of teaching students, launched an annual mathematical magazine for women in 1704 called the *Ladies' Diary*. Meyer – the most learned student of 'science for the fair sex' in our period – has written that 'Fontenelle, with some help from the Duchess of Newcastle, especially in her *Grounds of Natural Philosophy*, awakened in Englishwomen a dormant interest in sciences; afterwards, journalists like Dunton, Tipper and Philips fed that interest with natural history in general and with the findings of the microscope and telescope in particular.'[80] The key period is the two decades 1690–1710, for during this time women developed a new sense of themselves, despite Malebranche's verdict that they were biologically inferior to men. By 1710 English women were not merely reading more books; they also wanted to read books about 'male subjects', especially about science. Fontenelle catered to this need in *The Plurality of Worlds* (1715), a book ladies were reading at the same time they were scanning Pope's couplets about 'Belinda', and Fontenelle's lovely and inspiring Marchioness established the model of 'the scientific lady' for the eighteenth century. But even before this time male educators and philosophers emphasised that women ought to read science books. As early as 1677 Poulain de la Barre, a radical French priest, wrote in *The Women as Good as the Man* (which may have been translated into English by Anthony Le Grand), that young women should study Descartes's *Discourse on Method* and Rohault's *Cartesian Physics*. By the end of the century science books written primarily for women had appeared; in the 1690s John Dunton's *Athenian Mercury* invited scientific questions from ladies; then in 1704 Tipper's *Diary* beckoned ladies to submit their poetry for publication, yet printed material far more mathematical than poetic,[81] finally in 1718 Ambrose Philips's *Free-Thinker*, a semi-weekly sheet, introduced a paragon named Sophronia who had immersed herself in books about natural philosophy – our science books. By approximately 1740 any literate woman, of low or high class, could select from a very large number of printed works – almanacs, broadsheets, weeklies, magazines, manuals, books – to quench her thirst for science, a trend that continued throughout the century, with a pronounced resurgence at the end. She could also choose from any number of books like the anonymous *Female Physician* (1739; reprinted many times) which told her how to care for herself. Perhaps this trend accounts for the criticism of *Pamela* held by George Cheyne, the fashionable physician; Cheyne had corresponded with Richardson during actual composition, and upon reading the novel was surprised that Pamela had studied so little natural philosophy; 'a good Library of Natural Philosophy would be very proper for your Heroine, which you want and cannot otherwise procure I will help you.'[82] Surely Meyer is correct to notice that eighteenth-century ladies most certainly did not learn their science directly from the great scientists themselves – Newton, Huygens, Boyle, Hooke – or from their books: 'it was not to the primary works of these scientists that the ladies turned. Instead, professional expositors simplified and popularised for lay consumption the major scientific advances of the time.'[83] Yet the leading printers and booksellers of the day (about which more must be said) did not exploit this available market without injury to the fair sex. Although science as an activity designed for women as distinct from men was becoming increasingly popular in the early period from 1700 to 1720, there was a natural resistance that elicited a number of printed satires written by both men and women; however peripheral they are, these works deserve to be considered as 'science books' together with the popularisations described above.[84] If it is not altogether clear, as Meyer suggests, to what degree the English bluestockings were genuinely interested in science, they were nevertheless reading and buying books about this subject. They may have bought them and subscribed to journals and magazines without digesting the contents, but until further evidence is brought to light, it must be assumed that the sheer volume of popular scientific literature written explicitly for women was consumed by them because they possessed a vigorous interest in the subject.[85]

NEWTON AND 'THE TRADE IN BOOKS': THE SCIENTIFIC WITS

All these developments would have been inconceivable without the advent of a Newton. In an important essay already cited,[86] Simon Schaffer has gone so far as to argue that the rise of natural philosophy in the Restoration period was nothing other than 'Newtonian matter theory' and a whole series of responses to it, and there is certainly a sense in which this is the case. If the scientific movement of the seventeenth century was essentially Baconian, and if that great age of scientific discovery, encompassing Harvey's discovery of the circulation of the blood, Leeuwenhoek's invention of the microscope, Willis's brain theory, Boyle's discovery of the chemical laws in mechanics, Hooke's hydrostatics and so forth, owned its primal impulse to Bacon's teachings – hence its common designation as 'the Bacon-faced generation'[87] – the next three generations from 1690 to 1780 were indebted to Newton and perhaps ought to be known as 'the Newton-faced generations'. Surely this is not the appropriate place to argue fro Newton's rightful succession to Bacon; besides, others have already done so more adequately than I could.[88] But however excellent their argument, they have rarely been vigilant to the effect of this succession on the book trade and the readers to whom it catered. Abundant evidence exists to support the contention that from roughly 1680 to 1750 science, or natural philosophy, *meant* Newton. Whether agreeing or disagreeing with his physical laws;[89] whether delivering physico-theological sermons or university lectures or popular coffee-house talks to disseminate his ideas; whether discussing his ideas in recently-formed science clubs and reading groups; whether simplifying and expatiating on these Newtonian laws in almanacs, dictionaries and encyclopaedias that laymen could understand; or – in areas not yet touched upon – whether applying his laws to non-physical areas such as the disparate realms of civil government, painting, medicine, and so forth;[90] the student was always in effect talking or writing about Newton. In *Newton Demands the Muse* (1946) Marjorie Nicolson demonstrated how the *Opticks* were versified in dozens of poems throughout the century, and a similar book could be written for the *Principia*. Yet only recent compilations – such important works as Peter and Ruth Wallis's *Newton and Newtoniana 1672–1975* (1976) – have demonstrated to what an extent this 'Newton-faced generation' was writing and reading books about the man who had changed everyone's concept of the universe about him. When I. B. Cohen and Marjorie Nicolson wrote about Newton 20 or 30 years ago they wrote perceptively, but neither scholar could then have known to what degree the century 1680–1780, bibliographically considered, ought to have been called 'the century of Newton'.[91] That century itself provided numerous clues, not merely in the literally hundreds of popularisations to which I have been referring – works such as

Elizabeth Carter's *Newton's Philosophy Explain'd for the Use of Ladies* (1739) and the dozens of published and unpublished poems invoking Newton as their muse – but also in the less obvious patterns of book-reading and book-collecting that constituted such an important aspect of eighteenth-century life.[92] It ought not to be forgotten that early in the century Addison, an ardent Newtonian himself, had laid down in a popular periodical that 'Sir *Isaac Newton's* Works' ought to be found in every lady's library; indeed this is the only 'science book' required for 'Leonora's library'.[93] And if Leonora had to read Newton, then one should not be surprised to discover Elizabeth Carter and her 'Saturday-correspondent' Catherine Talbot, and many other women, reading him; perhaps not the mathematical sections, but surely the Scholia to the *Opticks* and other popular works Newton had written. As the eighteenth century wore on, writers of all types, high and low, serious and in jest, assumed that a vast readership existed in Britain that wanted information about the greatest scientific genius who had lived, and consequently poured forth printed material in unprecedented amounts – a cornucopia of printed works in different genres that has still to be taken into account in major discussion of the spread of secular knowledge in Western civilisation.[94]

Most of Newton's own writings, especially the *Principia* and *Opticks*, did not lie at the centre of this dissemination. They continued to be reprinted and reissued in small quantities, and translated into many languages,[95] but were read by relatively small numbers and in any case were not the books that support the above generalisations. Nor were the more easily grasped interpretations of Newton's system by Samuel Clarke, John Keill, Colin Maclaurin, Henry Pemberton, Cadwallader Colden, and many others; although these books attempted to present Newton's system to literate readers they were not always successful.[96] Far more influential were the outright popularisations – simple works in almanacs, dictionaries and poems – that made no pretension whatever to be faithful to their original. This is why two types of Newtonian books need to be distinguished: the interpretations, or reinterpretations, and the popularisations. The authors of the second type were often hacks and wrote for financial gain; yet, almost paradoxically, they were capable of effecting cultural secularisation by reaching large audiences. In this connection the 'scientific poem' begs to be considered. Upon occasion even Pope, Gray, Cowper, and later on Blake and Wordsworth, wrote in this vein; more commonly, though, its authors were poets of lower ranks, the Akensides, Armstrongs, Jagos, Beatties, and Erasmus Darwins who were far more sedulous in their aim of versify scientific ideas than to discover an original 'poetic voice'. Modern discussions of 'the scientific poem' in England have imposed judgements of value in the relation of aesthetic considerations to didactic aims that were foreign to the original poets involved.[97] A Thomson or Akenside – both 'best selling poets' in their day – naturally sought to write as well as possible and to whet the reader's imagination by a marriage of form and content, craft and ideas; but science, especially Newtonian science, had been the poet's original source of inspiration, and he generally considered it unlikely that his readers would not be as excited about his subject as he was. Precisely why Newtonian science should have seemed such a likely topic for versification when there were so many others from which to select is a matter to be debated by historians and sociologists of eighteenth-century English culture; what is important for the production of science books is the relation of these writers – and by implication their readers – to their chosen subject. In almost every instance it is apparent that 'science' meant Newtonian science, even among writers who understood little of the substance of the *Principia* and *Opticks*. It is as if 'Newton' were a vast region of the literate imagination even in the most remote of English shires. To write a quasi-popular 'science book' in 1720 to 1740 meant to write about Newton, even if the author's debt were oblique; and the science books of this period demonstrate a universal lag of time between original scientific idea and popular or semi-serious adoption. The universities and their presses, as well as certain houses in the English and Scottish metropolis, continued to publish science books such as William Jones's *An Essay on the First Principles of Natural Philosophy* (Oxford, 1762), and Robert Greene's *The Principles of the Philosophy of the Expensive and Contractive Forces* (Cambridge, 1727), which none but a handful of scholars could read or understand. But these, no more than the original *Principia* or *Opticks*, were not the books that changed the thinking of the common man.

The influential books were those that popularised Newtonianism. As we have seen, Newton's ideas were rapidly applied to domains other than the physical universe; moreover, authors unfamiliar with Newton's actual texts were often unaware of the debt they owed to these applicators. By the end of the seventeenth century, for example, books both endorsing and rebutting Locke swelled into a major industry; but many of these books were attempts to reconcile Locke's 'science of the mind' with Newton's physics.[98] The same was true, though to a lesser degree, of books by Cheyne, Whiston and Hartley later in the eighteenth century. Whatever else these writers may be called, they were first and foremost 'science writers' – that is, authors whose 'systems' had been radically changed by recent Newtonian science more than any other sphere of knowledge. Cheyne was a trained physician with a deep understanding of religion and art who grew increasingly mystical as he aged. The remarkable aspect of his career is the uncanny way in which he engaged the public's attention: almost anything he wrote or said was quickly 'consumed' by the public after knowledge spread of his unparalleled reduction of weight, especially as recounted in his autobiographical book *An Essay of Health and Long Life* (1724). For years after 1724, his books about diet and health permitted Cheyne to sustain a reading public enjoyed by few other English authors of the day.[99] Yet Newtonian science more than any other sphere of thought – even more than the theology to which he was so devoted – had inspired his most serious reflections, especially *Philosophical Principles of Religion* (1705), an extended meditation on God's place in a Newtonian cosmos. Similar generalisations can be made and supported for Hartley's brilliant *Observations on Man* (1749). This popular and didactic large book attempted to demonstrate that virtually all man's ideas are formed by 'mystical vibrations' conducted from the sensory and 'vibrating' organ to the brain; yet without the heritage of Newton and Locke, Hartley's system would have been impossible. Hartley, also a physician like Cheyne, probably read Newton's original works rather than one of the many available popularisations. The more significant point is that many readers in the 1750s learned their 'Newtonianism' from Hartley's *Observations* rather than from a book written by Newton himself. Much earlier the same had been true of Whiston's *Astronomical Principles of Religion* (1717), which Whiston, Newton's successor as Lucasian Professor of Mathematics at Cambridge, had written after being deprived of the Chair. Throughout the 1720s there are references to this work as a vast source of knowledge about Newtonianism, rendering it clear that Whiston had brought Newton's ideas to the common reader. In different ways all these books were responses to Newton, as well as meditations upon cosmology; in this sense they are similar to the more apparent 'translations' of Newton's system in more technical books such as Rohault's *System of Natural Philosophy* (1723) and other 'systems of natural philosophy' that do not pretend to simplify or interpret Newtonian hypotheses.[100] This variety of printed responses raises a further important question about the *diversity* of 'science books' in the first part of the eighteenth century: while many historians of science have recognised that most 'systems' of natural philosophy were directly or indirectly replying to Newton, it has not been evident to what an extent derivative 'science books' such as those by Cheyne, Whiston, and Hartley owed their origins to Newton. Such a theory of origins may not sit well with certain historians because it lays such great weight on Newton's ideas and minimises those of his forebears and contemporaries.[101] Yet if viewed from the vantage of the eighteenth-century common reader – an angle to which I have repeatedly returned in this essay – these derivative books are the ones capable of teaching what Newton's ideas actually were, and this is the reason such works are reprinted more frequently than Newton's own books.

NATURAL HISTORY: THE 'BOOKISH SPORT'

Yet if Newton was the spring of much, if not most, of the printed natural philosophy of the first half of the century, natural history – which, unlike natural philosophy, cannot be traced to a single monumental scientist – was the well for the second half. The reasons for this precise succession are immensely complicated and cannot be reduced to generalisations compressed into a few sentences.[102] Furthermore, the specific reasons proffered will depend on the analyst's view of the history of scientific thought in the Enlightenment, and whether he adopts a linear, dialectical, synchronic, or other approach. Necessarily complicated though this matter of attribution is, some aspects are plain, and those that directly apply to science books and their readers ought not to be omitted. By 1750 the Newtonians had won their war – certainly at home – and there was no longer any reason to continue to campaign vigorously. Further popularisations and interpretations for students could still be produced, indeed they were,[103] but it was no longer necessary to proselytise as frantically and sedulously as the army of Newtonians did in the first three decades of the century. Moreover, as natural history increasingly attracted readers from 1700 onwards, natural philosophy gradually lost them: it was not a swift progression, but as the one continued to lose, the other continued to gain. Jacques Roger, perhaps the most learned student of Enlightenment natural history, has attributed much of this inverse succession and discontinuity to gradual 'abandonment of the Cartesian philosophy' and to the literary abilities of natural history writers: in England John Ray, William Paley, William Derham, Oliver Goldsmith, Gilbert White;' on the Continent Pluche, Fabricius, Lesser, Buffon, Bonnet and many others.[104] Roger's two reasons are no doubt the right ones, especially if one realises to what an extent abandonment of Cartesianism meant adoption of Newtonianism and to what degree this adoption had succeeded in England by 1750; 'that whole corpus of [natural history] literature, whose success was enormous, has not yet been studied in its own right, perhaps because it has been considered as an accident in the history of science'.[105] 'Accident' it may or may not have been, but its effect nevertheless was to steal the very readers of natural philosophy who had been taught to crave knowledge about their teraqueous globe. The dynamic antagonism of natural philosophy and natural history, suspected by Roger, was never more evident than at mid-century. As Roger has written, because 'a "history of nature" is opposed to a "natural philosophy", which is the search for the causes of the phenomena',[106] it is consequently unthinkable that both could coexist at mid-century – or in any period – and be read by the same readers.

The triumph of natural history over natural philosophy was ultimately based, as Rogers has argued, on a set of theological, intellectual and literary assumptions, yet the victory also had marked effects upon book producing in the second half of the eighteenth century. In a certain simple sense the consequence is only arithemetical: all that was printed about natural philosophy before 1750 is transferred after 1750 to books about natural history. Stated in this way, though, the succession is deceptive insofar as it implies an absolute equality among both types of books. Yet the exchange was disparate, as Roger has suggested and I shall attempt briefly to develop, for natural-history books were far better written and much more beautifully produced than their predecessors in natural philosophy had been; and their success consequently derives more directly from a relationship between book and reader, as distinct from merely writer and reader, than had been the case for natural philosophy.[107] In a sense it is almost as if the British readership at mid-century was ready for an elegant *and* readable factual book which natural history could supply. Also, if one recollects that English natural-history books begin with John Ray, enjoy a number of sumptuous editions in the 1720s and 1730s, then pass on to popularisers such as Richard Brookes and Oliver Goldsmith, and culminate in Gilbert White's *Natural History of Selborne* (1789), then Roger's point as well as my own readily makes itself; for there are no books *qua* books in the other sphere – natural philosophy – that can vie with these, not even the 'scientific poems' (Thomson's *Seasons*, Blackmore's *The Creation*, Pope's *Essay on Man*, Darwin's *Botanic Garden* – which is a 'natural history' poem) which had, of course, a certain following in the eighteenth century.[108] Both types of books centred on the role of God in his creation, and on the 'natural' as distinct from supernatural or socio-economic aspects of their subject; but whereas books on natural philosophy were often technical, dry, repetitive, and written by authors who had no sense of literary form or style, those about natural history tended to be simple, elegant, beautifully illustrated (sometimes in colour) and often endowed with an architectural shape that enticed readers. But there were other differences as well, differences pertaining to innate elements of the subjects themselves. A 'natural history' of birds or insects, fossils or shells, genuinely lent itself to simplicity, elegance and illustration in ways that natural philosophy did not; such a 'history' naturally divided into symmetrical units which would seem artificial if imposed on natural philosophy. And natural history was blessed in attracting masterful prose stylists who were not discouraged by the technical rigours and difficult vocabulary of natural philosophy. A Goldsmith could master the best classification schemes of animals or birds in a relatively short time; it is doubtful whether Goldsmith would have understood Newton's fluxions (calculus) or physics well enough to write brilliantly about them (the historian ought never to assume that the common reader perusing popularisations of Newtonianism *c.* 1700 or 1720 understood them well enough to explicate them to others). Therefore while a dynamic approach to the triumph of natural-history books over those about natural philosophy is valid, these less recondite reasons existed to ensure the greater success of natural history with an audience composed of intelligent but non-scientific readers.[109]

Buffon's remarkable popularity in Britain and among English-speaking readers documents the point in a nutshell. His books on natural history began to appear in 1749 in French, and in 1762 in English, the year when *The Natural History of the Horse* was first published. No previous books on natural history were sought after so eagerly. Roger has argued that Buffon's popularity in France and elsewhere in the second half of the eighteenth century can be attributed to the common man's interest in global exploration – the sense Goldsmith had felt in *The Traveller* of a world composed of 'lakes, forests, cities, plains, extending wide'; and Foucalt has attributed the same widespread popularity to the zeal with which naturalists from Ray and Lhuyd onwards were then collecting and reclassifying the entire kingdom of nature as a consequence of a new separation of 'words' and 'things'.[110] Both these arguments are no doubt valid, but however metaphysical their concerns, something must have been owing to the average Englishman's Francophilia and to the public's reading habits. After all, natural history had flourished in the Renaissance, and there is no surer proof of its vitality then than the number of books about the symbolism of the natural world (animals, vegetables, minerals). In the Restoration period, the natural history of wonders was commonly found in children's books, as were the natural histories of countries just slightly later. What was lacking throughout this period from approximately 1600 to 1750 was a writer genuinely capable of engaging the attention of the common reader, and a readership sufficiently large in numbers and leisured in the economic sense, to benefit from such natural-history books. Buffon provided all this for a ready audience. He also appealed by the lucid style in which he wrote and the pictorial format his publishers used. It is not surprising that Buffon, who gave the world its most celebrated definition of style – the notion that style can never be anything less than 'the man himself'[111] – employed an elegant and polished style in French which his English translators admirably captured. Yet the success of Buffon's works in Britain, as Charles Waterton commented in his own *Essays on Natural History* (1838), owed as much to his publishers; for Buffon was not merely lucky, and the unparalleled enthusiasm for his books derived from a collaboration between author and bookseller-printer that veritably ensured his niche as a best-selling author for four decades. The anonymous writer of an early review in the *Monthly* has captured some of the dramatic excitement he sensed upon reading the first English translation of Buffon's *Natural History* by William Kenrick, the dramatist.

Likening the reader's experience in the *Natural History* to that of the reader in Swift, the reviewer argues that '[Buffon] has made a voyage through our system, seated in the magnetical steel chair of his countryman, *Cyrano de Bergerac*, the adventurous Prototype and Precursor of our Lemuel Gulliver'.[112] The exhilaration of 'taking the voyage' with Buffon was achieved by a rhetorical and pictorial eloquence that even Johnson noticed, according to Boswell, and which almost every reviewer in England discoursed upon in the eighteenth century. And while criticism of Buffon's scientific 'system' or 'knowledge of his subject' abounds, his excellence as a stylist remains a constant in the chorus of eighteenth-century approbation. In England the reader of weeklies and monthlies continued to hear about 'the beauty of the Author's style, which is always enchanting, even where it betrays marks of negligence'.[113] One reviewer of the Kenrick 1775 translation in three volumes went so far as to contend seriously that Buffon was a sorcerer capable of bewitching his readers: 'let the reader decide whether his fame be owing to the solidity of his investigations and discoveries, or to those sublime flights, and that magic power, energy, and grace of style, that have astonished and *bewitched* a considerable part of Europe'.[114]

If Buffon 'bewitched' his readers, he accomplished this by an attitude to nature as well as a glowing pen. In the words of another anonymous reviewer – actually Buffon's first reviewer in England – he mesmerised them 'not by contracting the sphere of nature within a narrow circle, but by extending it to immensity, so that we can obtain a true knowledge of her proceedings'. Linnaeus was a more venerated naturalist than Buffon during the period 1750–1800; he was the Newton of that age; yet he was utterly incapable of 'bewitching' reader anywhere. If Buffon's *Natural History* ranks as one of the ten most popular books in the second half of the eighteenth century,[115] nothing Linnaeus ever wrote attained the status of a 'natural-history bestseller'. Linnaeus's impact, especially his systems of bisexual classification and natural taxonomy which were altering the course of serious science, was more consequential than any 'system' generated by Buffon; yet Linnaeus was important as a *scientist* rather than as an author, and the distinction is crucial for an understanding of 'science books'.[116] No serious botanist or naturalist in England could afford to overlook Linnaeus at mid-century, no matter in which language he read Linnaeus's works; he would do very well, as Johnson assured Boswell, to overlook Buffon.[117] Johnson seems to be intimating the very distinction between 'scientific systems' and 'scientific thought' which I have been attempting to make throughout this essay. Ideas and books differ, and granting that here and there in Linnaeus are light touches of wit and humour and elegance, nevertheless there is none of the literary sophistication that pervades every page of Buffon.[118] This is why the English reviews of Linnaeus constitute an important collective documents: they demonstrate how Linnaeus came to be viewed as a 'genius' worthy to be ranked among the greatest scientists of all time, in the words of Raymond Hirons, one of Ralph Griffiths's frequent reviewers of books on natural history for the *Monthly*, a giant worthy of comparison with other giants: 'the same science [botany] which has been disgraced by a butterfly-catcher, or a hunter after cockle-shells, is immortalized by the labours of a Bacon, a Boyle, and a Linnaeus'.[119] Buffon is not in the list; he was never viewed as a major scientist; yet he, rather than Linnaeus, held sway among the many readers active in the cults of natural history. A point about language also needs to be made. Others than Hirons recognised that natural history was altering the English language far more than natural philosophy had done in the previous half century.[120] Hirons himself had attributed much of the change to the introduction of the Linaean system into England: 'Linnaeus has totally reformed the language of botany, and indeed, in great measure, introduced a new language into the science;.[121] But scientific genius such as Newton possessed and linguistic ripples such as Linnaeus clearly caused are distinct from the books the common reader then wanted to consume. If the demand for editions of Newton was small before 1750 in relation to the demand for books by his staunch popularisers, the same is true of Linnaeus after 1750. Benjamin Stillingfleet, Richard Pulteney, William Withering, John Hill, Erasmus Darwin – all popularisers of Linnaeus – are the authors whose books sold, not Linnaeus or the members of the 'Norfolk Linnaean group' whose books were too serious for popular consumption. Pulteney's *General View of the Writings of Linnaeus* (1781) proved so popular that his publisher Tom Payne claimed 'not to have a copy in his hands after the year 1785'. The point then is that whereas Linnaeus was more important than Buffon for the record of science, he could not write a single scientific book everyone could understand. He would not do what Freud decided to do at a major turning-point in his career: popularise the serious in at least one readable book.

Goldsmith's *History of the Earth and Animated Nature* stands somewhere between the enormous popularity of Buffon on the one hand and the serious scientific attitude of Linnaeus on the other. Goldsmith plundered both authors as models, Linnaeus for his system of sexual classification and Buffon for his descriptions of animals and birds, but with a difference. Goldsmith, unlike thee authors, was a native English writer whose refined prose style had been noted long before *Animated Nature* appeared on 1 July 1774, seven weeks after his death. The reader could expect a 'natural history' from his pen that was a prose classic, something not even the translators of Buffon (Kendrick *et al.*) ever hoped to achieve.[122] Furthermore, Goldsmith had already written successful poetry, drama and fiction, and he knew how to appeal to the taste of a broad cross-section of the populace, especially to its relatively recent demand in prose for authorial sentimentalism and continuous sympathy. When these qualities, everywhere engraved on the stamp of *Animated Nature*, were wedded to a description of the *whole* terraqueous globe that quenched the reader's own lust for travel and exploration in unknown places, the result was a popular set of 'science books' that ran to almost a dozen edition by 1800 and at least 22 editions by 1876.[123] The argument then of some recent critics, to the effect that *Animated Nature* is *not* an important work of Goldsmith's, needs to be re-evaluated. It was reprinted 'by popular demand' as often as *The Vicar of Wakefield* and *The Citizen of the World*, and was rivalled in the genre of popular science books only by English editions of Buffon. Paul Kaufman, mentioned earlier, has discovered that this popularity of Buffon and Goldsmith was as evident in cathedral libraries as in other lending libraries. No other writers on a scientific subject rivalled these two, if numbers of editions and documented borrowings are an appropriate yardstick.

But *Animated Nature* demonstrates another tension between scientific ideas and scientific books that has been a *leitmotif* of this essay. Indeed it casts light on the public's reading habits by demonstrating that a book like *Animated Nature* could be lionised *despite* its reviewers. So well did the public know what it wanted to read, and the booksellers – in this case John Nourse – how to feed that hunger. Edward Bancroft, a young American then resident in London with his growing family, considered *Animated nature* a poor performance in every way *save* one: 'Our Author has adopted no methodical arrangement'; 'his work is to be considered as an compilation'; 'his descriptions ... are almost wholly employed upon their more amusing properties'; and in his conclusion: 'however well qualified he was to excel in works of *imagination*, his talents appear to be ill-suited to those of *science*'.[124] There again lay the old dichotomy: *science* and *imagination*. But the novice reviewer had missed the larger truth: scientific 'bestsellers' had excelled in all ages for precisely this reason – that they had been 'works of *imagination*; – and not because they promulgated radical theories. This was as true in the Renaissance as it has been in our own century. Johnson in the capacity of a reviewer was far more perceptive: he chided Goldsmith for plundering Buffon so mercilessly and for propagating the old saw about cows shedding their horns at three years of age, but concluded that *Animated Nature* was nevertheless 'an excellent performance'.[125] Johnson knew that scientific 'bestsellers' are less comprehensive than scientific 'dissertations' and always borrow from serious writers. This is precisely what Goldsmith had done: he lifted from Buffon, while applying the best aspects of Linnaeus's theory. Yet Goldsmith, like smaller fry in Grub Street, knew what he was about in *Animated Nature*, he knew he wanted to write a popular as opposed to a scholarly book, a book, as Johnson might have said, that mankind would actually red. Goldsmith's assessment of Richard Brookes's *Natural History* (1763), to which Goldsmith himself contributed all the introductions, remains the best commentary on *Animated Nature*. 'We could wish,' he lamented, 'that he [Brooks] had thrown more life and variety into his manner, and imitated those Painters who, to give their pieces greater force, throw all their animals into action'.[126] Goldsmith had certainly ensured that 'the art of contrast' figured prominently in his narrative and for good reason: when he described animals in *Animated Nature*, he sentimentalised them by pervading his book with rhetorical personification, authorial sensibility, and human sympathy, especially when strong animals prey upon weak ones.[127] And he did so with prudence and taste in mind: by 1774 the cults of natural history abounded, as did books (by the dozens) on the subject. Anyone then could write about natural kingdoms, but only a few could capture everyone's attention.

Books about natural history shared a vast readership eager to learn about living creatures in nearby and remote places. This accounts for the eagerness of booksellers to enlist the best authors to write on the subject. Andrew Millar had paid Fielding the 'huge fortune' of £600 for exclusive rights to publish *Tom Jones*, and as recently as 1771 – only a few years before *Animated Nature* appeared – Smollett's publisher offered £210 for *Humphry Clinker*; but William Griffin paid Goldsmith £840 for *Animated Nature*, a very large sum by any measurement. Natural history was the universal British pastime, if not the national sport, by mid-century: Why else would the influential *Critical Review* claim in 1763 that 'Natural History is now, by a kind of *national establishment* . . . the favourite study of the time', and a few years later, in 1785, its rival, the *Monthly*, ask why 'all nations of refinement especially, should forward this species of knowledge [natural history] in their own tongue'?[128] Such an abundance of similar pronouncements is found that no reason exists to doubt their collective validity. Yet if the cults of natural history were expanding, so too was the 'book craze', as one commentator noted. As early as 1720 the naturalist William Sherard wrote to Richard Richardson, the Yorkshire scientist, that 'Natural History of all sorts is much in demand' (he meant 'the demand for books'), and in 1746 the Quaker merchant and naturalist Peter Collinson commented in a letter to Linnaeus that 'we [English] are very fond of all branches of Natural History; they sell the *best of any books* in England'.[129] This was a pre-Buffonian 1747; by 1760 or 1770 the public was caught, the time was ready for bestsellers among a readership thirsting for them: books almost everyone could understand and afford, not merely finely illustrated coffee-table productions for the fashionable and the rich. A steady 'progress' of natural-history books had appeared: in the 1740s George Edwards, one of the most gifted naturalists of the time, brought out a four-volume *History of Birds* that was sound and sold well,[130] and two decades later Thomas Pennant's *British Zoology* appeared, which the poet Gray owned and regularly consulted and which Goldsmith often read while writing *Animated Nature*.[131] In this continuum the trend was towards reasonably priced books written in a language and style the educated layman could understand.[132] This lineage culminated in such masterpieces of the mode as Erasmus Darwin's *Botanic Garden* (1789) and *Zoonomia: or the Laws of Organic Life* (1794–6), and Gilbert White's *Natural History of Selborne* (1789). Procrustean as it seems, it is nevertheless true to claim that by 1770 or so the natural-history book outdistanced all other science books, and did so irrespective of the reader's social class. The wealthy adorned their Chippendale tables with John Hill's magnificently illustrated *Vegetable System*, a 26-volume set that is aesthetically more pleasing than it is scientifically accurate; from 1759 onwards it usually appeared in biennial instalments at the price of £11 1s 6d per volume.[133] Those in less fortunate circumstances could buy Richard Brooke's set of *Natural History* in six volumes, which Newbery brought out cheaply in 1763 for only 18s;[134] those in the middle economic bracket could afford Goldsmith's set, though they might also read Buffon. And the booksellers and printers were not worried that after 1774 Buffon would spend another 14 years (1774–88) dressed in full-court costume composing 'natural history' eight hours every day and writing a total of 44 volumes![135] So secure had the English 'trade' in natural-history books grown. Immense as was Buffon's popularity in England, it was still *Animated Nature*, not Buffon's books, that the British public read and bought. There were, of course, other competing books: dozens, even hundreds, of books too long to list here,[136] but even if they were discussed one strain would be salient: the idea that natural history finally prevailed on the bestseller market. Goldsmith and Buffon shared this market, and to a lesser extent certain hacks, though neither writer was scientifically solid and though neither received complete academic approbation from his contemporaries.[137]

Although no other bookish 'sport' could compete with 'the natural history craze', a few tried. Marjorie Hope Nicolson has traced the growth of popular interest in books about the microscope and telescope in the Restoration period, and discovered how this enthusiasm shaped scientific activity then.[138] Books about microscopes and other instruments could not vie with books on natural philosophy and natural history, but the works of writers such as Henry Baker, author of the very popular *Microscope Made Easy* (1742) and *Employment for the Microscope* (1753),[139] and Benjamin Martin, the prolific instrument-maker who delivered lectures that required handbooks for attendance,[140] were runners-up. The reason for widespread enthusiasm about these books is not difficult to fathom: as natural philosophy and, later on, natural history made their impact on the layman's imagination, the notion spread that the naked eye could perceive the intricacies and fine details of 'Nature's kingdom' only with the aid of instruments such as the microscope and telescope. There was, of course, an ancillary and more abstract reason that Marjorie Hope Nicolson has isolated in connection with the ancients-moderns controversy:[141] that 'the moderns' paraded their interest in the new science by flaunting their mechanical instruments. But this penchant for modernity among scholars could not vie with the common man's need to see Nature's kingdom with his own eye. A similar state of affairs existed for books about other types of instruments and clocks. The search for the longitude, for example, had been important at least since 1676 when the first Astronomer Royal, Sir John Flamsteed, moved into Greenwich; after 1707, discovery of the longitude grew more pressing than it had been for the previous quarter of a century: ignorant of their longitude, a flotilla of nearly 2,000 British sailors ran aground in the Scilly Islands and was destroyed with heavy loss of life. The government became panic-stricken – especially in view of the importance of the navy to England's global might – and offered maximum rewards of £20,000 for proposals that would lead to discovery.[142] Schemes were devised and published,[143] and by the 1720s readers could find them in all sorts of libraries and purchase them at bookstalls, but books about the longitude never approached the number of books in natural history. The market for printed material about marine instruments was steady but small; perhaps the same was true of books about clocks, although these books were read by a more diverse audience. Early in the century William Derham's classic was popular and continued to be reprinted,[144] but clocks really came into their own *c.* 1760 when marine chronometers were being much discussed as the only instruments capable of accurately determining the longitude at sea.[145] All sorts of books about clocks appeared in this decade, books serious and staid as well as light and newfangled, books such as *The Clockmakers [sic] Outcry* (1760) which parodies *Tristram Shandy* and is not a 'scientific book' at all except in so far as it invokes clocks all the way through, and 'mechanical books' such as Benjamin Martin's *The Description and use of a Table Clock upon a New Construction, Going by a Weight Eight Days* (1770).[146] Technology in the home as well as on the road and at sea was then expanding; it was not yet, however, in a league to be compared with natural history, the 'Sunday pastime' of the British Isles.

SCIENCE BOOKS AND THE LITERARY IMAGINATION

The impact of so many concomitant scientific trends – philosophical, theological, technological, social – was reflected on the contemporary literary imagination. It could not have been otherwise. To the degree that this influence is attributable to science *books* as distinct from scientific *talk*, the literary man responded primarily in three ways: insofar as science books were shaping his general world view; in his impulse to collect these science books; and in the myriad ways in which he was invoking science in his own writing. The first – world view – is subtle and elusive, rarely perceived by writers on a conscious level, and even when perceived it is not always clear what different this angle of vision resulting in a *gestalt* makes to a writer's final text.[147] Most great English writers alive in the period from 1680 to 1740 were aware of the immense proliferation of science books. Certainly Swift and Pope and the Scriblerians were, and if they had been alive in the decade 1650–60, they might have compared this surge to the proliferation of books about alchemy and magic then. Furthermore, most of these English authors – not merely 'the greats' but lesser talents as well – took political positions and defended ideological stands, but certainly not all seem to have been aware of the degree to which science, and science books were shaping their *perception* of the whole world about them. Consciousness of alteration was probably more manifest on the level of collecting: whereas 'the ancients' had been collectable in 1680, now, in 1740, the party of 'the moderns' was winning on library shelves. Documentation of this phenomenon is not impossible for the assiduous researcher: abundant catalogues are extant for seventeenth- as well as eighteenth-century writers, and types of titles and numbers of volumes on shelves can be counted. The appendix (p. 000) provides some indication of the distribution of science books among the major English writers of the period, and it is important to notice how many different scientific books appear. No comparable pattern exists for English writers in the same period of the two previous centuries: 1480–1540 and 1580–1640. The last evidence of influence – explicit invocation of science – is, of course, a more uncertain terrain but certainly not a no-man's land. Donald Davie is on firm ground when he challenges Mallarmé and Professor R. L. Brett for objecting to early eighteenth-century poetry, for example, because this body of verse has paid *too much attention* to science and to the mechanical metaphors of a scientific world view.[148] What greater proof of evidence of influence could possibly exist? This is surely not the place to engage in polemics about the merits of eighteenth-century poetry except to notice that Davie – a consummate poet himself – has not been the only apologist for this body of literature. Other champions include Auden and Eliot and scholars such as Marjorie Nicolson, Maynard Mack, G. S. Fraser, I. A. Richards, Reuben Brower, and dozens of others who agree that such scientific metaphors and scientific language strengthened rather than weakened it.[149] But these metaphors and tropes would never have been invoked in the first place if science books had not proliferated in unprecedented numbers. Cause and effect are relatively clear in this instance, and it does not matter if some poets assimilated their knowledge from newspapers.[150] What counts for more is the undeniable fact that the largest upswing in the publishing history of scientific books had occurred.

Evidences of this fact were apparent everywhere, in high society as well as low, certainly throughout the adult world. Two instances – one among the rich, the other the young – may suffice to show how widely disseminated science books were becoming. Aristocratic librarians had collected some technical 'science books' throughout the Middle Ages, and continued throughout the Renaissance. The Knyvet brothers – Thomas and Sir Henry – collected so many scientific books at Ashwellthorpe in Norfolk at the end of the sixteenth century that 'the medical books alone occupy ten folio pages of the list'.[151] But the Knyvet brothers were exceptional: even in the next century one would expect to find some science books in every aristocratic library, yet these were rarely a large percentage of the whole, and what is more, it was uncommon to discover a large collection of current science books. In the eighteenth century this trend changed, and the librarians of such great Whig lords as the dukes of Devonshire, Newcastle, and Norfolk ardently bought up science books. No one to my knowledge has systematically studied the percentage of science books in aristocratic libraries 1500–1800, but it is fair to assume that such a survey would reveal a very marked increase for the eighteenth century.[152] Science books for children show an equally amazing increase in our period. When L. F. Gedike, a German schoolmaster who professed to have visited the Leipzig book fair 'every year of his adult life', returned home in 1787, he noted that 'no one other form of literary manufactor is so active as [scientific] book-making for young people of all grades and classes . . . a flooding tide [of] . . . geography for children, history for children, physics for children'.[153] Many of the books Gedike saw were written in English, and the British were publishing and exporting them as a consequence of the revolution in childhood then taking place, as well as an unprecedented craving for information among the young. All types of fables and allegories – by Aesop, Perrault, and others – were available for children in the Restoration period but 'science books', not matter how simplified, were almost impossible to find. Yet a notion persists among some historians of education that these books continued to be scarce until the late eighteenth century, a generalistion the facts will not support except for a few isolated cases such as military schools and remote colleges in outlying provinces.

As early as 1710 such a book appeared, published both in English and French: *A Short and Easie Method to Give Children an Idea or True Notion of Celestrial and Terrestial Beings.* Containing 38 plates reproducing many figures drawn from the heavens, it combined the educational principles of Comenius and Locke while attempting to inculcate both French and English in order 'to teach the arts and sciences, plants, fruits, and living creatures'.[154] Despite an emphasis on terrestrial as distinct from celestial life, these early books for children were spurred by a hunger for knowledge about astronomy. This may be why Isaac Watts, such a popular voice a decade later, would compose his very first science book for 'young adults' and call it *The Knowledge of the Heavens and the Earth Made Easy, or the First Principles of Geography and Astronomy Explained*, which appeared during the same year as *Gulliver's Travels* and which commented upon 'the heavens and the earth' in remote places, albeit not in Swiftian manner. By the 1730s Thomas Boreman initiated the 'publishing industry' that would make him a rich man: the printing a natural-history books for children.[155] This progression from natural philosophy to natural history paralleled the movement in 'adult science', and was a development Boreman exploited to its utmost. A decade later, in the 1740s, John Newbery, the printer about whom so much has been written and so little is yet known, began to issue *The Circles of Sciences . . . in Ten Volumes for Children*, books for children that went through many editions over the next three decades and which opened up the plentitudinous world of science.[156] Dedicating one volume to 'one science', Newbery published each book in octavo and each was written in the simplest language imaginable to educate the beginning reader. For three more decades he and his firm continued to write and publish children's books such as *Tom Telescope's Newtonian System of Philosophy*, being 'the substance of six lectures read to the Lilliputian Society', which first appeared as *The Philosophy of Tops and Balls*.[157] Populated by intriguing concantenations such as 'Master Telescope', 'Countess of Twilight' and 'Duke of Galazy', the book, which bears so many Goldsmithian traces that it could have been written by him, reveals the heavens to 'the little gentry' just as Fontenelle's dazzling Marchioness had opened them up to an adult audience in *A Plurality of Worlds*. This is not the place to explain how 'the aesthetics of the infinite' were communicated to children in the eighteenth century or the proper moment for a detailed history of science books for children, but such a survey would no doubt show how these books increased and how their quality improved with every decade.[158] Women such as Anna Barbauld and Sarah Trimmer wrote these books as did their male counterparts,[159] thereby paving the way for the Jane Marcets who would capture 'the hunger for science among children' in the next century.[160] Every decade brought new titles, the best reprinted over and again in new editions: in the 1780s John Aikin's *Calendar of Nature*[161] and the anonymous *Jack Dandy's Delight*; in the 1790s several English versions of Arnaud Berquin's *Looking-Glass for the Mind* – the popular English translation of a 'science books' for children written by a master of the genre who had been influenced by Rousseau's theories in *Emile*; at the turn of the century

Jeremiah Joyce's *Scientific Dialogues . . . in Six Volumes for Children*, a more serious science book than any of these others which covered the main branches: mechanics, astronomy, hydrostatics, pneumatics, optics, magnetism, electricity and galvanism.[162] If this survey had been extended beyond 1800, the trend towards better and cheaper books would be readily evident.[163] The fortunes of the book trade were not surprising in view the amazing growth of eighteenth-century science, conceived as activity, knowledge, and ideology. No matter what one's social class, every literate man, woman and child had thirsted by 1800 for knowledge about the natural world which the booksellers in England were quenching as Napoleon conquered Europe.

SCIENCE IN GRUB STREET

'The buzzing tribe', the class of booksellers, printers and publishers denounced by Pope in *The Dunciad*, was altogether aware of this new market and ready to exploit it greedily. Yet before the 'dunces' – usually 'depressed' writers – could compile saleable anthologies and compilations of 'science', the booksellers had to be ready and willing to print these books. The cause and effect of this state of affairs, and the internal structure of these 'publishing houses', are matters few scholars have been willing to tackle because so few actual facts are known. It may be that the right questions have not even been put as yet. The questions that fall into the domain of science books are fortunately specific and capable of partial solutions, although much more research will be needed before authoritative generalisations can be made.

Certainly by mid-century, the houses of Davis, Dilly, Hodges, Innys – a very large house that published many of the works of Newton and Boerhaave – Millar, Murray, Nourse, Rivington, and Robinson were printing large numbers of science books, large, that is, in relation to the number printed before *c*. 1720 and as a part of the total number of published titles. John Nourse's publishing establishment is especially germane here. He was one of the first scientific booksellers in England, specialising in popular treatises on mathematics, and mathematically-based books from hack writers who produced pot-boilers partly based on learned treatises and on the theories of others, just as Goldsmith did with natural history. But unlike Davis, Millar and Rivington, Nourse paid his hacks well, and he also cornered a whole segment of the technical-books market, as a consequence of which his enterprise was profitable.[164] On the other hand, equally large and prominent houses – Dodsley and Lintot – published so few science books that it may be said that certain houses were 'science houses' and other not. Why, then, were certain houses enthusiastic, and others most unreceptive? Plomer provides no clue, nor do more specialised recent studies.[165] The direction most houses took may reflect the interests of their proprietors, or it may have been based on chance: a book falling into the bookseller's lap, as it were. For every William Innys who cared about the new science, who had genuine interest in medicine and who had actually sought out Boerhaave,[166] dozens of other booksellers ranging from the notorious Curll to Pope's *bête noire*, Lintot of *Dunciad* fame, could not be persuaded unless there were assurance of a quick pound. Given the enthusiastic reception of natural philosophy in the early eighteenth century and the number and types of science books available by 1730 or 1740, it is reasonable to wonder why houses devoted exclusively to science books did not spring up. Could they not have made a good living from these books, perhaps printing as well as a few other types of titles? The questions, however, may be poor – may suffer from 'telescopic history' that glances backwards. Furthermore, so many documents (contracts, ledgers, bank accounts) have disappeared that it may not be possible to retrieve answers. In almost every specific case, only tentative generalisations are possible because of the disappearance of these documents. For example, Innys had published the first English translation of *Boerhaave's Aphorisms* in 1715, and continued to reissue it in 1722, 1724, and 1728. Sometime after 1728 Bettesworth bought the copyright from Innys and brought it out in 1735 in the same version.[167] Why did Innys sell? Purchase of such copyright from Innys is not commonplace in his period, and whereas it is impossible to learn how many copies Innys had sold in the four editions, the book itself must have ben among the dozen or so most popular science titles of the time. Can Bettesworth, who published few science books, have hoped to expand his own science titles and establish himself as something of a 'science bookseller'? Was Innys reaping small returns on the *Aphorisms* and only too glad to sell out to Bettesworth on London Bridge or any other bookseller? It is tempting, but risky, to hazardous guesses.

John Hill's publishing manoeuvres also pose a challenge to the student of eighteenth-century science books, although they shed much light on the history of publishing, because Hill catered to such a different reading audience from Boerhaave's and because Hill, unlike Boerhaave, was such a prolific popular-science writer.[168] So far as can be learned, Hill published 96 books with 29 different publishers during his lifetime.[169] The house of Baldwin at 47 Paternoster Row printed 25 of these titles, with Hill often holding total or partial copyright; Thomas and Mary Cooper printed nine titles, as did Thomas Osborne, and C. Davis five. All were well-known and prolific houses at mid-century. Yet the firm of Baldwin did not specialise in our modern sense – in science books, and published few scientific titles other than Hill's. The Baldwins primarily printed political literature, especially in pamphlets, and *c*. 1750 did not have the copyright for the works of any acclaimed scientists. Why, then, did Baldwin, as distinct from any number of other large, leading houses take on a quarter of Hill's books? And why did Hill, having established such rapport and effective arrangements with the Baldwins, offer the other three-quarters to other houses? Hill's correspondence reveals no particular friendship with any of these publishers,[170] and there is no value in speculating about Hill's reasons for approaching one bookseller over another. If this avenue of exploration were extended to many London booksellers, it would be evident that at mid-century there was no such publishing organism as a 'science house'. Until further evidence is brought to light, it is probably more accurate to believe that all houses considered suggested titles on the basis of economic (and sometimes political) success, and that the 'science list' in our current nomenclature – was not yet a category in which the booksellers and publishers cared or hoped to specialise.

The structure of prices is another matter, although less irksome to sort out and settle. During the sixteenth and seventeenth century the book trade was flooded with cheap handbooks on English science, hastily compiled by greedy booksellers eager to tap a steadily grown market. This state of affairs has deftly been summarised by R. F. Johnson: 'Most of these books were, like [Thomas] Hill's.[171] translations and compilations; but they were poorly printed, written in atrocious English style, and translated with so little regard for accuracy and clarity that no reliable knowledge – scientific or pseudo-scientific – could be derived form them.'[172] These books, moreover, were cheap, perhaps all selling for under a shilling, and were written by incompetent hacks who knew little if anything about science. After the establishment in the Restoration period of the Royal Society and the new spur given to natural philosophy, more knowledgeable writers of science began to compile these books, and their prices steadily rose. By 1720 a Londoner eager to own a science book would have been fortunate to discover one for under three or four shillings.[173] Yet 'as late as the third quarter of [seventeenth] century £12 a year were still thought sufficient for the year's pay of a schoolteachers, who of all others might be expected to show an interest in books',[174] and the average earning power did not climb so much in the next 50 years (1675–1725) to invalidate the point about the cost of science books in relation to annual earnings.[175] By 1730 or 1740 the disparity between the cost of a work of fiction, poetry or imaginative prose – a book of 'literature' – and a science book, whether theoretical or applied, was considerable. Whereas readers eager to purchase books could choose from any number of literary, historical or theological titles for a couple of shillings, most science books averaged from four to ten shillings and there were many that were much more expensive.[176] Dodsley's science books were cheaper than most others, yet even Dodsley offered no science book, other than the odd pamphlet on herbal medicine or a 'penny cure' for venereal disease, for less than two shillings.[177] When Dodsley brought out in translation Anthony Cocchi's popular *Pythagorean Diet* (1745), he charged 1s 6d, and Robert Douglas's more theoretical *Heat in Animals* (1747), a work 'of natural philosophy', cost 2s 6d: in relation to other science books these were inexpensive. The reader who wanted to purchase Rice Charleton's *Bath Waters* (1754) – a balneological work about the medicinal value of spas – could have bought it in 1754 for 1s 6d, but a year earlier would have had to pay 6s for Thomas Gataker's *Operations on Venereal Complaints* (1753; second edition) and 18s – almost a tenth of his annual salary if he had been a schoolteacher then – for Robert Hooke's *Micrographia* (1745).[178] After mid-century these books were gradually reduced in price, and by the 1770s and 1780s literary and science books cost roughly the same amount. While the precise reasons for this trend are not clear, they must have depended, in whatever degree, on the demand for these books as well as the copyright amount or other advance royalty paid to authors. So far as can be learned, authors were paid smaller sums to write and compile science books than were literary writers, and this was as true for renowned authors as well as unknowns in Grub Street. Joseph Johnson, the late eighteenth-century bookseller and publisher, issued over 2,700 imprints in the 48 active years of his career, many of which were scientific titles, including almost all the works of Joseph Priestly and Erasmus Darwin, and William Cowper's *Poems*, yet he said rather small sums to his science writers: only £40 to George Walker for a *Treatise on the Sphere* (1777)[179] and £150 – the largest amount I can discover – to William Nicholson for an *Introduction to Natural Philosophy* (1782),[180] an elementary science textbook that soon superseded John Rowning's *System of Natural Philosophy* (1735). Expectation of sales, the number of actual buyers, the price paid to the author – all these influenced the publishers of science books and contributed to the way in which the new demand for science books was altering the eighteenth-century book trade.[181]

Libraries of every type – circulating, non-circulating, proprietary, community, school, university, cathedral, and other public types – were the natural resort of those who could not afford to purchase science books at these prices. Two questions about these libraries especially need to be asked: first, precisely which titles and about how many science books could a borrower expect to find in a library in our period, and secondly, do the lists of borrowings of these libraries afford any clue about the popularity of certain titles over others? The answers are influenced by the fact that the nature and function of these libraries have only been distinguished in the last few decades and even the most recent research does not permit altogether satisfactory answers.[182] What is nevertheless clear is that all libraries irrespective of type contained *some* science books even at the very beginning of the century, and that some time in the decade of the 1770s these books (calculated as a percentage of the total number) dramatically began to increase.[183] Furthermore, even in their primitive way borrowing records demonstrate that although there was a steady demand for science books throughout the century, it was small in comparison to borrowings in history, *belles lettres* and travel until the 1770s; after this time, science books gradually began to outstrip others, theology and philosophy notwithstanding, and by the end of the century science (all branches) became the top competitor of literature and travel.[184] The materials thus far brought to light also indicate that these trends are as valid for Scotland and Ireland as for England and they apply, moreover, to provincial libraries as well as urban.[185] According to most historians, the eighteenth century marked the rise of 'book clubs'; yet when book-clubs lists are scrutinised it becomes patent that the trends just delineated for public and commercial libraries apply to them as well.[186] What is surprising is the extensiveness of science books in the small circulating libraries such as Philpotts's in Bristol: here the borrower would expect to discover mostly works of non-fiction, in accord with the general taste of the provincial reading public, but would he think to find in such a small row of books Buffon's *Natural History*, Goldsmith's *Animated Nature*, Patrick Boyne's *Tour Through Sicily and Malta* – a work containing serious scientific discussion of the antiquity of the crust of the earth – and Sir William Hamilton's geological *Observations on Mount Vesuvius*?[187]

The evidence is not yet sufficient to reply with any degree of certainty, for to answer in the affirmative may seem the deck in favour of science, and it may appear to some that the deck has already been sorted too far. Nor is the evidence positive enough to know if urban borrowers demanded more science titles than provincial, or if certain regions of the realm varied in their demand for this type of non-fiction; but one thing is clear: by the 1780s imaginative literature and history (including travel) had to share the limelight with all manner of science books, including the sciences of the globe as well as the even newer sciences of man. This fact in itself ought to reveal something significant about eighteenth-century intellectual pluralism: the way that knowledge was in a ferment and rapidly exploding; and it demonstrates beyond any shadow of doubt how the 'new science', as Boyle and his contemporaries called it a century earlier, was now directly 'translating' into the public's actual reading taste in its thirst for knowledge about the practical and applied – even the technological – aspects of theoretical science. Perhaps this is why the first catalogue of the London Library – the precursor of the present library – was prefaced with these words: 'It is intended that the London Library shall contain all those great works in *science and literature*, which it is difficult for individuals to procure . . .'[188] The charter of a new lending library formed in the

1680s, a century earlier, might have contained the same words, but the words themselves would have been gratuitous inasmuch as few of the borrowers would have possessed enough literacy to understand even the simple science books. There was nothing arbitrary 1772 about the pronouncement of Dr James Ramsay, the Enlightenment philanthropist who was so important in the abolition of slavery: 'Natural History is, at present, the favourite science all over Europe, and the progress that has been made in it will distinguish and characterise the eighteenth century in the annals of literature in the future.'[189]

Natural history had eventually carved that niche, yet less is known about the tides of reading tastes in natural history than is thought. As Paul Kaufman has written, 'few important areas of literary history, in the broad sense, remain so nebulous and undeveloped as the description of reading vogues',[190] and these vogues encompass science books as well as literary. If, then, any single direction for the future study of books in the eighteenth century emerges from this essay, it is the author's desideratum that these 'reading vogues' will eventually be studied in and of themselves, for they would no doubt shed much light on everything herein discussed. I suspect, furthermore, that a detailed exploration of eighteenth-century medical books would reveal many of the same patterns and trends noticed above. But the precise curve of the development of medical books – its intellectual domain as well as it social exigencies and reading vogues – must be told elsewhere.

APPENDIX: SCIENCE BOOKS IN THE LIBRARIES OF SOME EIGHTEENTH-CENTURY AUTHORS

The catalogue that follows provides information about libraries of a select group of prominent figures in Restoration and eighteenth-century England: Addison, Burke, Congreve, Defoe, Gibbon, Goldsmith, Johnson, Newton, Pope, Sterne and Swift. While the number of figures is necessarily small owing to limitations of space, there is no reason to believe that a larger sample would change the essential nature of the patterns found below or, indeed, any conclusions that may be drawn from this tabulation. Newton is included because recent scholars have shown that he was far more prominent in literary capacities than was previously thought, and also on the assumption that the contents of his admittedly large library would point to important contrasts with the holdings of the purely 'literary' figures of the time.

The uses here of the terms 'science' and 'scientific' are admittedly problematic, but if the above essay has shown anything it has demonstrated that no satisfactory definition of a 'science book' can be made, precisely because the age itself did not draw fine distinctions between 'science' and 'non-science' – or between 'science' and 'literature' – upon which subsequent centuries seem to have insisted. This definitional failure of the eighteenth-century notwithstanding, I have applied the same rule here that I applied in the essay, and have included – among other types of 'natural philosophy' – eighteenth-century books dealing with natural history, geography, and anthropology.

Furthermore, I have included pre-1700 scientific authors in the survey because they provide an excellent context in which to interpret the scientific libraries of these literary writers. By including ancient, medieval and Renaissance scientific books, one can see at a glance the prevalence of eighteenth-century scientists in the library of an individual writer; and students working on a particular author – whether Swift or Pope or Johnson – will eventually be grateful for learning, for example, that the works of Galen are to be found in only a few of the libraries studied here whereas the books of Thomas Burnet, the author of *The Sacred Theory of the Earth*, are found in almost every writer's library. Such information is often obscure and not easily gathered, but it is quickly found in the chart below. However, comparisons of this type would be impossible if the compilation had been limited to eighteenth-century scientific writers. The same reasons explain why the list of scientific authors is not restricted to English nationality.

Conclusions to be drawn from this compilation must await study elsewhere; but certain trends are already apparent, and their identification is essential for understanding the literary writer involved, as well as the particular scientist. For example, it is now patent that a large part of Sterne's library is devoted to medical books – especially French medical books – with a physiological bent, and that the same Sterne who wrote so brilliantly in *Tristram Shandy* about 'Tristram *Trismegistus*' never owned a copy of any of the works of 'Hermes Trismegistus' whose books were so often reprinted in Sterne's lifetime; that Luigi Cornaro's works – whose books on health underwent so many reprints and in so many languages in the eighteenth century – are found only in Newton's library whereas the writings of Dr George Cheyne, who is discussed so prominently in the above essay, are found in many libraries; that not one of our writers had a book by John Freke, the scientific author referred to many times in *Tom Jones*; that Johnson's library, which contained many science books, had no copy of the works of William Cadogan, the populariser of children's care, although Johnson's is the only library with any of Linnaeus's books, and that Defoe's library is the only one to possess any works by Athanasius Kircher whose *Mundus Subterraneus* inflamed Coleridge's imagination; that only Newton owned copies of the books of his most ardent populariser, J. T. Desaguliers; that the best-known writings of the Newtonians were widely represented in the libraries of literary men, and, finally, that Defoe had many science books whereas Pope owned few. These and many other interpretations will be drawn from the compilation.

Nevertheless, the limitations of such a list must not be minimised or neglected. The mere fact that a book appears in a library does not mean the owner has read the book, or even that he is aware of its contents. Moreover, the lack of information here about the numbers of these science books in relation to other types of books in any individual library, and the further lack of discussion about the circumstances in which the owner obtained the book, will inevitably raise questions about the validity of such an approach. But it has not been my purpose here to interpret the compilation or draw significant conclusions from its patterns. I shall be happy if others engage in that work.

Finally, a word about the sources – primary and secondary – upon which I have drawn. I have used sales catalogues, both published and unpublished, whenever possible, and in cases in which no authoritative sales catalogue was available I have attempted to verify that the books listed here were actually in the library of the owner. Such verification, however, has not been possible in every case, and questions are bound to be raised about the actual appearance of the book in a particular library. I have also drawn upon secondary studies of individual literary figures and their reading.

In the format of the compilation, the number in parentheses indicates the number of different titles by the scientific writer; i.e. Cheyne (4) means that the owner of the library had four different titles by Cheyne in his library. Duplicate copies of the same book are not listed. Thus an owner whose library contained three copies of Cheyne's *English Malady* and no other books by Cheyne is listed as owning merely 'Cheyne' – that is, one book only. An author's first name is given only if the Christian name is necessary to distinguish him or her from other scientific authors with the same name.

Acosta, J. de Newton
Adams, George Gibbon
Adanson, M. Gibbon
Aelianus, C. Gibbon (3)
Agricola, Georgius Johnson, Newton (3)
Albineus, Nathan Newton
Albinus, B. S. Gibbon (2)
Aldrovandus Goldsmith
Alimari Newton
Alipili Newton
Allen Sterne
Alpinus, Prosperus Defoe, Johnson

Amatus Newton
Anderson, Robert Newton (2)
Anville, J. B. Gibbon (12)
Apollonius Pergaeus Defoe
Apperley, Thomas Johnson
Arbuthnot, John Congreve, Gibbon, Newton (2), Sterne, Swift
Archimedes Defoe, Johnson, Newton (2)
Aretaeus of Cappadocia Johnson (2)
Astruc, Jean Gibbon, Goldsmith, Johnson, Sterne

Aurelianus, Caelius Gibbon
Avicenna Swift

Bacon, F. Addison, Goldsmith, Newton, Pope, Sterne, Swift
Baglivi, G. Defoe
Bailly, J. S. Gibbon (6)
Baker, Henry Gibbon (2)
Baker, Thomas Newton
Balam, Richard Newton
Barba, Alvardo Newton (2)
Barrow, Isaac Johnson (4), Newton (2)
Bates, T. Congreve
Bauhinus, C. Defoe
Bayle, P. Sterne
Baylies, W. Sterne
Béardé de L'Abbaye Johnson
Beaumont, J. Swift
Becher, Johann Johnson, Newton
Bedford, Arthur Gibbon (2)
Bellini, L. Defoe
Belon, P. Gibbon
Belot, Jean Newton
Bernoulli, Johann Newton (4)
Bidloo, Govard Newton (2)
Birch, Thomas Johnson
Birkenhout, J. Goldsmith
Bizot, Pierre Newton
Blackmore, R Defoe (2) Sterne
Blair, Patrick Newton
Blanckaert, S Defoe
Blegny, N. de Defoe
Bochart, S. Defoe
Boerhaave, H. Johnson (4), Newton (3), Sterne
Bolletti, G. G. Gibbon
Bonet, Théophile Sterne
Bonnet, Charles Gibbon
Borel, Pierre Newton
Bordelli, Giovanni Johnson, Newton (3)
Borrichivs, Olaus Newton (2)
Bossche, G. Defoe
Bowles, William Gibbon
Boyel, R. Addison, Burke, Defoe, Gibbon (3), Goldsmith, Johnson (5), Newton (22), Pope, Sterne, Swift

Bradley, Richard Defoe, Newton (4)
Brahe, T. Defoe
Brice, A. Sterne
Briet, P. Gibbon (2)
Brookes, R. Goldsmith, Sterne
Brown, G. Defoe
Brown, J. Defoe
Browne, John Newton
Browne, Thomas Congreve
Bruno, J. P. Defoe
Buchan, William Gibbon
Buffon, G. L. Burke, Gibbon (8), Goldsmith
Burnet, Thomas Addison, Congreve, Defoe, Gibbon (2), Johnson, Newton (4), Sterne, Swift
Burton, Robert Johnson, Sterne
Buesching, A. F. Gibbon (13)

Cadogan, William Gibbon
Caesalpinus, Andreas Johnson
Caesivs, Bernardus Newton
Caius, John Defoe, Johnson
Cardanus, H. Defoe, Johnson (10), Sterne, Swift
Carré, L. Defoe
Cawood, Francis Newton
Celsus, Aurelius Congreve, Defoe, Johnson, Newton, Swift
Chauvin, E. Defoe
Cheselden, William Johnson, Sterne
Cheyne, George Defoe (4), Johnson, Newton (4), Sterne (3)
Clarke, James Newton (2)
Cockburn, William Defoe
Cooper, William Newton
Cornaro, Luigi Newton
Cowper, William Newton
Croker, T. H. Goldsmith, Sterne
Cronstedt Johnson
Crousaz, J. P. de Defoe
Culpeper, N. Congreve
Dale, Samuel Congreve
Dale, Thomas Newton
Davisson, William Newton

Derham, William Goldsmith, Sterne
Desaguliers, J. T. Newton (2)
Descartes, R. Defoe, Newton (7)
Diemerbroeck, I. de Congreve, Defoe
Digby, Sir K. Goldsmith, Newton
Dionis, P. Sterne
Ditton, Humphrey Newton
Douglas, James Newton, Sterne
Dover, T. Sterne
Drake, J. Sterne
Du Chesne, Joseph Newton (3)
Duhamel, J. B. Johnson (2), Newton

Edwards, George Goldsmith (4)
Eschuid, Joanne Swift
Eschuid, Joannes Swift
Euclid Burke, Defoe, Gibbon (3), Newton (6), Sterne

Fabri, Honoratus Newton (2)
Fabricius, Hieronymus Johnson (2)
Fallopius, G. Johnson
Fatio de Duillier, N. Newton (3)
Ferguson, James Gibbon, Johnson (3), Sterne
Fermin Goldsmith
Fernel Swift
Fevillée, Louis Newton
Fida, Abu al Gibbon (2)
Flamsteed, John Newton
Floyer Sterne
Fontenelle, Bernard Addison, Congreve, Sterne
Foster, Samuel Newton
Fracastaro, Girolamo Congreve
François Marie, Père Newton
Freind, John Defoe, Newton
Frénicle de Bessy, B. Newton
Fréret, N. Gibbon
Fromondus Newton
Fuller, Thomas Newton, Sterne

Gale, T. Defoe, Sterne
Galen Burke, Defoe, Johnson (2), Swift
Galileo Newton

Gassendi Congreve, Defoe, Johnson, Newton
Gautruche Newton
Geer, Carl de Goldsmith
Gellibrand Newton
Geoffroy, E. F. Sterne
Gerard, J. Johnson, Sterne
Gerhard, Johanne, M. D. Newton
Gesner, J. Gibbon, Goldsmith, Johnson (3)
Gibson, T. Swift
Gilbert, William Johnson
Glaser, Christophe Newton
Godfrey, Ambrose Newton
Goodall, Charles Defoe
Gordon, George Newton (2)
Gouan Goldsmith
Graaf Defoe
Grandi, Guido Newton (4)
Gravesande Defoe, Gibbon (4), Johnson, Newton (9)
Greaves, John Defoe (2), Gibbon (2), Johnson (2), Newton (2)
Gregg, Hugh Newton
Gregory, David Newton (4)
Gregory, James Newton (4)
Gregory, John Sterne
Grew, N. Defoe
Guidott Defoe
Gumilla, J. Gibbon (3)
Gunter, E. Newton

Hales, Stephen Johnson (2)
Haller Gibbon (15), Johnson
Halley Newton (3)
Hamilton, Sir William Gibbon (2)
Harris, John Defoe (2), Newton (2)
Harris, Walter Defoe
Hartley, David Johnson (2)
Hartlib Newton
Hartmann, Johann Newton
Harvey, W. Burke
Hatton Johnson, Newton
Hauksbee, Francis Newton (3)
Heister, L. Defoe, Sterne
Helmont Johnson, Newton
Helsham Johnson
Hermann, Jacob Newton, Sterne

Hertzberg, E. F. Gibbon
Hill, John Gibbon (2), Goldsmith, Johnson (2), Sterne (2)
Hippocrates Burke, Congreve, Gibbon (3), Johnson, Swift
Hobbes Addison, Defoe, Newton, Sterne, Swift
Hoffmann, Friedrich Johnson, Sterne
Holbach Gibbon (2)
Hollandus, J. Newton
Hooke Defoe, Newton (4)
Horrocks, Jeremiah Newton
Huxam Johnson

James, Robert Johnson (4)
Jones, William Newton (2)
Johnston, John Newton

Kaempfer Gibbon
Keill, John Defoe, Gibbon (3), Newton (4), Sterne
Kemp, W Newton
Kendal Newton
Kennedy, Peter Defoe, Newton (2)
Kepler, J. Defoe (2)
Kersey, John Newton
Kircher Defoe (2)
Kirwan Gibbon
Klein, Jacob Goldsmith
Knowles, G. Defoe, Sterne
Koenig, Emanuel Newton

La Hire Defoe, Newton (2), Sterne
Lawrence, Thomas Johnson
Leadbetter Sterne
Le Clerc, Daniel Defoe, Johnson, Newton
Ledran Sterne
Lémery, Nicolas Defoe (2), Newton (2)
Lemort Newton
Leow, J. F. Newton
Lesser, F. C. Gibbon (2)
Lewis, William Johnson
Lhuyd, Edward Newton, Sterne
Lilly, William Congreve
Linnaeus Goldsmith (3)

Lister, Martin Defoe, Newton
Long, Roger Sterne
Lowe, Peter Sterne
Lull, R. Defoe

Maclaurin, Colin Gibbon, Newton (3), Sterne
Macquer Johnson (3)
Malcolm Johnson (3)
Manilius Congreve, Defoe, Gibbon (4), Johnson, Pope
Marggraf Johnson
Martin, Benjamin Sterne
Maubray Sterne (2)
Maupertius Johnson
Mayerne, T. T. de Defoe
Mead, Richard Gibbon (3), Johnson (2), Newton (2), Sterne
Melanchthon Sterne, Swift
Menuret de Chambaud Sterne
Michelotti, P. A. Newton
Miller, P. Sterne
Moffett, T. Defoe
Moivre Newton (7)
Molyneaux, W. Defoe
Monro, A Burke, Sterne
Moore, Jonas Newton (4)
Morgagni Newton
Morin, J. Swift
Moxon, J. Defoe, Newton, Swift (2)
Mudge, John Johnson (2)
Musschenbroek Johnson, Newton (3), Sterne
Mylius, J Newton
Mynsicht Newton

Newton Addison, Burke, Congreve, Defoe, Gibbon, Johnson (5), Newton (21), Pope, Sterne, Swift
Norwood Newton (2)

Oughtred Defoe
Owen, Charles Goldsmith

Palfyn Newton
Paracelsus Defoe, Johnson (3), Newton (5), Swift

Paré Defoe
Parkinson, John Newton
Paxton, P. Sterne
Payne, William Johnson (4)
Pemberton Congreve, Sterne
Pennant Gibbon (9), Goldsmith (4)
Petiver Newton (4)
Pitcairne, A. Defoe, Johnson, Newton
Place, E. Newton
Plat, H. Defoe
Pliny the Elder Congreve, Defoe (2), Gibbon (20), Goldsmith, Johnson (3), Sterne
Plot Newton
Pluche Goldsmith (4), Sterne (2)
Porterfield, R. Sterne
Pott, Percivall Johnson (3)
Pringle, John Sterne
Pysegur, A. M. J. Gibbon

Quincy, John Burke, Congreve

Randolph, G. Sterne
Ray Congreve, Gibbon (2), Goldsmith, Newton (2)
Réaumur Gibbon (20), Johnson
Ripley Newton (3)
Riverius, L. Johnson
Robinson, N. Defoe
Rohault, J. Gibbon, Newton, Sterne
Rudd, Thomas Newton
Ruysch Newton, Sterne
Rzaczynski, G Johnson

Salmon, William Congreve, Defoe, Johnson, Newton (2)
Sanctorius Congreve
Schelhamer(us) Johnson, Newton
Scultetus, J. Defoe
Sénac, J. B. Gibbon (2)
Sendivogius Newton (3)
Spallanzani Gibbon (4)

Sprat, Thomas Gibbon, Newton, Sterne
Stahl Johnson
Starkey Newton (8)
Stephens, J. Sterne
Stone, E. Newton
Strother Defoe
Stuart, Alexander Newton
Stukeley Sterne
Swammerdam Goldsmith, Sterne
Sydenham Congreve, Defoe (2), Sterne

Templeman, P. Sterne
Tolet, F. Defoe
Tournefort Goldsmith
Towne, R. Defoe
Trembley, A. Gibbon, Goldsmith
Tunstall Johnson
Turner, Daniel Sterne

Valsalva, A. M. Defoe
Verheyen Sterne
Vesalius Defoe, Johnson

Wallis, John Defoe (2), Newton (2)
Ward, Seth Newton
Watson, Richard Johnson (3)
Webster, John Newton
Whiston, William Addison, Defoe (2), Newton (5), Sterne
Whytt Burke
Wilkins, J. Congreve, Sterne
Willis, Thomas Congreve, Defoe, Johnson, Newton, Sterne
Willoughby, Francis Goldsmith
Wingate, Edmund Newton (3), Sterne
Winslow, J. B. Gibbon (4)
Wintringham, C. Sterne
Wolfus Johnson (11)
Woodward, John Johnson, Newton (3), Swift

NOTES

I am grateful to Dr Roger Hambridge, as well as Dr J. P. Feather and Dr I. Rivers, for commenting on this essay at various stages of its preparation, although I myself am responsible for any gaps the reader may discern. I must also thank Mrs Leila Brownfield for her unrelenting energy in ferreting out some of the obscure details in the appendix.

1 Quoted in E. L. Eisenstein, *The Printing Press as an Agent of Change* (1979), II, 691. Madame du Bocage's remark is found in her *Letters Concerning England*, 2 vols. (1770), I, 34, written while she was a visitor in London.

2 As evident in Richard Helsham, M.D., *A Course of Lectures in Natural Philosophy* (1743). See also A. Ferguson (ed.), *Natural Philosophy through the Eighteenth Century* (1972).

3 See R. K. Merton, 'Puritanism, pietism, and science', in *Social Theory and Social Structure* (New York, 1957), 574–606; idem, *Science, Technology and Society in Seventeeth Century England* (1970); C. Webster, *The Great Instauration* (1975); S. B. Barnes, *Scientific Knowledge and Sociological Theory* (1974); S. Shapin and S. B. Barnes (eds.), *Natural Order* (1979). For deism, see N. Torrey, *Voltaire and the English Diests* (New Haven, 1930) and P. Gay, *Deism: an Anthology* (Princeton, 1968).

4 C. Hill, *Puritanism and Revolution* (1958); idem, *Society and Puritanism in Pre-Revolutionary England* (1964); idem, 'Mechanic preachers and the mechanical philosophy', in *The World Turned Upside Down: Radical Ideas during the English Revolution* (1972), 287–205; K. Thomas, *Religion and the Decline of Magic* (1971); P. M. Mathias (ed.), *Science and Society 1600–1900* (1972); B. Easlea, *Witch Hunting, Magic and the New Philosophy* (1980).

5 S. Schaffer, 'Natural philosophy', in G. S. Rousseau and Roy Porter (eds.), *The Ferment of Knowledge: Studies in the Historiography of Eighteenth-Century Science* (1980), 55–92.

6 For example M. Foucault, *The Archaeology of Knowledge* (1972) and *The Order of Things* (1970); see also Schaffer, op. cit., who discusses this matter at some length.

7 See the discussion of 'Baconianism' in M. Purver, *The Royal Society: Concept and Creation* (Cambridge, 1967), 20–62, and R. Frank, *Harvey and the Oxford Physiologists* (Berkeley, 1981).

8 See W. Houghton, 'The English virtuoso in the seventeenth century', *Journal of the History of Ideas* III (1942), 51–73, 190–219; R. Porter, 'Gentlemen and geology: the emergence of scientific career, 1660–1920', *Historical Journal*, XXI (1978), 809–36; S. F. Cannon, *Science in Culture* (1978).

9 L. Stephens's *History of English Thought in the Eighteenth Century*, 2 vols. (1876) remains the classic study of deism in England during this period.

10 The common man in the period 1660–1720, so far as one can generalise, would hear the 'argument from the limitations of reason', in biblical, allegorical, moral and psychological versions, every Sunday from the pulpit and read about it in printed sermons.

11 See the subsequent two pages and J. M. Levine, *Dr Woodward's Shield: History, Science and Satire in Augustan England* (Los Angeles, 1977).

12 The extent of this awakening has been shown by M. H. Nicolson in *Pepys' Diary and the New Science* (Charlottesville, 1965); Evelyn, Dryden, Butler, Waller, Denham, Cowley, the Duchess of Newcastle and many other literary figures, male and female, are treated in addition to Pepys.

13 G. Sherburn (ed.), *The Correspondence of Alexander Pope*, 5 vols. (1956), II, 264. Pope had a great deal to say, of course, about this 'class'; see, for example, the *Dunciad*, IV, which expatiates on these *virtuosi*-scientists who 'Impale a Glow-worm, or Vertu profess, / Shine in the dignity of F. R. S.' (569–70).

14 See D. Stimson, *Scientists and Amateurs: a History of the Royal Society* (New York, 1948); L. S. Feuer, *The Scientific Intellectual* (New York, 1963); D. S. L. Cardwell, *The Organisation of Science in England* (rev. ed., 1972); M. Berman, ' "Hegemony" and the amateur tradition in British science', *Journal of Social History* VIII (1975), 30–50, who studies aspects of class-structure and fashion.

15 And even in histories of natural philosophy. Bishop Thomas Sprat – the historian of the Royal Society, high churchman, Bishop of Rochester and 'believer' in the new science of natural philosophy – explains in his *History of the Royal Society* (1667) how a new '*Systeme of Natural Philosophy*' (327) had been built up: 'one great Man [Bacon], who had the true Imagination of the whole extent of this Enterprize [the new science], as it is now set on foot' (35). By 'this Enterprize' Sprat means the Royal Society, as it had developed before 1667, but his monolithic attribution to one man of the whole development of 'new science' in the seventeenth century is of course no longer credited. Yet credit to Bacon thunders down throughout the eighteenth century in books about the development of the Royal Society and in biographies of Bacon; see D. Mallet, *The Life of Francis Bacon* (1740); T. Birch, *The Life of the Hon. Robert Boyle* (1744).

16 See Purver, op. cit.; the useful introduction by J. I. Cope and H. Whitmore Jones to *Sprat's History of the Royal Society* (St Louis, 1966); Stimson, op. cit.; H. Lyons, *The Royal Society, 1660–1940* (1944); Frank, op. cit.

17 One hundred years later Samuel Johnson selected the same motto as the epigraph for *The Rambler* (1752), almost ironically and without any of the resistance to language felt by the early Fellows of the Royal Society.

18 Thomas Hobbes, *Behemoth* (1679), 155, quoted in Purver, op. cit., 73.

19 For analysis of the authors of these papers, see H. Lyons, *The Royal Society* (1944), 126; M. Hunter, 'The social basis and changing fortunes of an early scientific institution: an analysis of the membership of the Royal Society, 1660–1685', *Notes and Records of the Royal Society of London*, XXXI (1976), 9–114; idem, *Science and Society in Restoration England* (1981).

20 The point is supported further by material gathered by P. Laslett in 'The foundation of the Royal Society and the medical profession in England', *British Medical Journal*, II (1960), 167. Sir G. Clark, *A History of the Royal College of Physicians of London*, 3 vols. (1962–71), II, 435–6, discusses the income of physicians and the style of life they enjoyed; so far as I am aware the libraries of physicians 1660–1760 had not been studied in any depth except for those of a few physicians of note, such as Richard Mead, who amassed about 30,000 volumes, and the surgeon John Hunter, whose vast library formed the basis of the Hunterian Collection in Glasgow University. See also P. G. M. Dickson, *The Financial Revolution in England* (1967), for some discussion of the incomes of physicians in relation to other incomes then, and J. F. Fulton, *The Great Medical Bibliographers* (1951), for a study of these physicians as collectors.

21 See the section on 'ladies' philosophy' below and, for a witty and early treatment of women and 'natural philosophy', Nicolson, *Pepys*, ' "Mad Madge" and the "The Wits" ', 101–76.

22 The internal ideological wars of the Fellows and of outsiders in the book trade with whom they quarrelled also played a role, but delineation of this matter lies beyond the scope of this essay; see R. F. Jones, *Ancients and Moderns; a Study of the Rise of the Scientific Movement in Seventeenth-Century England* (St Louis, 1961, rev. ed.). Joseph Priestley, the so-called 'father of chemistry' and ingenious student of rhetoric, writing at the end of the eighteenth century thought in retrospect that the early Royal Society had done more to foster publication in science books and pamphlets than any other group'; see *The Theological Repository*, 2 vols. (1769), I, vii–viii.

23 See K. Mannheim, *Ideology and Utopia* (New York, 1936), 61; a fine discussion of Bacon's concept appears on 61–2.

24 E.g., William Derham, *Physico-theology, or a Demonstration of the Being and Attributes of God from his Works of Creation* (1713), which went through 12 editions by 1754.
25 M. C. Jacob, *The Newtonians and the English Revolution 1689–1720* (1976), 33.
26 No account of science books in the eighteenth century can omit this factor and hope to treat the subject adequately, yet even J. L. Thornton and R. I. J. Tully are silent on literacy in *Scientific Books, Libraries and Collectors: a Study of Bibliography and the Book Trade in Relation to Science* (3rd ed., rev. 1971). For important background, see also L. Rostenberg, *Literary, Political, Scientific, Religious and Legal Publishing, Printing and Bookselling in England 1551–1700* (New York, 1965).
27 From the Restoration down through the end of the eighteenth century the perception is found in the diaries of foreigners as well as natives: see S. Sorbière, *Voyage to England* (1664), together with T. Sprat's *Comments upon S. Sorbière's Voyage* (1665), and P. J. Grosley's *Tour to London*, 2 vols. (1772), I, 185–90.
28 As D. J. Greene has shrewdly argued in 'Swift: some caveats', *Studies in the Eighteenth Century*, II (Sydney, 1973), 354–8.
29 *The Tatler*, No. 236 (1710) containing the 'Will of a Virtuoso', Addison's most extended satire on *virtuosi*. See also *Tatler*, 119, 216, and 221 for other attacks. Only M. C. Jacob (*op. cit.*) has commented, albeit briefly, on the political and ideological aspect of Addison's response to science.
30 See D. Bond (ed.), *The Spectator*, 5 vols. (1965), I, 191; further discussion of Addison's attacks on the *virtuosi* and other natural philosophers is found in M. H. Nicolson, *Science and Imagination* (Ithaca, 1962), 174–6.
31 Much of the criticism was that naturalist-philosophers were not Baconian *enough*, and had allowed themselves to be deflected from their calling, i.e. the discovery of techniques with clear social utility. See Feuer, *op. cit.*, 44–6. But literary men were not the only group to attack science. John Sergeant's (1622–1707) *Method of Science* (1696) objects to the philosophy of 'natural philosophy' on the grounds that it is incapable of establishing 'the truth' about reality and that it is therefore an altogether limited form of knowledge; approaching both truth and knowledge from an Aristotelian realist position, Sergeant, who was vigorously opposed to Locke's views, argues that the truth produced by 'natural philosophy' is incommensurate with innate ideas; see my forthcoming essay 'Sergeant's neo-Aristotelian *Method of Science* (1696) and "the way to truth"'. Commentaries on Malebranche and prefaces to his works are also revealing for the relation of 'natural philosophy' to 'truth': see, for example, the 1694 English edition of the *Treatise Concerning the Search after Truth* . . . (1694).
32 Actual ridicule has been surveyed in Nicolson, *Science*, and in P. Rattansi, 'Satire on science in the seventeenth and eighteenth centuries' (Ph.D. thesis, University of London, 1971), but not the extent to which the literati embraced aspects of the new latitudinarian ideology.
33 See F. Watson, *The Beginnings of the Teaching of Modern Subjects in England* (1909), 288; see also M. L. Clarke, *Classical Education in Britain 1500–1900* (1959); A. M. d'I Oakeshott, 'English grammar schools 1660–1714 (Ph.D. thesis, University of London, 1969); W. A. L. Vincent, *The Grammar Schools 1660–1714* (1969); M. Seaborne, *The English School* (1971); R. S. Thompson, *Classics or Charity? the Dilemma of the Eighteenth-Century Grammar School* (Manchester, 1971). Useful comments about the lack of 'science books' are also found in contemporary works on education: see Gilbert Burnet, *Thoughts on Education* (1668, repr. 1761); N. Carlisle, *A Concise Description of the Endowed Grammar Schools in England*, 2 vols. (1881).
34 Perhaps Daniel Defoe's *Complete English Tradesman* (1726) ought to mark the dawn of a new era for the stress laid on 'natural philosophy' in a 'young gentleman's education'; see also E. C. Mack, *Public Schools and British Opinion 1780–1860* (New York, 1939), 57; and G. C. Brauer, *The Education of a Gentleman: Theories of Gentlemanly Education in England 1660–1775* (New York, 1959).

Science books and their readership 313

35 Quoted in Vincent, *op. cit.*, 215, whose source is E. H. Pearce, *Annals of Christ's Hospital* (1908).

36 J. Sargeant, *Annals of Westminster School* (1898), 135. By 1751 Barrow's edition of Euclid had gone through six editions in Latin and at least four in English, and other scholars prepared other editions, totalling more than 50 by 1780.

37 A. C. Babenroth, *English Childhood* (New York, 1922), 173. D. E. Allen notes in *The Naturalist in Britain: a Social History* (1976), 16, that Banks could not find books about botany even when he went up to Oxford.

38 A. K. Cook, *About Winchester College* (1917), 316. A. C. Benson, *Fasti Etonenses* (1899), makes no mention of science education at Eton.

39 Quoted in W. Birch, *The School Master; a Tribute to the Memory of Thomas James . . . with a Short Memoir* (1829), 14.

40 Bentham's *Chrestomathia* (1816), his philosophy of education, has no eighteenth-century equivalent; Locke's *Thoughts Concerning Education* (1693) is the closest approximation but is remarkably silent on the matter of textbooks, and however 'modern' Aubrey's *Idea of Education of Young Gentlemen* claimed to be in the 1690s, it nevertheless placed little emphasis on 'natural philosophy'.

41 See D. M. Simpkins, 'Early editions of Euclid in England', *Annals of Science*, XXII (1966), 225–49.

42 We tend to forget for how long 'science' was excluded from the curriculum of English schools, whether endowed or unendowed: as late as the Regency Charles Darwin claims to have been incapable of learning anything about 'science' at Dr Butler's school in Shrewsbury and equally unable to obtain science textbooks there; see C. Darwin, *His Life Told in an Autobiographical Chapter* (New York, 1893), 9, and P. Kaufman, *Libraries and their Users: Collected Papers in Library History* (1969), 132–3, who confirms Darwin's statement with hard evidence.

43 I discuss the economics of this state of affairs below. For dissenting academies see chap. 6 above and J. W. A Smith, *The Birth of Modern Education: the Contribution of the Dissenting Academies 1600–1800* (1954).

44 C. Wordsworth, *The Undergraduate* (1928), 13.

45 *Ibid.*, 13; Wordsworth is quoting here from *Scholae academicae* (1877, repr. 1968), 189–90.

46 R. W. T. Gunther, *Early Science in Cambridge* (1937), 226.

47 *Ibid.*, 226; presumably Gunther means a student of superior talent. Gunther notes in *Early Science in Oxford*, 14 vols. (1945), XIV, 3–4, that science teaching there was no better, yet he maintains that 'mathematics and natural philosophy are so generally . . . understood, that more than 20 in every year of the Candidates for a Bachelor of Arts Degree [at mid-century] are able to demonstrate the principal Propositions in the *Principia*'; see *Early Science in Cambridge*, 59.

48 See A. Friedman (ed.), *Collected Works of Oliver Goldsmith*, 5 vols. (1966), I, 334, and for the passage cited below, I, 463; the passages quoted are respectively from *An Enquiry into the Present State of Polite Learning in Europe* (1759) and *The Bee*, No. 6, 'On education'.

49 C. Wordsworth, *Scholae academicae* (1877; repr. 1968), 78–81, 248–9, 330–7; in his other appendices Wordsworth gives reading lists, the names of textbooks required on examinations, the reading lists of the schools, the lecture notes of several tutors, etc.; see also D. A. Winstanley, *Unreformed Cambridge* (1935); A. D. Godley, *Oxford in the Eighteenth Century* (1908); Gunther, *Early Science in Oxford*; and W. A. Pantin, *Oxford Life in Oxford Archives* (1972). The situation at Oxford was similar to that at Cambridge, with the exception that 'natural philosophy' was adopted even more slowly there, and this may be why the Oxford University Press printed so few

'science books' between 1700 and 1800, as H. G. Carter, *A History of the Oxford University Press* (1979), has shown.

50 Wordsworth, *Scholae academicae*, 69.

51 The spread of Newtonianism in the early eighteenth century, and the means by which it was adopted, has acquired an ocean of scholarship that cannot be treated here. What is of interest is the degree to which the adoption depended on printed books.

52 A large percentage of these 'books' are merely printed versions of the lecturer's notes, but this fact does not alter the matter of origins.

53 See J. Crellin, 'Chemistry and eighteenth century British medical education', *Clio Medica*, IX (1974), 21, and the discussion below of natural history books.

54 Especially T. S. Kuhn, *The Structure of Scientific Revolutions* (1962; rev. ed. 1970) and the various books of M. C. and J. R. Jacob, especially 'The Anglican origins of modern science', *Isis*, LXXI (1980), 251–67.

55 'The Navy', in A. S. Turberville (ed.), *Johnson's England*, 2 vols. (1933), I, 57.

56 Quoted in Sir J. Smyth, *Sandhurst* (1961), 33.

57 F. G. Guggisberg, *'The Shop': the Story of the Royal Military Academy* (1900), 26.

58 John Muller's publishing career demonstrates the point: he was one of the first to complain about the need for military science textbooks (see p. 000 above) and then exploited the argument of need to persuade London booksellers to continue to print his textbooks: works such as *A Treatise of Fortification* (1746) – the books Sandham was copying and which went through three editions by 1774; *A Treatise Containing the Practical Part of Fortification* (1755), which also went through several editions in a few years; *A Treatise of Artillery* (1768); *A New System of Mathematicks [for the use of Military Students]* (1769); *The Attack and Defence of Fortified Places* (1770) – many of which were assigned in his classrooms at the Royal Academy at Woolwich.

59 See H. Barnard, *Military Schools and Courses of Instruction in the Science and Art of War . . .* (New York, 1969), 525–8, 559–65.

60 *Ibid.*, 628–9.

61 I have presented some of the evidence in an essay entitled 'Immortal Doctor Cheyne and the "Scientific Wits": aspects of science and millenarianism in the eighteenth century', in R. H. Popkin (ed.), *Messianism and Millenarianism in Eighteenth-Century England* (Berkeley and Los Angeles, 1987) (chap. 4 above).

62 N. Hans, *New Trends in Education in the Eighteenth Century* (1966), on whose book I have heavily drawn in this section.

63 J. T. Desaguliers, *Physico-Mechanical Lectures* (1717), i.

64 As Hans notes (*op. cit.*, 141) many well-known persons attended these lectures, including famous poets and writers, financiers, politicians, physicians, courtiers, foreigners, and other dignitaries. If as diverse a group bought these science books as attended the coffee-house lectures, then the dissemination of these books indeed penetrated almost all segments of society. Attendance at these courses was then as fashionable as it was educational, and purchase of the lecturer's manual was a status symbol.

65 Desaguliers, Watts and Worster worked in collaboration with the Innys brothers, the successful family of booksellers whose premises were in the north-west alley behind St Paul's Church. See also below, p. 231.

66 In 1713–14 Pope attended Whiston's lectures at Button's coffee-house and bought the course-manual, *A Course of Mechanical, Optical . . . Experiments . . . the Explanatory Lectures read by William Whiston* – for only 6s; in 1731 Shaw's lecture course cost 5 guineas and the printed lecture notes 6s.

Science books and their readership 315

67 J. Lawson and H. Silver, *A Social History of Education in England* (1973), 219; italics mine. Passages discussed below are also found on p. 219.

68 No one has undertaken a prosopographical study of the English coffee-houses in which those who are known to have attended and their backgrounds are studied.

69 A failure to distinguish the various professions of the audience of the early coffee-houses is the chief weakness of A. Ellis's otherwise informative *The Penny Universities: a History of the Coffee Houses* (1956).

70 The matter is further complicated by the relation of these itinerant lecturers and their 'science books' to Grub Street, a relation which no one seems to have studied in any detail; even our finest student of Grub Street – Pat Rogers – is silent on the matter in *Grub Street: Studies in a Subculture* (1972), as were his predecessors: A. S. Collins, K. Hornbeak, E. E. Kent, and E. A. Bloom.

71 And it is crucial to note about this readership that no similar books were then being prepared for 'courses' or 'demonstrations' on other subjects, i.e., history, literature, politics.

72 The contrasts discussed in this paragraph derive from study of a representative group of each type of book, including but not limited to: F. Hauksbee, *Physico-Mechanical Experiments* (1709); W. Whiston, *A Course of Natural Philosophy* (1715?); W. Vream, *A Description of the Air Pump* (1717); Desaguliers, *Lectures*; B. Worster, *The Principles of Natural Philosophy* (1722); W. Whiston, *Astronomical Lectures* (1728); P. Shaw and F. Hauksbee, *Proposals for a Course of Chemical Experiments* (1731); and then, later on, J. Ferguson, *Astronomy* (1756); *idem*, *Lectures on Select Subjects* (1760); *idem*, *An Epitome of Natural and Experimental Philosophy* (1769); A. Walker, *An Analysis of a Course of Lectures* (1770); J. Banks, *An Epitome of a Course of Lectures* (1775). A complete list of the Boyle Lectures from their inception in 1691 to their demise in the 1730s is found in G. S. Rousseau, 'Science', in P. Rogers (ed.), *The Contexts of English Literature: the Eighteenth Century* (1978), 160.

73 A fifth edition was brought out by 1736 'with very considerable additions ... from later discoveries in mathematicks and philosophy'. In 1744 a supplement appeared as prepared 'by a Society of gentlemen who were studying science'. Perhaps David Abercromby's compendium ought to be added to this list: see *Academia Scientarum: or the Academy of Sciences, being a Short and Easie Introduction to the Knowledge of the ... Sciences* (1687). Trained as a physician, Abercromby also wrote a crisp *Discourse on Wit* (1686) and was praised by the prestigious and influential Haller for anticipating the ideas of the Scottish school of Common-Sense Philosophy.

74 W. T. Starnes and G. E. Noyes, *The English Dictionary from Cawdrey to Johnson 1604–1755* (Chapel Hill, 1946); M. Segar, 'Dictionary making in the early eighteenth century', *Review of English Studies*, VII (1931), 230–8; and several articles studying the impact of Bayle's dictionary. See also M. H. Nicolson, 'English almanacs and the new astronomy', *Annals of Science*, IV (1939), 1–33; A. Hughes, 'Science in English encyclopaedias, 1704–1875', *Annals of Science*, V (1940), 220–39; D. A. Kronick, *A History of Scientific and Technical Periodicals: the Origins and Development of the Scientific and Technologic Press, 1665–1790* (New York, 1962); D. Layton, 'Diction and dictionaries in the diffusion of scientific knowledge: an aspect of the popularization of science', *British Journal of the History of Science*, II (1965), 221–34.

75 See the Strahan ledgers, 1775–90.

76 A small sample shows that at mid-century these books averaged £1 1s so the common man could not readily afford them; but aristocrats, libraries, colleges, and other institutions bought them. It may not be going too far to affirm that 'dictionary-reading' became a national pastime among the new leisured class; see also J. H. Plumb, *The Commercialisation of Leisure in Eighteenth-Century England* (1973).

77 E.g., S. Blankaart, *The Physical Dictionary* (1702); E. Stone, *New Mathematical Dictionary* (1726); W. Hooson, *The Miner's Dictionary* (1747); John Hill, *A New Astronomical*

Dictionary (1768); a re-issue of his earlier *Urania* (1754); Richard Brookes, *General Gazeteer, or a Compendious Geographical Dictionary* (1762); G. Smith, *A Universal Military Dictionary . . . Explanation of the Technical Terms* (1779). Stone was a populariser of mathematics, but many of the authors of these dictionaries were physicians who had the leisure and cash to prepare such compendious works.

78   J. E. McClellan III, 'The scientific press in transition: Rozier's journal and the scientific societies of the 1700s', *Annals of Science*, XXXVI (1979), 425–49.

79   Nicolson, *Pepys*, 101–75 and *idem, Conway Letters* (New Haven, 1930); see also S. Mintz, 'The Duchess of Newcastle's visit to the Royal Society', *Journal of English and German Philology*, LI (1952), 168–76.

80   G. D. Meyer, *The Scientific Lady in England 1650–1760* (Berkeley, 1955), 49. See also Hans, *op. cit.*, 'Education of women', 202–7.

81   From the start Tipper's strategy contributed to the financial success of the periodical: women were invited to submit their scientific questions 'in verse'; which Tipper promised to publish; yet he printed only a fraction of the poems he received, reserving most available space for his *own* scientific questions. Nevertheless, the enticement impelled his female readers (were there any males?) to buy each issue, thereby ensuring the solvency of the *Diary*.

82   C. F. Mullett (ed.), *Letters of Dr George Cheyne to Samuel Richardson (1733–1743)* (Columbia, 1943), 70. Hans (*op. cit.*) lists the relevant printed works but does not speculate about the causes for the revival; yet it may be that the scientific individualism displayed by courageous women from the time of 'Mad Madge' and Fontenelle was now sentimentalised by a new generation of young women in the 1790s who had read or heard about these female pioneers. Valuable comparisons can be made between scientific women c. 1690, 1790, and in the Regency, e.g., Jane Marcet's *Conversations on Natural Philosophy* (1819) and her many other popular science books for ladies.

83   Meyer, *op. cit.*, viii.

84   Meyer concentrates on Susannah Centlivre's *The Bassett-Table* (1705), Thomas Wright's *The Female Vertuosos* (1705), and James Miller's *The Humours of Oxford* (1730), but others should also be considered. In *A Treatise on Female Nervous, Hysterical . . . Diseases* (1780), William Rowley considered excessive application to 'natural philosophy' a cause of madness and eventual suicide. A further detailed study of this subject needs to be made.

85   Their interest in natural philosophy c. 1690 was related, of course, to the widespread secularisation of culture about which so much has been written; yet the part played by scientific women in this emancipation of the religious and philosophic imagination has been overlooked. There are many books about female writers before 1800; see M. R. Mahl and H. Koon (eds.), *The Female Spectator: English Women Writers before 1800* (Bloomington, 1977); but none on scientific women.

86   Schaffer in *The Ferment of Knowledge*, 58–71.

87   See Jones, *op. cit.*, 'The "Bacon-faced generation" ', 237–67.

88   Among cultural historians see: Feuer, *op. cit.*, 411–19; P. Gay, *The Enlightenment*, 2 vols. (1966–9); Kuhn, *op. cit.*; G. Buchdahl, *The Image of Newton and Locke in the Age of Reason* (1961); F. Manuel, *A Portrait of Isaac Newton* (1968). Among historians of science: A. Thackray, *Atoms and Powers* (1970); M. C. Jacob, *op. cit.*; and the secondary works discussed by Schaffer in *The Ferment of Knowledge*, 58–71. Indispensable also are G. J. Gray, *A Bibliography of the Works of . . . Newton* (1888) and P. and R. Wallis, *Newton and Newtoniana 1672–1975* (1977).

89   The disagreements were studied by H. Guerlac, 'Where the statue stood; divergent loyalties to Newton in the eighteenth century', in E. Wasserman (ed.), *Aspects of the Eighteenth Century* (Baltimore, 1965), 317–34.

90   Examples include: (for government) J. T. Desaguliers, *The Newtonian System of the*

Science books and their readership 317

World, the Best Model of Government (1728); (for painting) anon., *The Geometry of Landskips and Painting . . . useful to Line Limners in Drawing and Gentlemen in Choosing Pictures* (1735); (for medicine) N. Robinson, *A New Theory of Physick and Diseases*, founded upon the Principles of the Newtonian Philosophy (1725).

91 I. B. Cohen, *Franklin and Newton* (1956) and M. H. Nicolson, *Newton Demands the Muse* (1946).

92 Especially in the new age of leisure as J. H. Plumb has shown in *Commercialization of Leisure* and *Georgian Delights* (1979).

93 Addison, *Spectator* (37 (cf. p. 165, above) D. Bond comments in his Clarendon edition, 5 vols. (1965), I, 154, n. 6, that Addison may have been thinking of Newton's recently published *Arithmetica Universalis* (1707); but 'young ladies' could have learned simple arithmetic from any number of available books in 1711, especially those by Isaac Barrow, and did not need this work. Probably Addison was thinking of the *Opticks* and *Principia*. Culpeper's *Directory for Midwives* (1651), reissued many times by 1711, is also mentioned in Addison's list.

94 D. Knight has rightly stressed the cultural importance of science books in the development of western civilisation, in *Natural Science Books in English 1600–1900* (1972) and in *Sources for the History of Science 1660–1914* (1975), but has not enquired into the spur Newton gave to the book trade.

95 See P. and R. Wallis, *op. cit., passim*.

96 These Newtonians are studied by Schaffer in *The Ferment of Knowledge*, 58–71; Manuel, *op. cit.*; and in various papers by Peter Heimann.

97 See R. Cohen, *The Art of Discrimination* (1964); a sounder treatment, more faithful to the original intentions of these poets, is found in A. D. McKillop, *The Background of Thomson's Seasons* (Minneapolis, 1942).

98 See Buchdahl, *op. cit.*; and an important series of articles by G. A. J. Rogers demonstrating the various ways in which Locke's works ought to be interpreted as a direct response to Newton and vice versa. Of especial importance is Rogers's 'Locke's *Essay* and Newton's *Principia*' *Journal of the History of Ideas*, XXXIX (1978), 217–32. It would be interesting to learn whether the banning of Locke's *Essay* in Oxford University in 1703 had any effect on either the reading or sales of Locke's book then.

99 The precise size of this readership has never been and perhaps cannot be determined, but few books sold as well as Cheyne's *Essay of Health and Long Life* (1724) which went through four editions in just the first year, and his *English Malady* (1733) which, although not as sought after as its predecessor, nevertheless went through several reissues by mid-century. Even the more concrete *Observations on the Gout* (1720) reached a fifth edition by 1723, and was in a 'ninth corrected edition' by 1738, destined to go through many subsequent editions. *Philosophical Principles of Religion* (1705) was in a 'fifth corrected' edition in 1736. Both Cheyne and his publishers – principally the house of Strahan – reaped vast profits from these reprints. See chap. 4 above.

100 Many other similar titles are discussed by Schaffer in *The Ferment of Knowledge*, 58–71.

101 It also suggests that scientific revolutions – and, more generally, revolutions in knowledge – occur as the result of one man's work rather than a whole change in world view or socio-economic organisation; see Kuhn, *op. cit.*, 'Revolutions as changes of world view', 111–35. Aspects of Cheyne's and Whiston's debt to Newton have been treated by H. Metzger in *Attraction universelle et religion naturelle chez quelques commentateurs anglais de Newton* (Paris, 1938).

102 Important discussions of the problem include: J. Roger, *Les Sciences de la vie* (1963) and *idem, The Ferment of Knowledge*, 'The triumph of natural history', 263–70; M. Foucault, *Les Mots et les choses* (Paris, 1966), translated into English as *The Order of Things*; W. Lepenies, *Das Ende der Naturgeschichte* (Munich, 1975); for the social and

economic aspects see J. M. Chalmers-Hunt, *Natural History Auctions 1700–1792* (1976). An important contemporary book that provides still another point of view is William Smellie's *Philosophy of Natural History* (1790).

103 Lists of these books and secondary studies of them are found in P. and R. Wallis, *op. cit.*, *passim*.

104 See Roger, *Ferment of Knowledge*, 263–4.

105 *Ibid.*, 263.

106 *Ibid.*, 264.

107 *Ibid.*, 263. Roger does not state that the reason for the success is 'literary', but this is his implication, especially when he urgently sends the reader to C. V. Doane, 'Un succès littéraire du XVIIe siècle: le *Spectacle de la nature de l'abbé Pluche*' (dissertation, University of Paris, 1957). See also my own discussion below.

108 The poems were often reprinted and reissued and were clear money-makers. Their contents have been analysed by several scholars (A. D. McKillop, M. H. Nicolson, R. Cohen, *et al.*), but I am unaware of any study that examines them from the reader's point of view, that is, from his expectations while consulting a didactic-scientific poem.

109 I certainly do not wish to imply that after 1750 books about natural philosophy ceased to be published, or that readers of natural history books were pre-eminently amateurish; there is a certain overlap in both directions, but for the sake of generalisation I have epitomised the evolution of each province of book – natural philosophy and natural history – in extreme terms. D. E. Allens' approach to the rise of natural history is social; see *The Naturalist in Britain*. Yet, excellent as this study is, it says little about the reasons for the sudden craze for natural history in the eighteenth century, except that the phenomenon expressed itself 'in social activities' from the beginning.

110 J. Roger, *Buffon: les époques de la nature* (Paris, 1962) and *Ferment of Knowledge*, 262–8; M. Foucault in *The Order of Things* (1970).

111 See Buffon, *Discours sur le style* (Paris, 1753).

112 *Monthly Rev.*, LII (1775), 616, a review of the three-volume *Natural History of . . . Animals, Vegetables, Minerals* (1775) translated by William Kenrick. This edition was followed in 1781 by a nine-volume set translated by William Smellie, the Scottish naturalist and printer who must not be confused with the Scottish physician who lived at the same time. Containing 260 plates, the edition is a landmark in the evolution of eighteenth-century science books for its remarkably low price of £3 12s. The three-volume Kenrick set (1775) had cost twice this, and the current (i.e. 1781) edition of Buffon in French was available in 16 volumes for 16 guineas. Smellie's firm, known as 'Creech and Smellie', became printers to the University of Edinburgh and published many scientific and medical books from 1770 to the end of the century, including William Buchan's *Domestic Medicine*, the first edition of the *Encyclopaedia Britannica* (1771), a popular *Philosophy of Natural History*, 2 vols. (1790), sometimes referred to as Smellie's *Natural History*, a reissue of Benjamin Stillingfleet's *Tracts Relating to Natural History* (1762) and a now scarce 1778 edition), the works of Blair, Beattie, George Campbell, William Cullen, John Gregory – all important 'scientific works' – and the translations of Buffon. Information about Buffon in England is found in O. E. Fellows and S. F. Milliken, *Buffon* (New York, 1972), much of which is based on S. F. Milliken, 'Buffon and the British', (Ph.D. thesis, Columbia University, 1965).

113 *Monthly Review*, LXII (1780), 400. Nangle was unable to identify this reviewer or the one discussed in n. 116.

114 *Monthly Review*, LXI (1780), 531. Many similar comments can be compiled in the literally dozens of reviews that appeared from 1762 onwards, when Ralph Griffiths, editor of the *Monthly Review*, brought out the *Natural History of . . . the Horse* (1762).

Science books and their readership 319

115 By the 1790s English readers could choose from at least five sets of Buffon's works in translation.
115 See Kaufman, *Libraries*, 156, where the point is documented for Scotland; there is no reason to believe it is not equally valid for England.
116 The distinction was clearly recognised in Linnaeus's own lifetime or shortly after his death in 1778; see Thornton and Tully, *op. cit.*, 192–5; B. Henrey, *British Botanical and Horticultural Literature before 1800*, 3 vols. (1975), II, 650–2.
117 See L. F. Powell (ed.), *Boswell's Life of Johnson*, 6 vols. (1934), III, 84.
118 F. A. Stafleu, for example, writes in *Linnaeus and the Linnaeans* (Utrecht, 1971), 123, of Linnaeus's 'sardonic humour', and others have noticed a witty vein, but these qualities are minimal in the whole *oeuvre*.
119 *Monthly Review*, XLV (1772), 258.
120 This fascinating subject is not treated in D. Davie's *Science and Literature 1700–1740* (1964) which terminates before the 'natural history' movement had advanced very far; for some discussion see T. Savory, *The Language of Science* (1953).
121 Raymond Hirons, *Monthly Review*, XLV (1772), 259.
122 As early as 1762, before anyone had read Buffon in English, William Rider announced in *An Historical and Critical Account of the Lives and Writings of the Living Authors of Great Britain* (p. 14) that while Goldsmith 'is surpassed by few of his Contemporaries with Regard to the Matter which his Writings contain, he is superior to most of them in Style, having happily found out the Secret to unite Elevation with Ease, a Perfection in Language, which few Writers of our Nation have attained to'. I have learned much from J. H. Pitman's *Goldsmith's Animated Nature* (New Haven, 1924) and from mounds of notes I collected in the British Library while compiling *Goldsmith: the Critical Heritage* (1972), where the Rider passage is cited on p. 158.
123 Pitman, *op. cit.*; since Pitman published his study several other editions of *Animated Nature* have come to light.
124 *Monthly Review*, LII (1775), 310–14; most of Bancroft's reviews in 1775 were harsh, for example, his savage attack on the English translation of Bienville's *Treatise on Nymphomania*. Light on Goldsmith's relation to the book trade and on his reviewers is found in E. E. Kent, *Goldsmith and his Booksellers* (Ithaca, 1933).
125 Powell (ed.), *op. cit.*, III, 84.
126 *Monthly Review*, XXIX (1763), 285; elsewhere on the same page Goldsmith chastises Brookes for unrelentingly 'dry descriptions' and extols 'the flowing manner of the French Naturalists' – a 'manner' he himself invoked in *Animated Nature*.
127 One of the literary triumphs of *Animated Nature* is enticement of the reader into believing he is reading about outlandish animals in distant exotic places; while engaging the reader by this device, Goldsmith at the same time loses no opportunity to philosophise and extemporise about *man*. The reader thus gains the impression (given nowhere in Linnaeus or Buffon) that 'natural history' is a province of knowledge about *familiar mankind* as well as animals.
128 See *Criticial Review*, XVI (1763), 312 and *Monthly Review*, LXXII (1785), 403.
129 Quoted without reference to Allen, *op. cit.* 36.
130 The first three volumes cost 2 guineas apiece, and the fourth may have sold for the same amount; there was also a cheap pocket edition in two volumes for 6s. Three more volumes appeared from 1758 to 1764 priced at a total sum of £6 6s. These prices are inexpensive compared to the prices of many other natural-history books.
131 *British Zoology* began to appear in 1766 at a price of £2 2s 6d and continued to be regularly issued until 1812; for Gray's reading of Pennant, see P. Toynbee and L. Whibley, *The Correspondence of Thomas Gray* 3 vols. (1935), III, 1163ff. and C. E. Norton, *The Poet Gray as a Naturalist* (Boston, 1903), 27. Goldsmith's indebtedness to Pennant is discussed in Pitman, *op. cit.*, 138–9.

132 Especially 'how-to-do-it' books for those who were 'Sunday naturalists', but as these books often appeared in the form of chapbooks rather than science books, I have not discussed them in any detail here.

133 An entire set of 26 volumes cost 38 guineas 'with 1600 copper plain plates' and 160 guineas 'with coloured plates and 26,000 figures' in 1775 when the last volume appeared. Volume 1 (1759) with copper plates cost £1 11s 6d; volume 2 (1761) £2 12s 6d, thereafter, the volumes continued to cost £1 11s 6d for the next two decades. Curiously the first three volumes fetched only 17s 6d in 1794 at the auction of the library of the Earl of Bute (Hill's patron), and 'an elegant copy, the plates beautifully coloured, 27 volumes' fetched £79. See the *Sales Catalogue of the . . . Rt. Hon. the Earl of Bute 1794* (BL. 1255. c. 15.1–3). Hill's *Essays in Natural History and Philosophy* (1752), a cheaper book, is worth consulting for contemporary reflections on the popularity of natural history as distinct from natural philosophy. After 1760 the market was flooded with books on the *Florae Britannicae*.

134 A second edition appeared in 1772 for the same price. The copper plates in Brookes's volumes are inferior to Hill's in the *Vegetable System*, yet illustration in the 1760s was an important aspect of the public's attraction to natural-history books. See W. Blunt, *The Art of Botanical Illustration* (1951), and D. Knight, *Zoological Illustration* (1977). Knight maintains that Hooke's depiction of specimens seen through the microscope led the way to the beautiful illustrations of the eighteenth century, but he does not comment on the social history of interest in science.

135 For editions of Buffon in England 1750–1800 and Buffon in relation to the book trade, see Milliken, thesis.

136 The reader who is curious about the extensiveness of these books should consult Henrey, *op. cit.*, III: *the Eighteenth Century: Bibliography* (1975). There was even a big trade in children's natural-history books and demand for books such as *Jack Dandy's History of Birds and Beasts* (c. 1781).

137 For example, Sir William Watson, a leading Fellow of the Royal Society at this time, judged Linnaeus's *Systema Naturae* 'the masterpeice of the most compleat naturalist the world has seen' (*Gentleman's Magazine*, XXIV (1754), 558), and by comparison considered the natural histories of Buffon and Goldsmith the products of amateurs. Yet Watson judged as a scientist without any interest in these men as writers of books. A study such as S. E. Hyman's *The Tangled Bank: Darwin, Marx, Frazer and Freud as Imaginative Writers* (New York, 1962) has curiously not as yet been written for these 'imaginative writers' who moulded the taste of their century.

138 'The telescope and imagination', *Modern Philology*, XXXII (1935), 233–60 and *The Miscroscope and English Imagination* (Northampton, 1935).

139 *The Microscope Made Easy* went through 5 editions by 1769, its brilliant success owing something to the simple manner in which Baker explained how to prepare and observe specimens. For the popularity of this book see also G. L'E. Turner, 'Henry Baker, F.R.S.', *Notes and Records of the Royal Society of London*, XXIX(1974), 53–79.

140 Martin wrote dozens of books and tracts about instruments that have now been studied by J. R. Millburn in *Benjamin Martin, Author, Instrument-maker, and 'Country Showman'* (Leyden, 1976), Millburn has appended a useful bibliography of these works on pages 193–207; of interest are *Micrographica Nova . . . a New Treatise on the Microscope* (1742); *All Sorts of Philosophical, Optical, and Mathematical Instruments* (1756); *An Essay on Visual Glasses* (1756); *Horologia Nova* (1770), all of which were published in England and went through frequent reprints. The elder George Adams's *Micrographia Illustrata* (1746) underwent 4 editions by 1771.

141 The controversy is worthy of reference because of the light its sheds on other types of 'scientific' books and their readers. See Nicolson, *Science*, 200–1.

142 The prize was awarded in 1753 when it was shared by John Harrison, Tobias Mayer

Science books and their readership 321

and others. See E. G. Forbes, *Tobias Mayer's claim for the longitude prize'*, *Journal of Navigation*, XXVIII (1975), 77–90; H. Quill, *John Harrison* (1966).

143 In great numbers, for example: Whiston and Ditton, *A New Method of Discovering the Longitude* (1714); an anonymous *Longitude to be found out with a New Invented Instrument* (1715); numerous satires on the Whiston-Ditton book by the 'Club of Scriblerians': Arbuthnot's satire *To the Right Honourable the Mayor and Aldermen of the City of London* (1716), in which the author ironically asserts that the reward for the discovery of the longitude has had beneficial effect on 'common folk' because it has obliged 'Cooks, and Cook-maids to study Opticks and Astronomy', and to 'throw the whole Art of cookery into the Hands of Astronomers and Glassgrinders'.

144 *The Artificial Clockmaker, or a Treatise of Watch and Clock Work, Showing to the Meanest Capacities the Art of Calculating Numbers to All Sorts of Movements . . . with the Ancient and Modern History of Clockwork* (1696) was in a fourth edition by 1734, each time having been reprinted by 'the author', after which time it began to appear in 'corrected editions'.

145 For discussion of clocks and chronometers see Millburn, *op. cit.*; H. Michel, *Scientific Instruments in Art and History* (1967); C. M. Cipolla, *Clocks and Culture* (1967); M. Dumas, *Scientific Instruments of the . . . Eighteenth Century* (1972); O. Kurz, *European Clocks and Watches in the Near East* (1975); S. L. Macey, *Clocks and Cosmos* (Hamden, 1980).

146 To the reader it must have seemed that interest in clocks was peaking in this decade; as much is announced in the preface of A. Cumming's *Elements of Clock and Watchwork* (1766).

147 On this subject – especially the role of ideology in literature – see M. Bradbury, *The Social Context of Modern English Literature* (1971), and although it deals with a period just after ours, R. D. Altick, *The English Common Reader: a Social History of the Mass Reading Public* (Chicago, 1957); M. C. Jacob, *op. cit.*; J. R. Jacob, *Robert Boyle and the English Revolution: a Study in Social and IntellectualChange* (New York, 1977).But also see Barnes, *Scientific Knowledge: idem, Interest and the Growth of Knowledge* (1977).

148 D. Davie, *The Language of Science and the Language of Literature 1700–1740* (1963), 2–3.

149 The attack on this literature in our time has been made by neo-Romantic critics such as H. Bloom and G. Hartman who argue that this body of verse possesses *no* tropes, only *topoi* that have been substituted for tropes. But the assessment must not be ambiguous: Auden, Eliot *et al.* have praised these writers – Pope, Swift, Gray, Thomson, Johnson, Churchill – for their use of scientific metaphor, but *not* because they are 'scientific poets' or 'scientific writers'. A poet must invoke more than 'scientific' language and mechanical tropes before he becomes 'a scientific poet'; for this he must at least versify scientific topics as a large part of his programme. Whereas Akenside, Armstrong and Thomson are 'scientific poets', Pope and Churchill are not. *An Essay on Man* is an exception.

150 See n. 74 above for the dissemination of science in newspapers.

151 Quoted in A. Maclean, *Humanism and the Rise of Science in Tudor England* (New York, 1972), 220. See also D. McKitterick, *Sir Thomas Knyvet* (1977).

152 Here one wonders if any light on this matter will be shed by the studies of subscriptions to books on which Peter Wallis and his team at Newcastle are engaged. It may also be that the percentage change is also evident in other types of libraries, such as those of curates and naturalists; see J. Salter, 'The books of an early eighteenth-century curate', *The Library*, XXXIII (1978), 33–46 and R. A. Harvey, 'The private library of Henry Cavendish', *The Library*, II (1980), 281–92. Valuable material is also found in M. Plant, *The English Book Trade* (1939), 55–64.

153 Quoted without source in P. Muir, *English Children's Books 1600 to 1900* (New York, 1954), 67.

154 This rare work is No. 261 in William Sloane's checklist of *Children's Books in England* (New York, 1955); for further discussion see M. F. Thwaite, *From Primer to Pleasure* (1963), 208.

155 Almost nothing is known about Boreman and the financial records of his firm seem to have disappeared.

156 In accord with the *utile-dulce* ethic of the age, Newbery intended this series to be both instructional and delightful, as the first advertisement I have been able to locate states: 'the Whole will seem rather an Amusement than a Task'; see *Penny London Post*, 18 Jan. 1745. C. Welsh, *On Some of the Books for Children of the Last Century* (1886) and *A Bookseller of the Last Century* (1885) has been replaced by S. Roscoe's *Provisional Check-list of Books . . . Issued under the Imprints of John Newbery* (1966) and *John Newbery and his Successors* (1973).

157 Essentially a popularisation of Newtonian science, this is also a children's book and succeeds by understanding the structure of curiosity of a child's astronomical imagination. From 1761 to 1794 it went through seven editions.

158 Improvement in illustration is described by J. Whalley in *Cobwebs to Catch Flies: Illustrated Books for the Nursery . . . 1700–1900* (1974).

159 E.g., Sarah Trimmer's *An Easy Introduction to the Knowledge of Nature* (1780) went through 10 editions by the turn of the century, having been composed 'to supply the *lack* of natural history knowledge' in Anna Barbauld's *Lessons for Children* (1778).

160 Begun in 1806, Miss Marcet's series was successful in its rhetorical habit of casting each book in the form of 'conversations': *Conversations on Chemistry* (1806) reached 16 editions by 1853, *Conversations on Natural Philosophy* (1819) reached 20 by 1880, and so forth.

161 The same John Aikin who wrote the more 'adult' *Essay on the Application of Natural History to Poetry* (1777) and who was the only serious scientist then who approved of Goldsmith's *Animated Nautre*.

162 Comparison of this book published at the end of the century with *A Short and Easie Method* (1710), discussed above, shows how much more information about science children were now expected to possess.

163 J. A. Paris's *Philosophy in Sport Made Science in Earnest* (1827) was another bestseller after 1800. See also C. Meigs, 'Roots in the past up to 1840', in *A Critical History of Children's Literature* (1953, rev. ed. 1969), 105–10.

164 For Nourse, see J. P. Feather, 'John Nourse and his authors', *Studies in Bibliography*, XXXIV (1981), 205–26. J. P. Feather and Giles Barber have in preparation a full-length biography of Nourse.

165 Such studies as R. Straus, *Dodsley: Poet, Publisher and Playwright* (1910); A. Briggs (ed.), *Essays in the History of Publishing. . . . the House of Longman, 1724–1974* (1974); P. Hernlund, 'William Strahan's ledgers 1738–1785', *Studies in Bibliography*, XX (1967), 89–111; I. Maxted, *The London Book Trades 1775–1800* (1977).

166 See G. A. Lindeboom, *Boerhaave and Great Britain* (Leyden, 1974), 19.

167 But Innys bought back the copyright, and in 1742 he issued it again in an identical version to that of the 1735 Bettesworth edition; C. Hitch was a partner in both the Innys and Bettesworth editions.

168 Some idea of this audience is gained by examining my 'Chronological checklist of the works of John Hill', in D. J. Greene (ed.), *The Renaissance Man in the Eighteenth Century* (Los Angeles, 1978), 107–29.

169 Based on the findings of the above checklist; the *DNB* list is highly incomplete.

170 See G. S. Rousseau (ed.), *The Letters and Private Papers of Sir John Hill* (New York, 1982), introduction, *passim*.

171 Not to be confused with John Hill, just discussed. See F. R. Johnson, 'Thomas Hill, an Elizabethan Huxley', *Huntington Library Quarterly*, VII (1944), 329–51.

172 Ibid., 341.
173 The price applies to popular handbooks as well as theoretical science books by Boyle, Locke or Newton, all which types had greatly improved since the times of Pepys. An English translation of Burnet's *Sacred Theory of the Earth* cost 12s. in the early eighteenth century; Fontenelle's 1757 edition in English of the *Plurality of Worlds* costs 6s.
174 Plant, op. cit. (1974 rev. ed.), 54.
175 See E. S. Gilboy, *Wages in Eighteenth-century England* (Cambridge, 1934). Some information is also found in M. S. Rosen, 'Authors and publishers: 1750–1830', *Science and Society*, XXXII (1908), 218–32.
176 Based on a survey I have made of representative science books for each decade of the eighteenth century.
177 The prices on Dodsley's lists at mid-century divide between theoretical science books and practical handbooks for the home. Books on the nursing of children or the suppressing of hoemorrhages are usually under 2s and cheap in contrast to science books without practical application.
178 Some of Dodsley's prices at mid-century are quoted by T. F. Dibdin in *Bibliomania; or Book Madness* (1876, rev. ed.).
179 According to George Walker in *Essays on Various Subjects*, 2 vols. (1809), I, lxxx; see also L. Chard, 'Bookseller to publisher: Joseph Johnson and the English Booktrade, 1760 to 1810', *The Library*, XXXII (1977), 138–54.
180 See Henry Crabbe Robinson, *Diary*, 2 vols. (1872), I, 194–5.
181 One other consideration ought to be mentioned: the practice, common by about 1720, of booksellers and printers of keeping in stock current pills and potions for sale. Scholars have continued to be perplexed at the origin of this practice, but it seems to be clearly related to the distribution of goods in the period. Because well-established networks provided a ready-made distribution system, some publishers – notably Newbery – bought into medicine patents to maximise their profits: see J. Alden, 'Pills and publishing; some notes on the English book trade, 1660–1715', *The Library*, VII (1952), 21, and P. S. Brown, 'Medicines advertised in eighteenth-century Bath newspapers', *Medical History*, XX (1976), 152–68. The traditional explanation has been that the practice worked to the profit of both parties: the booksellers (e.g. Newbery) received a commission on all sales in return for advertising the panaceas in their newspapers and selling them in their shops; the medical-pharmaceutical profession gained free advertisement in newspapers and had low overhead costs. Booksellers also carried in stock a large number of scientific instruments, especially those sold in coffee-houses by itinerant lecturers. The booksellers Jonah Bowyer and Jonas Brown carried tickets as well as instruments used for Desaguliers' lecture-demonstrations which began on 14 December 1713 at the Bedford coffee-house. This date is written in an eighteenth-century hand in the BL copy (551.d. 19.81) of J. T. Desaguliers, *A Catalogue of the Experiments in Mr Desaguliers Course* (1713). I greatly profited from J. P. Feather's 1981 lecture at the University of California, which dealt with some of this material.
182 See H. M. Hamlyn, 'Eighteenth-century circulating libraries in England', *The Library*, 5th ser. I (1946–7), 197–222; F. Beckwith, 'The eighteenth-century proprietary library in England'. *Journal of Documentation*, III (1947), 81–98; T. Kelly, *Early Public Libraries: a History of Public Libraries in Great Britain before 1850* (1966); C. Parish, *History of the Birmingham Library: an Eighteenth-Century Proprietary Library* (1966); Kaufman, *Libraries*, and 'Readers and their reading in eighteenth-century Lichfield', *The Library*, 5th ser., XXVII (1973), 108–15, dealing with cathedral libraries.
183 In the catalogues and borrowers' lists surveyed by Kaufman, 'Readers', the crucial decade is the 1770s.
184 Kaufman goes even further when he observes that by the 1780s 'there is a preponder-

ance of scientific works'; see P. Kaufman, 'The Westminster library', *The Library*, 5th ser. XXI (1966), 243.

185 Kaufman, *Libraries*, 'Scotland as the home of community libraries [in the eighteenth century]', 134–47 and R. C. Cole, 'Community lending libraries in eighteenth-century Ireland', *Library Quarterly*, XLIV (1974), 111–23.

186 See P. Kaufman, 'English book clubs and their role in social history', *Libri*, XIV (1964), 1–31, and Beckwith, n. 184, 94–5.

187 Kaufman, *Libraries*, 31–2; Goldsmith's *Animated Nature* continued to be one of 'the ten most popular books' in public libraries down to the end of the century.

188 *A Catalogue of the Books of the London Library Instituted in the Year 1785*, a manuscript quoted by Kaufman in *Libraries*, 19.

189 *Scots Magazine*, XXXIV (April 1772), 136.

190 Kaufman, *Libraries*, 28.

13

Wicked Whiston and the English wits

Some projects, it seems to me, actually suffer from a higher theoretical profile than they require. Even in this post-disciplinary age, when universities and other research institutions feverishly ask themselves hard questions about the relevance of the humanities to ordinary daily life, certain topics require much more historicist reinvigoration than theoretical speculation. A case in point is William Whiston, the English astronomer-mathematician who tried to discover a reliable means of plotting longitude. His transformation from a quasi-scientific Newtonian into the ardent millenarian and inspired messianic figure he later became, was such a topic.

Whiston was notorious in his time for religious heresy as well as radical millenarianism. He was also a shrewd entrepreneur who could survive fierce bouts of poverty. As Newton's successor at Cambridge University, by the early years of the eighteenth century he was already widely known as an astronomer and mathematician. But his unorthodox religious and cosmological views transformed him into an object of ridicule, especially when attacked by the sharp quills of the Scriblerus Club, that group of wits in the age of Queen Anne which included Pope, Swift, Arbuthnot, Thomas Parnell, and Robert Harley, the Earl of Oxford.

The point I developed here was again that about discourse, scientific in relation to literary, and the truths and ideologies of each. In this instance, the collective literary mind – the Scriblerians' – triumphed over the scientific, as the wits ruthlessly exposed in their satires how foolish and unscientific – how pathetically unempirical – were Whiston's prophecies of universal cataclysm. If I were rewriting the essay I would have dwelled at greater length on the ideologies implicit in Whiston's prophetic strain, and would have shown that under certain circumstances the muses are often more scientific than their alleged cousins, the sciences.

> Who can the Comet's wond'rous Journey tell;
> Seats not unaptly deem'd the Place of Hell.
> Now burning in the Sun's immediate Beams;
> More frigid now than Greenland's frozen Streams.
> Of all God's Works, our Reason Nothing shows,
> So fitly form'd for Torments and for Woes.
>   (quoted in Benjamin Martin, *The Young Gentleman and Lady's Philosophy*
>   (1756), dialogue xv, 'Of the comets')

The Scriblerians were less interested in Halley's comet, or its implications for Newtonian astronomy and celestial mechanics, than in William Whiston's prophecies based on comets.[1] The wits had no reason to suspect Halley's mathematical authority or astronomical competence, and Halley had prudently confined himself to the field (celestial mechanics) he knew best rather than applying these physical laws to historical events. But Whiston's millenarian predictions were of another order, had more immediate social relevance than Halley's tables and periodicities: indeed his predictions provided the wits with ready-made fuel for satire and derision. For them Whiston was a comic, if trifling, figure who became a natural butt for their Scriblerian satire, as he read significance into every motion of the stars and tremor of the earth. In time his public image was transformed into that of the worst type of 'ancient': supernaturally interpreting signs above and beneath the earth as proof that the hour of apocalypse was imminent but constantly changing his mind about the day. For this ridiculous behavior the wits enthroned him as one of the archdunces of the age, if not accorded a place in Pope's *Dunciad* nevertheless the explicit target of others of their satires. Newton and Halley were – in this sense – also of the party of the 'ancients', yet they were less objectionable for their more modest millenarian activities. But 'wicked Whiston', as Pope eventually denounced him, was incorrigible: an Arian heretic; a lapsed Cambridge professor; a social pariah; an eccentric millenarian, quack doctor, and fool-like character whose strengths could not redeem his foibles. The Scriblerian wits assume the role of 'moderns' in adoption of this position against irrationalism and supernaturalism, Whiston and his millenarian cohorts, the ancients; yet time was needed before the Scriblerians could asses his prophesies and penetrate through his mumbo-jumbo. The story of their response to cometary Whiston is my subject here.

Long before 1682, comets had captured the visual and literary imagination.[2] The comet's immense size, tail, blaze, and hairiness, inflamed the imagination of poets and provided them with new sources of imagery, new senses of space and magnitude, new ways to relate the supernatural to the natural world, new empires of concrete color on which to draw, as in Donne's 'vagrant transitory Comets', and Milton's dozens of poetic references to comets. Long before 1682, Shakespeare had written that 'Stars with trains of fire and dews of blood' reflected 'disasters in the sun', and it is accurate to note that by 1600 the comet was the single most accessible image to writers succumbing to magic and superstition. As Horatio had admonished Hamlet: 'The moist star Under whose influence Neptune's empire stands Was sick almost to doomsday with eclipse'; and Gloucester had forewarned his peers that 'these late eclipses in the sun and moon portend no good to us . . . We have seen the best of our times'.[3] A century later the Scriblerians would have agreed with Gloucester, but as the early eighteenth century wore on, the wits' attitude to comets gradually shifted from high seriousness to comic levity. Popular interest in comets after 1680 enhanced this transition to levity, as did increasing secularism and deism. By 1682–3 the popular imagination was saturated with all types of astronomical speculation, as well as fantastic hypotheses regarding prior collisions of comets with the earth, cometary winters as the primary cause for the death of the legendary giants, monsters and dinosaurs, and astro-theological predictions, as in Christopher Nesse's astrological treatises from which prints and engravings were made.[4] In the 1680s and 1690s comets elicited every type of religious response, especially supernatural ones about doomsday and the New Jerusalem.

By 1684 John Edwards, a Calvinist minister who believed that Socinianism and John Locke's philosophy had been the two worst developments of his lifetime, could write a three-hundred page treatise surveying the history of comets from the Greeks and Romans to the present time. Here Edward's explained that comets were the ultimate source of vital animal spirits that gave motion to all matter, that the Romans believed comets to be the souls of deceased heroes, and demonstrated that astrology as well as human health depended on the periodicity of comets.[5] Edwards's *Cometomantia* was not a Restoration equivalent of Eaton Barrett's best-selling Victorian book *The Comet Mania* (1857), published when Halley's comet was to return once again, but it made larger claims for comets than any book written in English before 1684. As the dire political events of 1685 evolved, and four years later, when William and Mary were crowned, prophets increasingly cited comets as prognostications and proof of these abrupt political changes. 1689 was an important year, politically as well as philosophically, noteworthy for crownings, acts of toleration, and Lockean philosophy. It also printed works and engravings noting the conjunction of astronomical, astrological, and theological realms, as in a caricature labelled 'The Dutch Stockhouse of the Jesuit Father Peter: The Great Three-Horned Doctor, a great naturalist',[6] who, as a Jesuit priest and oppressor of Protestant dissenters, collaborates with demons to enslave Protestants, and searches with these dissenters for the periodicity of Halley's comet. But nothing cometary written in English in the 1680s could rival the appearance, in 1696, of Whiston's *New Theory of the Earth*. This treatise written by a *bona fide* Newtonian who appeared to have the highest scientific credentials was ostensibly a reply to Thomas Burnet's *Sacred Theory of the Earth*, but struck some contemporary readers as outrageous and as more extravagant than Burnet.[7] Here Whiston related how the earth itself began its existence as a comet that evolved into a planet. A second comet with 'little or no atmosphere' struck the Equator and set the Earth spinning. Then human sinfulness – he maintained – brought a third and punitive comet that passed in front of the orbiting earth at noon,

Peking time, on Monday 2 December 2926 B.C. The tidal force cracked the earth, and the tail of the comet drenched it with water six miles deep: this was the great Noachian deluge that Whiston labelled 'the Choc of a Comet'.

Whiston seems to have borrowed his theory of 'a Choc' directly from Halley, and his absorption may have involved plagiarism. In 1694 Halley had read two papers before the Royal Society in which he suggested as a natural cause for the Flood, the 'Choc of a Comet'.[8] Either Halley or the editor of the *Philosophical Transactions* suspected Whiston of plagiarism, since the editor's note concludes with a bizarre comment about Whiston's *New Theory of the Earth* having appeared a year and a half *after* Halley read his two papers. The innuendo of plagiarism from Halley appears not to have harmed Whiston immediately, not until the publication in 1708 of a second edition of *A New Theory*. The appearance of this edition coincided with the first charges of antitrinitarian heresy read into Whiston's Boyle Lectures, and proved that Halley's caution about not wanting to publish Whiston's papers postulating the comet as a cause for the flood was justified.[9] The Burnet controversy raged in the 1690s; after 1696 it took a new turn as Halley's comet and the flood were widely debated. The Scriblerians had not yet assembled, and in the 1690s Pope was still a boy, but after 1708 the wits would have had to exist in a tunnel not to have heard the public debates about Whiston's heresy and his cometary floods. There was no need to apprise themselves: talk about Whistonism was pervasive. After the appearance of Whiston's second edition of a *New Theory of the Earth*, Whiston's notoriety spread: in 1708 he was formally charged with heresy; two years later he was banished from the Lucasian Chair at Cambridge on grounds that he was an arian. During 1708–10 he earned a reputation as a heretic who had to be removed from university office for the protection of the Church of England. When the young Pope wrote to Cromwell that summer he referred, in verse, to 'the wicked Works of Whiston', alluding to Whiston's heretical Boyle Lectures.

Throughout 1709 Whiston remained in seclusion. After his banishment from Cambridge in 1710 he claimed to be impoverished, unable to feed his children, and migrated to London. With the help of Addison and Steele he began to offer coffee-house lectures on science and theology.[10] It is unknown to what degree they personally knew him, yet there is evidence that on several occasions in 1713–14 Steele presided over the coffee-drinkers and introduced his speaker. At Button's, Whiston's subject was the new astronomy with an emphasis on eclipses and comets. By August 1713 Pope had heard Whiston on at least one of these occasions, as Marjorie Nicolson and I demonstrated in *'This Long Disease, my Life'*; and on 14 August, Pope wrote to John Caryll:

You can't wonder my thoughts are scarce consistent, when I tell you how they are distracted. This minute, perhaps, I am above the stars, with a thousand systems round about me, looking forward into the vast abyss of eternity, and losing my whole comprehension in the boundless spaces of the extended Creation, in dialogues with W[histon] and the astronomers; the next moment I am below all trifles. . . .[11]

Pope's attitude to 'wicked Whiston' was never so positive, but how did he reconcile heresy with Whiston's spellbinding performance in the coffee-house? If Pope was transported by Whiston's coffeehouse 'dialogues,' as he told Caryll, especially by Whiston's 'Scheme of the Solar System' based on Halley's 'Table of Comets',[12] he would be less charitable about cometary Whiston on other occasions. Just two months after Pope heard Whiston at Button's he advanced a plan to Swift to form a 'Club of Scriblerus' that would study the works of the learned. From this project emerged *The Memoirs of Martinus Scriblerus* a decade later, which refers to Whiston and his comets. Yet before Martinus's memoirs saw the light of day, something happened to persuade Pope and the other Scriblerians that among the 'falsely learned', cometary Whiston, unlike Halley, ranked in the front line.

Chief among these offences (provocation of the wits would be too strong a term) was Whiston's publication, just a single sheet, entitled *The Cause of the Deluge Demonstrated*, which appeared in February 1714 as Whiston's indicters were pressing charges against him for arian heresy.[13] Although no comet had been seen in England for almost a generation, Whiston claimed in this publication the idea of a cometary start and cometary end of the world. As a type of eighteenth-century Immanuel Velikovsky, he identified the millenarian comet as Halley's comet of 1682, and claimed that his earlier hypothesis (i.e. the flood's origination in a comet) was empirical fact. This was the same comet, he asserted, seen in 44 B.C., a year after the assassination of Julius Caesar. Most alarming, Whiston claimed, this comet was the only one listed in Halley's tables that would reduce the whole earth to flames and ashes, for it was the only one that 'can come near enough to our Earth in the comets' Ascent from the Sun to cause the great Conflagration'.[14]

To the wits such certain knowledge seemed ludicrous. It was one matter for the modest and cautious Halley to compile tables and espouse hypotheses about celestial mechanics; quite another for the defrocked, dechaired, banished pauper to identify Halley's comet as the cause of a new deluge that would suddenly bring the world to an end. The Scriblerians lost no time before exposing their target: as they made grandiose plans to satirise false learning in the midsummer of 1714, they focused on Whiston and his cometary theory in an artificial letter from Martinus written 'to the most amiable Lindamira' in chapter 14, expressing Whiston's fear that the 'Ages must be numbred [sic], nay perhaps some Comet may vitrify this Globe on which we tred'.[15] Among the 'Discoveries and Works of the Great Scriblerus . . . written and to be written, known and unknown', is a list of 'Tide-Tables, for a Comet, that is to approximate towards the Earth'.[16] Martinus assembles these tide-tables in preparation for the approaching comet, which, Whiston contended, was the same one as that of 1682. His fictive tables are Scriblerian parodies of those Whiston is thought to have filched from Halley.

To the Scriblerians Whiston's theory of the millenarian deluge was derisory: all things *began* with a comet – so they shall *end* with a comet. This was the view that had to be challenged, and the astronomical events of the next three years – 1715, 1716, 1717 – provided them with plenty of ammunition. John Gay, like Pope, had heard Whiston at Button's and knew what a spellbinder he proved to his captivated audiences. Inspired by one of these occasions but not setting quill to paper for over a decade, Gay assumed the persona of a London draper named 'J. Baker, Knight', who was a regular attendant at Whiston's Tuesday lectures and was probably the author of a delightful prose satire called *A True and Faithful Narrative of What passed in London during the general Consternation of all Ranks and Degrees of Mankind*, probably composed in 1731–2 and printed anonymously in the Swift-Pope *Miscellanies*.[17]

The work is supposedly a true account of Whiston's lecture near the Royal Exchange on Tuesday 13 October 1714. Gay (assuming he was the author) fabricates a name for each of the fourteen regular auditors (tradesmen and craftsmen, not the learned), and recounts both the lecture and the course of Whiston's predictions. Millenarians, world-enders, and doomsdayers are addressed in particular: 'Friends and Fellow Citizens,' the narrator remembers Whiston saying prophetically, 'all speculative Science is at an end; the Period of all things is at Hand; on *Friday* next this World shall be no more. Put not your confidence in me, Brethren, for tomorrow Morning, five Minutes after Five, the Truth will be Evident; in that instant the Comet shall appear. . . . As ye have heard, believe. Go hence and prepare your Wives, your Families and Friends for the universal Change.'[18] Nothing more convinced the draper that Whiston believed his own prophecy than Whiston's returning 'a shilling apiece to the Youths that had been disappointed of their Lecture'. Then the draper describes how word of the prediction spread like fire throughout London, how initial scepticsm was replaced by awesome credulity and with what consternation the dreaded comet appeared 'three minutes after five'. Here Gay takes a dig at Whiston for his earlier error in predicting the time of the total eclipse as two minutes *later* than it actually appeared.[19] According to the draper, the learned in London thought they were to watch an eclipse, but instead they watched a comet. Yet if Whiston proved, as an astronomer, to be exact to the minute, he was much less successful as a millenarian for, according to the narrator, the comet came and passed, 'Friday came, and the People covered all the streets . . . they Drank, they Whor'd, they Swore, they L'yd'.[20] If the millenarian event had demonstrated a deeper side of human nature than evidenced in normal times, London itself soon returned to normalcy, Londoners to their old habits and vices. Whiston's prediction had proved, like his apostasy, to be a fraud. He was, it now seemed, its false prophet, false apostle.

I have already suggested why Whiston was associated with the cometary end of the world in 1714 (especially in his *Cause of the Deluge Demonstrated*), even though no comet had been seen in England for some time. It remains to be shown how Whiston's unrelenting prophesies continued to irk and arouse the Scriblerians after 1714. In the year of Queen Anne's death there had been no *total* eclipse of the sun over England for 575 years. According to Halley's computations, the last total solar eclipse had occurred in 1140;[21] the last partial solar eclipse had occurred on 13 September 1699, and was barely observed in England because of inclement weather. But a total solar eclipse was more engrossing and great preparations were being made for it, not only in the Royal Society but widely among laymen and the learned and in such poems as an anonymous work called *The Eclipse* which none of the Scriblerians wrote. It was to occur on 22 April 1715, in the preparation for which Whiston was much involved. Claiming that he and Halley had predicted it would arrive in April, Whiston devised an instrument for the calculation of solar eclipses called 'The Copernicus'.[22] By March he was selling his Copernicus and manuals of instruction for its use in the coffee-houses; Pope and Gay may have bought one and become proficient in elliptical calculations through its use, as they both told Caryll in their letters. They and other Scriblerians eagerly awaited the total solar eclipse, especially as a millenarian event, but it came and went without excitement.

Two months after the April eclipse the second nova known to appear and disappear periodically appeared in England in June, which astronomers recognised as a reappearance of the one first described in 1686.[23] First observed on 15 June, it so increased in visibility during the summer that by August it could be seen by the naked eye.[24] Many laymen considered it a comet because of its periodicity, and Whiston did nothing in the coffee-houses to disabuse them. Encouraged by Whiston, the printed ephemera that summer and autumn suggested that these continuing strange portents in the heavens were presages that would direly affect mankind. The next winter of 1716–17 was phenomenally cold; so cold that the *Historical Register* considered the frost to be among the 'most memorable occurrences and events' in English history and described how the Thames was frozen over, with whole oxen and sheep roasted on the ice.[25] All this caused the public, and especially the Scriblerians, to wonder about Whiston's role as a prophet.

On 6 March another celestial phenomenon appeared, vividly described by Halley. 'This Phenomenon', he wrote in the *Philosophical Transactions*, 'found all those that are skill'd in the Observation of the Heavens unprepared.'[26] Calling it an aurora borealis rather than a nova or comet, Halley despaired that he would ever again see what he saw that night but he was silent about its mellenarian significance. Amateurs like the Scriblerians and other laymen thought the aurora to be a comet with a spectacular tail, just as they had believed eclipses and other celestial phenomena to be comets, suggesting to what terrific degree comets then had a stranglehold on their imagination. Whiston was in the meantime feeding the popular imagination about the end of the world in the London coffee-houses and, more particularly, in Steele's Censorium: claiming that these continuing appearances in the skies were unequivocal warnings to mankind sinful humanity. The newspapers brimmed with speculation that a comet had just passed or was about to pass, and even Steele's *Chit-Chat* proclaimed that 'all this Degree of Light' in the skies is proof of a comet soon to collide with the earth and destroy the world we know by flames, flood and conflagration.[27]

Four days after the 6 March aurora appeared, the unsigned epilogue written for Addison's play *The Drummer* linked Pope and Whiston in transformative images suggestive of the fourth canto of *The Rape of the Lock*:

> If any Briton in this Place appears,
> A slave to Priests, or superstitious Fears,
> Let these odd Scenes reform his Brainsick Notions,
> Or BYFIELD'S[28] ready-to apply his Potions.
> Those Wits excepted, who appear'd so wise,
> To conjure Spectres from the vap'ry Skies.
> A very POPE (I'm told may be afraid,
> And tremble at the Monsters, which he made.)
> From dark mishapen Clouds* of many a Dye.
> A different Object rose to every Eye:
> And the same Vapour, as your Fancies ran,
> Appear'd a *Monarch*, or a *Warming-Pan*.
> Well has Friend WHISTON every Scene apply'd
> And drawn th' unmeaning Meteor to our Side.
>   *The late *Meteor*

Only a few weeks earlier Pope had described Minerva's descent in his just published translation of the *Iliad* as 'like a comet', a description associated with further Jacobite prophecies of uprisings,[29] and on 18 February an article in the *Weekly Packet* had called attention to the cometary passage in the *Iliad* translation. This article commented that the cryptic message concealed in Pope's note accompanying the *Iliad* passage was not surprising in view of Pope being a Catholic himself.[30] A few weeks later, Dr Arbuthnot tackled Whiston, who was still prophesying the end of the world in the coffee-houses, in another Scriblerian paper: a satire on the many sorts of projects with which civic authorities were deluged. This was a pamphlet addressed *To the Right Honourable The Mayor and Alderman of the City of London* and subtitled *The Humble Petition of Colliers, Cooks, Cook-Maids, Blacksmiths, Jackmakers, Braziers, and Others.*[31] Satirically claiming that a whole class of virtuosi who call themselves 'Catoprical victuallers, by gathering, breaking, folding, and bundling up the Sun-Beams, by the help of certain Glasses', planned to procure a monopoly for 'cooking' in the future, and 'oblige Cooks, and Cookmaids to study Opticks and Astronomy', Arbuthnot charged Whiston, among others, with participating in the scheme, just as Whiston had earlier attempted to discover the longitude. Then, early in November, yet another Scriblerian satire on Whiston appeared, originally printed on both sides of a single sheet. Sherburn and Ault both attributed it to Pope, but Nicolson and Rousseau made what still seems to me a stronger case for Gay, especially if Gay had been the author of *A True and Faithful Narrative*, since it follows as a natural sequence to that prose satire.[32]

As Swift had parodied astrological prognostications in the Bickerstaff Papers, so the Scriblerian *God's Revenge* satirised astronomical predictions, and whether one or another of the Scriblerians wrote it the satire it contains is unmistakably theirs. Looking back over the calamities of English Restoration history, Gay (if it was his) recalls the Great Plague and the Great Fire sent to chastise a sinful people. He recounts how the nation had scarcely recovered from these disasters when 'other Abominations rose up in the land', which not even the abrupt political discontinuities of the 1680s could correct.[33] For 'still the Nation so greatly offended, that Socinianism, Arianism, and *Whistonism* triumph'd in our Streets',[34] and '*Whistonism*' is invoked here with such resonance that one can speculate with what frequency it was a term bandied about in the town. Yet considering the profound effect Whiston's lectures had on Pope's imagination for at least two years now, it is strange that only a year later – in 1716 – Pope should return to his earlier position about the 'wicked Works of Whiston' and acerbically denounce '*Whistonism*' as the natural culmination of Socinianism and Arianism. Gay, on the other hand, or possibly Swift, had already had his fun with cometary Whiston in the earlier satire, and I believe that he (Gay) was providing a sequel to the earlier work. But whether written by Gay, Swift or Pope, or some combination of the Scriblerians, the author at this point launches into an account of recent astronomical phenomena: two real, one exaggerated or invented. 'And yet still, after all these Visitations [in the skies],' the satirist genially satirises millenarian Whiston, 'it has pleased Heaven to visit us with a Contagion more Epidemical, and of consequences more Fatal [i.e. namely Punning]: This was foretold to Us, By that unparallel'd Eclipse of 1714 . . . [1715].[35] Secondly, By the dreadful Coruscations in the Air this present Year [1716]: and thirdly, by the Nine Comets seen at once over Soho-Square, by Mrs. Katherine Wadlington . . .' The eclipse and aurora we recognise, but to what do the satirist's nine comets refer? Most likely they are the author's hyperbole for other celestial phenomena that appeared during that astronomical *annus mirabilis* of 1715–16, especially the nova that had been visible to the naked eye and which so many laymen (including all the Scriblerians) thought to be yet another comet. But the nova returned in 1716, just as the author of the 1715 article in the *Philosophical Transactions* predicted, and shone most brightly around 10 September, a few weeks before *God's Revenge* appeared.[36]

God's Revenge inspired a pseudonymous 'E. Parker, Philomath' to compose a series of prophecies for the New Year 1717, which appeared in December 1716 as *Mr. Joanidion Fielding. His True and Faithful Account of the Strange and Miraculous Comet Which was Seen by the Mufti at Constantinople, As appears by the Daily Courant of the Month.*[37] Whoever E. Parker was, he unequivocally attributed *God's Revenge* to Alexander Pope, whose name is included among the prophecies for 1717: 'If the Pope be not Eat up by Pun-aises, for Anathema's that he never denounc'd, he shall at least be Tickled to Death, or receive a Phillip from St. Ambrose' – that is, Pope's enemy Ambrose Phillips. Whiston's name also appears in close proximity to Pope's: 'Mr. Whiston's Scheme of Primitive Christianity', with its prediction of the return of the flood by collision of the earth with a comet, 'shall not prevail this Year [1717]; so that it will be more for [Whiston's] Profit', the satirist's bathos continues, 'to meddle only with Inspiration of the Air-Pump at White Hall.'[38] The appearance a few weeks earlier of Whiston's *Astronomical Principles of Religion*, again announcing that the earth was originally a comet and that Halley's comet of 1682 had been the cause of the great Noachian deluge, fed into Parker's aim. Parker's point of departure – 'this Strange and Miraculous Comet . . . seen by the Mufti at Constantinople' – was a real comet described by the *Daily Courant* for Thursday 15 November:

> Paris . . . They write from Malta, that an English ship arrived there from the Isles of the Archipelago and reports that there has been seen at Constantinople for eight Days, a Comet, hairy with a long Tail; which appeared soon after Sun-rising, and extended itself from North to South. This has

very much frighted those People [the Turks], who are not used to such Appearances in the Heavens.[39]

Parker visualises the Mufti of Constantinople as he sits pensively musing on the banks of the Hellespont. His 'Favorite Muse', he narrates, came to him prophetically crying out: 'Behold, thou High Priest of Allah, how the light streameth out of Darkness, and the Firmament blazeth as a Topaz.'

> At this the High-Priest of Mahomet lifted his Eyes from the Ground, and lo! he beheld a palish Light, like the Crescent of the Imperial Turbant; it soon shot itself forth into the Form of a Comet, whose Body appeared two Degree Diameter; and its Tail in the Form of a Paraboloid, shot up within 20 Degrees of the Zenith; so that it appeared like the One-ey'd Polyphemus, in a Full-bottom'd Periwig.[40]

The pseudonymous Parker also recounts that consternation throughout the land continued unabated, together with various attempts to explain the mysterious comet and anticipate its possible malign or benign effects. The explanation offered by the Sannadrin, we are told, was the one that satisfied the Ottoman Emperor: for 'it [the comet] only denoted the Arrival [in Constantinople] of a British Ambassadress of marvellous Beauty, with a long Train of Attendants'.[41] Wortley Montagu had become Ambassador to Turkey in August 1716, three months before the pseudonymous Parker composed his delightful satire, the satirist's 'British Ambassadress' being none other, of course, than Lady Mary Wortley Montagu, although the Montagus arrived in Constantinople too late to see the comet.[42] Yet even more crucial than the topical prophecy about Pope's Lady Mary, which someone as informed as John Gay would have understood in view of her relation to Pope at just this time (1716), are the references to Whiston, and I believe that they clue us in to the real author. 'Mr. Joanidion Fielding' playfully writes: '[while viewing the comet] I did not see the late Coruscations, which Mr. Whiston rightly judges to be the Fire-works of Aerial Spirits . . .'[43] A few paragraphs later: 'a Friend of mine did see the Nine Comets at Mrs. Wadlington's in Soho-Square, when he was going to celebrate a Geocentrick Conjunction', adding perhaps naughtily, 'I did see Mr. Halley (whom indeed I never had a good Opinion of) flying a fiery Chariot over Greenwich-Park'.[44]

This chronological survey brings us to the beginning of 1717 when Gay, or one of the other Scriblerians, had now had his fun with Whiston at least twice, in equally playful sequels. If we continued to survey the satires of the wits in this chronological vein, and if we followed other diversions of the original Scriblerians, we would see that their fascination and vexation with cometary Whiston tapered off after 1717, as the skies became more peaceful. Furthermore, by 1717–18 cometary Whiston himself had become old hat to Londoners who had, like Pope, actually heard him, or, like Swift, at least heard about his spell-binding coffee-house presentations for almost a decade. But the Scriblerians, as well as Addison and Steele, were not through with cometary jests arising from their cosmic wonder and cometary imagination. A decade later Swift would, of course, make the possible collision of the comet with the earth one of the memorable events of Lemuel's travels through Laputa, about which so much Swiftiana has now been written that it would belabour the point to say anything further about it here.[45] And, looking further ahead, it may have been the recent death of Halley in 1742 that prompted Pope to compare his third group of dullards in the *New Dunciad* published in October 1743, to a comet approaching perihelion.[46] Pope's cometary wit in the *New Dunciad* cannot be traced to a single source, nor need it be since he had grown up with comets, Halleian and Whistonian, and he need not have read anything to insert them wittily into mock-epics and burlesques about 'wicked Whiston'. But Pope was too infirm during the winter of 1744 – his last winter – to notice that yet another comet threatened to visit England. Its imminence did not prevent Whiston, now of a great age, writing memoirs, and contemplating religious conversions, from returning to millenarian prophecies, as his *Memoirs* relate, and lesser poets than Pope and Swift from exercising their own millenarian imaginations. Even the *Gentleman's Magazine* devoted a portion of each issue in 1742–3 to discussion of the new threatening comet, with the figure of Whiston lingering in the background and often mentioned in these accounts.[47]

By 1758 Whiston and all the Scriblerians were dead, unable to survey the town to see if Halley's comet really did return. Of course it did in 1759 and was first glimpsed near Dresden, and then widely viewed throughout Europe and America; but by now, exactly seventy-six years after the comet of 1682, comets had lost some (certainly not all) of their millenarian force. The comet glimpsed in 1758 reappeared early in the spring of 1759, and permitted Laurence Sterne, among others, who was then composing the first two books of *Tristram Shandy*, to be piqued enough to remember the cometary wit of the Scriblerians and the clever way they had identified Whiston as their favourite target. In Sterne's witty version Dr Slop is likened to the sight of a besmirched Obadiah who is worse than 'the worst of Whiston's comets'.[48] Whiston is thus yoked to sentimental fiction of the 1750s but without the lengthy treatment he had received five decades earlier. During that spring of 1759 Horace Walpole played a card game at Strawberry Hill called 'comets', with a pack of cards he had used seventeen years earlier in 1742.[49] The rules of the game elude me but Walpole's attitude does not: he was eager to learn whether his friend in Italy, Horace Mann, had seen or heard of the new comet, as eager to know as Madame Tussaud was in 1986 to make a waxwork model of Edmond Halley. Yet Walpole, who knew almost nothing about astronomy, doubted that the comet of 1758–9 was Halley's comet, and as he interrogated Mann it is clear that his reason was historical rather than millenarian. Walpole had read somewhere, or heard about the events in the skies in the years immediately preceding his birth, and he wondered if another cometary *annus mirabilis* was on the way, as it had been in 1715–16. 'In short,' Walpole wrote sardonically to Mann, 'like pineapples and gold pheasants, these comets of Halley will grow so common as to be sold in Covent Garden market' for a shilling.[50] But during that autumn of 1759 terrific heat struck England, and as the reclusive Walpole fried in a torrid Strawberry Hill, he was not entirely willing to surrender the possibility of supernatural powers of comets. He complained, once again, to Horace Mann that all this conflagration of heat is no doubt the work of the spring's comet, and if only George II had died a few months earlier, the returning comet would have been blamed no doubt for his death too. Meanwhile, the notorious Madame Sevigny was ailing in France, and was 'not well [enough]', according to Walpole, to sit up until the early hours of the morning awaiting the comet, for which purpose 'she had appointed [her own personal] astronomer to bring his telescopes to the President Henault's, where a large party had gathered to view it'.[51] On both sides of the Channel interest in comets was as great as ever, but no one seems to have replaced Whiston in prophesying their millenarian consequence. By the 1760s comets (but not Halley's) had come and gone for decades, some falsely called Halley's comet, others not, and the new generation of Walpole and Sterne was as eager to view comets as the Scriblerians had been. Now the ladies assumed a more active role than they had earlier in the century: all sorts of astronomical books and manuals composed at mid-century taught them how to locate comets and to plot their periodicities, and if there was no exact equivalent for females of Whiston's 'Copernicus' instrument, there were nevertheless dozens of similar devices, made by Benjamin Martin and other instrument makers in anticipation of the first expected return of Halley's comet. When Caroline Herschel, for example, read her account of a new comet before the Royal Society on 9 November 1786, long after Halley's comet returned, she represented the very best of a tradition for ladies now decades old.

But despite Halley's achievement in identifying the comet of 1682 and bearing in mind the inevitable growth of astronomical knowledge during the eighteenth century, a fundamental fabric of the social domain had nevertheless altered since the Interregnum and Restoration: especially the gradual decline of superstition. When the young gentlewoman Euphrosyne and her vacationing undergraduate brother Cleonicus discuss comets in Benjamin Martin's 1755 survey of the physical sciences,[52] a question arises about the purpose of these 'wondrous Bodies'. Cleonicus, recollecting Whiston, answers that each comet is a type of hell – 'Seats not unaptly deem'd the Place of Hell' – and cites the poetic passage found as an epigraph to this essay; but sister Euphrosyne dismisses the absurd notion that comets portend extraordinary events, raising various sensible objections to 'a Plurality of Hells' and to the idea that comets possess any supernatural powers. Brother Cleonicus concurs, reminding his sister how all these superstitious ideas are 'Nothing but Conjecture', the worst type of conjecture at that. But from the vantage of the more general historian today who can see beyond the gaze of Euphrosyne or Cleonicus it is unthinkable, given the climate of religious opinion before the eighteenth century, that a group of writers equivalent to the Scriblerians could have thought of deriding cometary prophecy as they did. And even the satiric author of *Hudibras*, who died just as the comet of 1680 was approaching England and as John Hill, a London physician and astrologer, was issuing an *Allarm to Europe by a late prodigious Comet seen* (1680), has Sidrophel interpret a lit-up kite flying in the sky as a supernatural omen only once:

> This Sidrophel by chance espy'd,
> And with amazement staring wide,
> Bless us! (quoth he), what dreadful wonder
> Is that, appears in heaven yonder?
> A comet, and without a beard!
> Or star that ne'er before appear'd?
> ...
> It must be supernatural,
> Unless it be that cannon-ball,
> That shot i' th' air point-blank upright,
> Was borne to that prodigious height,
> That learn'd philosophers maintain,
> It ne'er came backwards down again.[53]

If 'wicked Whiston' would have agreed with Shakespeare's superstitious Gloucester that 'these late eclipses in the sun and moon portend no good to us. . . . We have seen the best of our times', the more modern Scriblerians, like Euphrosyne, begged to differ. They were not erudite in matters mathematical or astronomic, but their intuition informed them that in every movement in the skies there could not be portents of imminent catastrophe. In this sense the Scriblerians were of the party of the moderns – more discriminating, less superstitious – and eventually recognised to what degree cometary Whiston was no more prophetic than they were. For them the best of times was just beginning.

NOTES

1. For Whiston see: J. E. Force, *William Whiston Honest Newtonian* (Cambridge, 1985); M. Farell, *William Whiston* (New York, 1981); M. H. Nicolson and G. S. Rousseau, *'This Long Disease, My Life': Alexander Pope and the Sciences* (Princeton, 1968), hereafter cited as NR.

2. For the comet in literature before 1680 see: K. B. Collier, *Cosmogonies of our Fathers* (New York, 1934); J. C. Greene, *The Death of Adam* (Ames, 1959); Nigel Calder, *The Comet is Coming!* (New York, 1980).

3. Shakespeare, *King Lear*, I. ii. 98ff.

4. Christopher Nesse, *A True Account of this present Blasing [sic] Star. Presenting itself to the view of the World. This August 1682*, a single-sheet folio. Nesse or Ness (1621–1705) was a non-conformist divine who wrote astrological and theological discourses about Halley's comet in 1682. His *Astrological and Theological Discourse . . . of the Great Comet* (1682, a broadside in the British Library), in which Biblical predictions are fulfilled according to the zodiac, is reproduced here.

5. John Edwards (1637–1716), *Cometomantia: A Discourse of Comets: Shewing their Original, Substance, Place, Time . . . and, more especially, their Prognosticks, Significations and Presages* (1684).

6. No. 1209 in F. G. Stephens, *Catalogue of Political and Personal Satires in the British Museum, vol III, part I* (1877), 738, where the cometary references in the engraving are explained.

7. The most thorough treatment of the Burnet controversy is found in M. H. Nicolson's *Mountain Gloom and Mountain Glory*; for 1689 and Locke's philosophy see John Yolton, *Perceptual Acquaintance from Descartes to Reid* (Minneapolis, 1984).

8. Both papers were published thirty years later in *Philosophical Transactions*, XXXIII, No. 383 (May-June 1724), 118–25; see NR, 141.

9. Halley gave as his reason: 'he being sensible that he might have adventured *ultra crepidam*: and apprehensive lest by some unguarded Expression he might incur the Censur of the Sacred Order'. See NR, 141.

10. James Force (n. 1, 163) recognises Addison's acquaintanceship with Whiston at this time but fails to explain, as we did (NR, 170–2), why Addison continued to satirise Whiston from 1713 to the late 1720s.

11. G. Sherburn (ed.), *The Correspondence of Alexander Pope*, 5 vols. (Oxford, 1956), I, 185.

12. NR, 145–8, 167, where the 'Scheme' is described.

13. NR, 180–1.

14. Whiston, 'The Cause of the Deluge Demonstrated', appended to Whiston's *Astronomical Principles of Religion* (1717), 6.

15. C. Kerby-Miller (ed.), *Memoirs of . . . Martinus Scriblerus* (New Haven, 1950), 149.

16. *Ibid.*, 167.

17. As the fourth volume of the set which appeared in 1732. Authorship of the prose satire remains problematic: it has been attributed at one time or another to Gay, Swift, Pope, and Arbuthnot, all of whom have a claim; but whichever of these authors actually composed it, it clearly remains a Scriblerian probject. Composition and authorship of the work has been discussed by L. D. Peterson in 'Jonathan Swift and a Prose "Day of Judgment" ', *Modern Philology*, LXXXI (1984), 401–6. I have used the copy in the Clark Library on whose title-page is written in an eighteenth-century hand 'by Mr John Gay'.

18. *A True and Faithful Narrative*, 257.

19. *Ibid.*, 260–1.

20 Ibid., 275–6.
21 E. Halley, 'Observations of the late Total Eclipse of the Sun on the 22nd of April last past, made before the Royal Society', *Philosophical Transactions* XIX, No. 303 (March–May 1715), 245–62.
22 Whiston, *The Copernicus Explain'd: Or a Brief Account of the Nature and Use of an Universal Astronomical Instrument, For the Calculation and Exhibition of the New and Full Moon, and of the Eclipses, both Solar and Lunar: with the Places Heliocentrical and Geocentrical of all the Planets, Primary and Secondary, & c.* (n.d.); discussed and described in NR, 158–60.
23 E. Halley, 'A Short History of the several New-Stars that have appear'd within these 150 years: with an Account of the Return of that in Collo Cygni', *Philosophical Transactions* XVIII No. 346 (Nov. 1715).
24 NR, 162.
25 *Historical Register* (1716), 115.
26 *Philosophical Transactions* XIX, No. 347 (Jan.–March 1716), 406.
27 See Rae Blanchard, *Richard Steele's Periodical Journalism* (Oxford, 1942), 261–2.
28 Timothy Byfield, a physician who had turned radical millenarian, joined Whiston, and later linked up with the French prophets from the Cevennes. The text cited here is from the copy in the Yale University Library entitled 'An Epilogue written for the late celebrated New Play called the Drummer, but not spoke', and did not appear with editions of Addison's play until 1744.
29 Pope, *The Iliad of Homer*, 95–106.
30 Quoted in the *Weekly Packet*, 18 Feb. 1716, section on 'letters from Paris [that] tell of earthquakes and meteors in Italy, which scientists regard as portents of war'.
31 *Miscellanies . . . by Dr. Arbuthnot, Mr. Pope, and Mr. Gay* (1742), III, 176–80, discussed in NR, 176.
32 For date, attribution, and contents, see NR, 182–3.
33 *God's Revenge*, quoted in NR, 183.
34 NR, 183.
35 The author writes 'in 1714' but the date must have been a typographical error, since neither Gay nor Pope could have made a mistake about so recent a phenomenon.
36 See n. 23 above.
37 Sherburn mentions the satire in passing in his *Early Career*, 161, 182–83; C. Kerby-Miller considers it the work of the Scriblerians in *The Memoirs of Martinus Scriblerus*, 46, n. 126; NR discuss it in detail on pp. 184–6 but without coming to any conclusions about authorship.
38 NR, 185.
39 *Daily Courant*, No. 4703, 15 Nov. 1716, n.p.
40 NR, 185.
41 NR, 186.
42 They arrived in Adrianople, then the capital of Turkey, on 24 March 1717, and travelled to Constantinople late in May 1717; see Robert Halsband, *The Life of Lady Mary Wortley Montagu* (Oxford, 1956), 68–73.
43 NR, 186.
44 NR, 186.
45 Not only by M. H. Nicolson but the many studies that have appeared since her classic article, 'The Scientific Background of Swift's *Voyage to Laputa*', *Annals of Science* II (1937), 299–334.

46 *The Dunciad*, book, 81–90 and the lengthy accompanying note by Pope and Warburton.
47 Signing some of the articles merely 'Cometographus', the *GM* also excerpted passages about comets from Akenside's poem *Pleasures of Imagination*; commented on Cotton Mather's *Essay on Comets* (Boston, 1744); listed another anonymous millenarian work entitled *The Language of Comets, a lively Call to Repentance for National Sins* (1744), which some believed to have been by Whiston; and reprinted extracts about the life of Halley. A fine study of the public's anticipation of the return of Halley's comet after 1755 is found in C. B. Waff, 'Comet Halley's First Expected Return: English Public Apprehensions, 1755–58', *Journal for the History of Astronomy* SVII (1986): 1–37.
48 *Tristram Shandy*, II, chap. 9. This passage about 'Whiston's comets' is doubly interesting in so far as it contains the only use of the word 'NUCLEUS' [*sic*] in a major novel of the period.
49 W. S. Lewis (ed.), *The Yale Edition of Horace Walpole's Correspondence*, 48 vols. (New Haven, 1937–83), XVIII, 36.
50 *Ibid.*, 294. A few years earlier Mann wrote to Walpole from Italy saying that young women there distinguished themselves from older women by a cometary hairstyle, 'so that the preceding mode a la comete is only fit for Madame Swares and such antiquated beauties' (*ibid.*, XX, 148).
51 *Ibid.*, 249 and 290.
52 Benjamin Martin, 'Of the Comets', dialogue 15, 'The Young Gentleman and Lady's Philosophy', in *General Magazine of Arts and Sciences* (1755–6), 99–106. Martin's cometary views are much less supernatural than Whiston's; see, for example, his *Theory of Comets Illustrated . . . and exemplified in the orbit of the comet of the year 1682, whose return is now near at hand* (London, 1757).
53 Samuel Butler, *Hudibras*, 3 vols. (1819), II, 199–200.

INDEX

Aberdeen, Scotland, 81–2, 85, 88
Abrams, M. H., 64–5
Adams Parson, 79
Addison, Joseph, 47
Addison, Thomas, 134, 269–70, 283, 303, 328
Adorno, Theodor, 145
Adrastus, 56–7
Aeschylus, 42
Aesop, 296
aesthetics, 9, 35, 47, 123
AIDS, 37, 47–8
Aikin, John, *Calendar of Nature*, 297
Akenside, Mark, 240, 283–4
alchemy, 203
Allen, D. C., 203
Allen, Dr. John, 127
Allen, Ralph, 191
Althusser, Louis, 145, 258–9
America, 33, 69, 130, 151–2, 158, 165, 202, 214, 221, 224, 253, 336
 and Americans, 15, 46, 56–60, 64–5, 68, 70, 72–3, 151, 161, 203, 208–9, 238, 255, 258–61, 291
 America Dissected, see Berkeley, George, McSparran, James.
Amsterdam, 85, 151
anatomy, 69, 124, 159, 183, 185
animism, 105
Annals of Scholarship, 238
anthropologists, 79
anthropology, 31, 44, 55, 78, 221, 237, 256, 304
Antigone, 36
antiquarianism, 135
anxieties, 10, 16, 45
Apollo, 51
Arbuthnot, Dr John, 92, 100–1, 126, 164, 182, 325, 333
archaeology, 31, 255, 265
archetypes, 41–9
architecture, 9, 135, 264
Arianism, 86, 103, 154, 333
Arion, 57
Aristotle, 73, 80, 120, 136, 177, 179–80, 223, 265
 Poetics, 35

Arkhangelsky, P. A., *The Treatment of Cholera*, 31
Armstrong, John, 283
 and *The Art of Preserving Health*, 191
Arnaud, George, *Dissertation on Hermaphrodites*, 188
Arnold, Matthew, 203, 220, 247
artefacts, 32
Aschenbach, Gustav von, 2
asthma, 47, 150, 168
astrology, 203
astronomy, 3–4, 123, 125, 137, 182, 203, 273, 279, 325–7, 333, 338
atheism, 214
Atwood, Margaret, 247
Auden, W. S., 295
Ault, Donald, 333
autobiographies, 10, 13, 83–4, 285
Ayre, John, *Arithmetick: a Treatise Designed for the Use . . . of Tradesmen*, 276

Bachelard, Gaston, 15, 67, 15, 223–4, 252–62, 265
 and *The New Scientific Spirit*, 253–4
 and *The Poetics of Reverie*, 253–4
 and *The Poetics of Space*, 253
 and *The Psychoanalysis of Fire*, 253–4, 256
Bacon, Francis, 51, 120, 123, 127, 214, 237, 248, 265, 267, 282, 290
Baglivi, G., 61, 121
Baker, Augustine, *Sancta Sophia*, 88
Baker, Henry, *Employment for the Microscope*, 293
 and *Microscope Made Easy*, 293
Bakhtin, Mikhail M., 145
Baldwin, printing house in London, 299–300
Balzac, Honoré de, 17
 and *The Country Doctor*, 4
Bancroft, Edward, 291
Banks, Joseph, 271
Barbados, 158, 189
Barbauld, Anna, 297
Barbon, Nicholas, 90
Baroque, the, 3
Barre, Poulain de la, *The Women as Good as the Man*, 281

Index

Barrett, Eaton, *The Comet Mania*, 327
Barrow, Isaac, 271–2
Barthes, Roland, 13, 15, 67, 223
Bartholin, Thomas, 184
Bath, 40, 66, 78, 86, 88, 91, 95, 97–8, 131
Bathurst, Henry, 2d Earl of, 190
Battie, Dr William, *Treatise on Madness*, 7, 30
Baudrillard, Jean, 262
Beattie, James, 283
Beddoes, Thomas Lovell, 31
Bedford Coffee House, 276
Behmenist, 103
Bekett, W., *Chirurgical Tracts*, 133
  and *Practical Surgery Illustrated*, 133
Bellet, Isaac, *Letters on the Force of the Imagination in Pregnant Women*, 185
Benson, Donald R., 246
Bentham, Jeremy, 271
Bentinck, Hans Willem, 1st Earl of Portland, 190
Bergson, Henri, 255
Berkeley, George, 132, 137, 145–71, 185, 215
  and *A Chair of Philosophical Reflexions and Inquiries concerning the Virtues of Tar Water*, 147
  and *Farther Thoughts on Tar-Water*, 151
  and *A Letter to Thomas Prior*, 151
  and *A Second Letter from the author of Siris to Thomas Prior*, 151
  as Bishop of Cloyne, 145, 151, 153–5, 158, 164–5, 168, 170
Berkshire, 190
Berlin, 69
Bermuda, 145–6, 161, 165
Bernouilli, Johann, 102
Berquin, Arnauld, *Looking-Glass for the Mind*, 297
Bethel, Hugh, 166
Bettesworth, Thomas, publisher, 299
Bible, 129, 264
Bichat, Marie François Xavier, 68
biographies, 10, 63, 80–1, 147–8, 185, 194, 223, 247, 255, 257
biology, 51, 125, 131, 136, 194, 224, 227, 240, 278
Bird, Robert Montgomery, 31, 34, 42
Birmingham, 131
birth, 14, 16, 228
Blackmore, Richard, 126
  and *The Creation*, 287
blacks, 28
Blake, William, 105, 170, 217, 220, 242–6, 283
  and *The Marriage of Heaven and Hell*, 243

Blome, Richard, author of *The Gentlemans [sic] Recreation* (1686), 278
Blondel, James Augustus, 178, 183–6
Blount, Martha, friend of Alexander Pope, 190
Bocage, Madame du, prolific commentator on England, 269
body, the, 17
Boehme, Jakob, 88–9, 91, 96–8, 101, 103
Boerhaave, Hermann, 34, 60–1, 121, 126–7, 130, 159, 183, 298–9
  and *Academical Lectures on the Theory of Physic*, 122
  and *Aphorisms Concerning the Knowledge and Cure of Diseases*, 124
  and *The Influence of the Sun and Moon upon Human Bodies*, 127
Bonnet, Charles, 286
Booth, Wayne, 64
Bordeu, Théophile de (1722–1776), Montpellier physician
  and medical vitalist and philosopher, 128
Bordieu, Pierre, xii-xiii
Boreman, Thomas, 297
Boston, 145–6
Boswell, James, 43, 80, 289
  and *Life of Johnson*, 120, 274
botany, 69, 189, 194, 273, 278, 290
Bourignon, Antoinette, 89
Bowles, Geoffrey, 85, 91, 103
Boyle, Robert, 62, 84, 88, 119, 121, 123, 137, 154, 159, 184, 215, 217, 223, 265–6, 268–9, 281–2, 290, 302, 328
  and *The Christian Virtuoso*, 266
Boyne, Patrick, *Tour Through Sicily and Malta*, 302
Bradley, Richard, 189–90
brain, 68, 151, 183, 282
Brayton, Alice, 146
Brazil, 189
Brescia, province in northern Italy, 166
Brett, R. L., 295
Bridges, Robert, 12, 14, 30–1, 34, 42
Bright, Richard, 134
Bristol, 89, 91, 131, 302
Britain, 82, 130, 137, 283, 288
British, 57, 61, 103, 124, 136, 153, 159, 220, 227, 259, 264, 267, 269, 280, 287, 292–4, 296, 335
Bronowski, Jacob, 238, 242
Brontë, Charlotte, *The Professor*, 9
Brookes, Richard, 279, 287
  and *Natural History*, 292–3
Brower, Reuben, 295
Brown, John, 121, 129, 133–4
Brown University, 206

Index

Brown, William Cullen, 133
Browne, Sir Thomas, 17, 30, 37–9, 51, 98, 239
 and *Religio Medici*, 39
Buffon, G. L. L., 286, 288, 290–1
 and *Natural History*, 288–9, 302
Bunyan, John, 48
Burgess, Anthony, *The Doctor is Sick*, 5
Burgess, Thomas, *The Physiology of Blushing*, 31
Burke, Edmund, 164, 170, 303
Burlington, Lord Richard Boyle, 3rd Earl of, 190
Burnet, Thomas, 83, 328
 and *Sacred Theory of the Earth*, 82, 304, 327
Burton, J., *Treatise on the Non-Naturals Mechanically Accounted For*, 132
Burton, Robert, 13, 51
 and *Anatomy of Melancholy*, 3–4
Butler, Samuel, 12, 240
 and *Erewhon*, 11
Butterfield, Herbert, 40, 239
Buttons, coffee-house, 277, 330
Byfield, Thomas, 91
Byrom, John, 91, 220

Cabanis, George, 30
Cadogan, William, 63, 204
Caesar, Julius, 329
Cajal, Ramón y, 18
calculus, 82, 124, 216
California, University of
 at Berkeley, 78, 206, 236, 241, 248
 at Irvine, 58
 at Los Angeles, 206
 at Santa Cruz, 58
 see also literature and medicine; literature and science
Cambridge, 40, 86, 104, 159, 182, 215, 223, 237, 268, 271–3
 University of, 189, 285, 325–6
Camisards, 84
Campion, Thomas, 12–14, 30
cancer, 6, 11–12, 65
Canguilhem, Georges, 145, 257, 265
Canterbury, 162
Carlyle, Thomas, 68, 137, 220, 247
Caroline, Princess, 151
Carter, Elizabeth, 167, 283
 and *Newton's Philosophy Explain'd for the Use of Ladies*, 283
Caryll, John, a friend of Alexander Pope, 181, 328–9, 331
Cassirer, Ernst, 248
catarrh, 47

catharsis, 35–9, 42
Catholicism, 333
the 'Catholicon', a panacea claimed to be a universal medicine, 159–61
Céline, Louis Ferdinand, 47
Celsus, 135
Cérisy, Colloque de, 260
Cévennes, the, France, 84, 103
Chambers, Ephraim, 192, 279
Chargaff, Erwin, *Heraclitean Fire*, 13
the Charité Hospital, Paris, 69
Charles II, 266
Charleton, Rice, *Bath Waters*, 301
Chaucer, Geoffrey, 4, 203, 224
Chekhov, Anton, 17, 30–1, 37–8, 42, 46, 51
chemistry, 3, 15, 127, 215, 217, 221, 269, 272–3, 275
Cheselden, William, 39, 101
 and *Anatomy of the Human Body*, 133
Chester, Bishop of, 182
Cheyne, Dr George, 61, 78–106, 121, 126, 133, 281, 284–5, 304
 and *The English Malady*, 78, 80, 92–5, 100, 305
 and *Essay Concerning the Nature of Ailments*, 92
 and *Essay on Health and Long Life*, 91
 and *Essay on Regimen*, 96, 133
 and neo-Pythagoreanism, 103
 and neo-stoicism, 103
 and *Philosophical Principles of Natural Religion*, 84–6, 88–9, 91
 and Wesley's *Primitive Physick*, 80
chiliasm, *see* millenarianism, 82, 102–3
China, 104
Christ, 103
Christianity, 33, 46–7, 62, 79, 86, 88, 92, 95, 99, 103–4, 161, 243–4, 266, 268, 304, 334
Christmas, 86
civilisation, modern, 69
civilisation, Western, 12, 17, 73, 208, 222, 244, 283
Claremont College, 206
Clarke, Samuel, 283
classicism, age of, 213
Claudian, 135
Cloyne, town in County Cork, Ireland, 148, 157
Cocchi, Anthony, *Pythagorean Diet*, 301
Coetlogen, Dennis de, 130–1
 and *Universal Dictionary of Arts and Sciences*, 130
Cohen, I. Bernard, 282
Cohen, Ralph, 206, 260

Colden, Cadwallader, 151-2, 283
Coleridge, Samuel Taylor, 35, 105, 169, 206, 219, 240, 247, 304
Colie, Rosalie, 206
Collinson, Peter, 292
Columbia University, 145, 202
Comenius, John Amos, 296
compassion, 10, 35
Comte, Auguste, 220
Congreve, William, 303
Conrad, Joseph, 224
Constantinople, 334-5
constructionism, 213
  see also externalists; internalists; literature and science
Conway, Viscountess Anne, neo-Platonist philosopher and proponent of medical research, 280
Cooper, Mary, printer and bookseller, 299
Copernicus, Nicholas, 216
Cornaro, Luigi, 91, 95, 304
  see also Lessius, Leonardus
cosmology, 17, 256, 274
Cour, Monsieur Le, 189
Cousins, Norman, 45
Cowper, William, 169, 283, 301
Craig, John, *Theologiae Christianae Principia Mathematica*, 82, 84
criticism, 15, 32, 37, 41, 58-9, 64-5, 70-1, 92, 194, 205, 221, 224, 226-7, 239, 253-4, 256
Cromwell, Oliver, 189, 328
Cullen, William, 121, 128, 133
Culpepper, Nicholas, 177, 180
  and *Directory for Midwives*, 179, 192
  and *The English Physician*, 192
cultures, 30-1, 40, 57, 70-1
Cuninghame, James, 88-9
Cyprianus, Charles, 184

Dagognet, François, 257
Danckerts, Dutch artist, 188
d'Annunzio, Gabriele, 224
Dante Alighieri, 30, 224
Darwin, Charles, 18, 43, 220, 222-3
  and *Origin of Species*, 247
Darwin, Erasmus, 29, 39-40, 240, 283, 287, 290, 301
  and *Botanic Garden*, 293
  and *Zoonomia: or the Laws of Organic Life*, 293
Davie, Donald, 295
Davis, C., publisher, 298-9
death, 14, 16, 36, 48, 85-6, 105, 228
Decker, Sir Matthew, 189-90
deconstruction, 41, 51, 228

Defoe, Daniel, 18, 30, 48, 51, 303-5
  and *Journal of the Plague Year*, 47, 162
deism, 327
Deleuze, Gilles, 15, 58, 223-4
demonism, 28-9
Deptford, borough south-east of London, 159
Derham, William, 286, 294
Derrida, Jacques, xi, 58, 223, 258
  see also discourse theory; literature and medicine; literature and science
Desaguliers, J. T., 275-7, 304
Descartes, René, 206, 220, 255, 265
  and *Discourse on Method*, 281
determinism, 95
Devonshire, 296
Dickens, Charles, 4, 9, 17, 30
  and *Great Expectations*, 170
Diderot, Denis, 243, 247
Digby, Sir Kenelm, 184
Digby, Robert, 165, 184
Dilly, publisher, 298
Dilthey, Wilhelm, 237-8, 241
  and *Geisteswissenschaften*, 236, 241
  and *Naturwissenschaften*, 236, 241
Dionysius, 152
diphtheria, 47
directionalism, 5
disciplinarity, xii-xiv
  and disciplines, ix-xiii
discourses (and discourse theory)
  and practice, 2-53
  and theory, 2-53
  of literature and medicine, 2-51
  of literature and science, 202-50
disease, 11-12, 41, 43-4, 47, 105, 123, 130-1, 133, 136, 162-3, 168, 176, 192, 279
dissenting academies, 271-4
Dodsley, Isaac, 191, 298, 300-1
Donne, John, 170, 204, 326
Dostoyevsky, Feodor, 47
Douglas, Dr James, 188
Douglas, Robert, *Heat in Animals*, 301
Drape, Captain, 150, 153
Dresden, Poland, 336
Dryden, John, 4, 40, 71-2, 244, 269
dualism, 241, 245
dualists, 227-9
  see also metaphors and metaphor theory
Dublin, 147, 149, 155, 159, 164, 170
  society, 150
Dubois, Paul, *Psychic Treatment of Nervous Disorders*, 31
Duncombe, William, 151
Dunton, John, *Athenian Mercury*, 281

Index

Dutens, Louis, 135
Dysart, Alan, 3
dystentery, 148
dyspepsia, 92

Easter, Proclamation on, 83
East Indies, 151
economics, 15, 51, 72, 123, 137, 224, 227, 267, 287, 300
Edinburgh, 80–1, 83, 131, 159
  University of, 133
Edwards, George, 292
  and *Cometomantia,* 327
  and *History of Birds,* 293
Egmont, Lady, 151
eighteenth century, the, 3, 6, 12–13, 17, 29, 33, 39–40, 45, 48, 49, 55–6, 61, 63, 65–6, 69, 72, 78–9, 83–4, 89, 93–4, 98, 100, 102–4, 118–38, 146–8, 169, 176–9, 189–90, 192–4, 206, 216–19, 239, 241, 265, 267, 270–1, 273–6, 279, 281, 283–5, 287–9, 295–9, 301–5, 327, 329, 337–8
Einstein, Albert, 206, 223
Elector of Hanover, 189
Eliot, George, 3, 18, 30–1
Eliot, T. S., 4, 69, 243, 253, 295
embryology, 125, 136, 178
Emerson, Ralph Waldo, 46, 56, 220
  and Emersonian values, 56
emotions, 11
Empedocles, 135
empiricism, 118–38, 223–5, 227
England, 39–40, 42, 46–7, 55, 62, 67, 72, 79–80, 87–9, 91–2, 95–6, 102, 104–5, 124, 131, 135, 138, 151–2, 161–2, 165–6, 169, 177–8, 183, 188–90, 223, 246, 264, 266, 283, 286, 289–90, 292–4, 298, 303, 329, 331, 338
Enlightenment, the, 4, 16, 33–4, 40–1, 45, 47, 49, 55, 57–8, 60, 63, 79–80, 93, 98, 100–1, 118–38, 216–17, 219, 243, 247, 264–305
  and Enlightenment studies, xiii–xiv
Entralgo, P. Lain, 44
ephebi in the disciplines, 55–73
Epicurus, 165
epidemics, 48
epigoni in the disciplines, 55
Episcopalians, 81–2
epistemology, 258–9, 261
erotica, 44
Essex, county in south-east England, 43
Ethiopians, 152
Eton, 270–1

Europe, 8, 61, 69, 82, 124, 151, 178, 187, 214, 223, 289, 298, 303, 336
  and Europeans, 3, 32, 48, 68, 70, 238, 248, 260, 262
Evelyn, John, 189, 194, 269
externalists, 5, 7–8, 14–15
  see also internalists

fanaticism, 181
fantasies, 6, 11, 13, 46–7, 56
Faraday, Michael, 220
Fatio de Duillier, 86–7, 89, 104
Faulkner, George, 152
fear, 10, 35–6
feminists, 51
Feyerabend, Paul, 224, 238, 260
Feyjoo (Feijoo), Benito, 126
  and *Uncertainty of the Physick,* 133
Fielding, Henry, 30, 72, 79, 94, 160, 168, 292
  and *Tom Jones,* 193
Flamsteed, Sir John, 294
Flaubert, Gustave, 30–1, 46–7
Fogle, French, 206
folklore, 256
Fontenelle, Bernard le Bovier de, 297
  and *The Plurality of Worlds,* 280
Foote, Samuel, 30
fornacalia, 55–71
Fortescue, William, 190
Foucault, Michel, xiii, 13, 15–16, 40, 42, 48, 55–6, 58, 61, 65, 68–71, 118, 145, 208–9, 223–4, 226–7, 240, 253, 257–8, 260–2, 265, 288
  and *The Archaeology of Knowledge,* 15
  and *The Birth of the Clinic,* 15–16, 67–8
  and *Discipline and Punishment,* 15
  and *The History of Sexuality,* 15
  and *Madness and Civilisation: a History of Insanity in the Age of Reason,* 15
  and *The Order of Things,* 15
Fox, George, 72
France, 15, 39, 67, 162, 220, 223, 248, 253, 255, 257–8, 260, 288, 337
Franklin, Benjamin, x
Fraser, Alexander Campbell, 156, 159
  and *The Works of George Berkeley,* 147
Fraser, G. S., 295
Freind, Dr John, 82, 124
  and *History of Physick,* 124
Freke, John, 304
Freud, Sigmund, 6, 19, 30–2, 35, 206, 209, 223, 255–7, 290
Frye, Northrop, 253, 256
Funkenstein, Amos, 104

Gainsborough, Thomas, 190
Galen, 30, 61, 80, 126, 130, 135–6, 179, 304
Galileo, 137, 216, 222–3
Garden, George, 82, 103
Garrison, Fielding, 62–3, 65–71
   and *History of Medicine*, 60
Garth, Samuel, 162–3
Gataker, Thomas, *Operations on Venereal Complaints*, 301
Gay, John, 79, 164, 331
Gay, Peter, 206, 330, 333, 335
Gedlike, L. F., 296
gender, 145
geography, 2, 130, 271, 274, 279, 296, 304
geometry, 82, 125, 270, 273–4
George I, King of England, 182
George II, King of England, 337
Germans, 89, 100, 214, 221, 223, 225, 257, 296
Gibbon, Edward, 303
Gide, André, 47
Giles, John, *Ananas, a Treatise on the Pine Apple*, 177
Gilson, Etienne, 253
Glauber, Johann Rudolf, 159
God, 90, 99, 101, 105, 126, 169–71, 183, 191, 214, 220, 266, 285, 287, 325
Godalming, borough south-west of London in West Surrey, 180–1
Goethe, Johann Wolfgang von, x, 3, 224
Goldsmith, Oliver, 5, 30, 38, 98, 170, 192, 272–3, 279, 286–8, 290–3, 297–8, 303
   and *A Citizen of the World*, 161
   and *History of the Earth and Animated Nature*, 290–3, 302
Göttingen, 69
Graunt, John, 123
Gray, Thomas, 72, 80, 164, 283, 293
Greece, 12, 56–7
Greeks, the, 4, 35, 37, 42, 60, 126, 152, 179, 205, 214, 257, 327
Greenberg, Mark L., 242, 246
Greene, D. J., 169–70
Greene, Robert, *The Principles of the Philosophy of the Expensive and Contractive Forces*, 284
Greenhill, A. G., commentator on Dr George Cheyne, 103
Greenland, 130
Greenwich, 294
Gregory, David, 83, 86
Gregory, Richard L., 245
Griffin, William, 292
Griffith, Ralph, 290
Guattari, Felix, psychoanalyst and collaborator of Gilles Deleuze, 223–4

Guggisberg, F. G., 274
Guildford, borough in West Surrey, 181
Guinea, 150
Gunther, R. W. T., 272
Guyon, Mme Jeanne Marie Bouvier de la Motte, 89, 103

Habermas, Jürgen, xi
Hales, Stephen, 156, 159–60
   and *An Account of Some Experiments and Observations on Tar-Water*, 158
   and *A Description of Ventilators*, 130
   and *Haemastaticks*, 159
   and *Philosophical Experiments*, 133
   and *Statical Experiments on the Sap in Vegetables*, 132
   and *Vegetable Staticks*, 159
Haller, Albrecht von, 12, 121, 128
Halley, Edmund, 123–4, 326–32, 334–7
Hamilton, Sir William, *Observations on Mount Vesuvius*, 302
Hancock, John, 30, 156, 162
Harley, Robert, Earl of Oxford, 268, 271–2, 325
Harris, John, 192, 279
   and *Lexicon Technicum*, 278
Harris, Walter, 123
Harrison, Frederic, 220
Hartley, David, 101, 126, 193, 284–5
   and *Ten Cases of Persons who have taken Mrs. Stephen's Medicines for the Stone*, 132
Harvard University, 118
Harvey, Gideon, 32
   and *Vanities of Physick and Philosophy*, 30
Harvey, William, 40, 119–20, 124, 134, 282
Hawes, William, 163–4
   and *An Examination of Mr John Wesley's Primitive Physick*, 163
Hawksbee, Francis, 276
Hawthorne, Nathaniel, 46
health, 11, 14, 17, 26, 32, 43, 46, 48, 51, 83, 91, 94–6, 166, 169
Hebus, town in Lancashire, 186
hedonism, 85
Hegel, Georg Wilhelm Friedrich, 255
Heisenberg, Werner, 206
Heliodorus, 184
Helmont, Jan Baptista van, 159
Henault, President Charles, 337
Henderson, G. D., 103
hermaphrodites, 180, 187–8
heroes, 16, 43, 50
heroism, 17
Herring, Archbishop Thomas, 151
Herschel, Caroline, 337

Index

Hervey, Lord John, 46, 80
Hesiod, 184
Hesse, Mary, 226
Hill, John, 80, 119, 186, 290, 299–300, 338
  and *The Vegetable System*, 293
Hillary, Dr William, 128
  and *A Rational and Mechanical Essay on the Small Pox*, 126, 133
Hilton, Nelson, 242, 245, 248
Hippocrates, 61, 80, 120, 135, 154
Hirons, Raymond, 290
historians, 3–20, 44, 48, 59, 61, 63, 66–73, 93, 102–3, 118–22, 124, 127, 135–6, 138, 145, 156, 186, 202–7, 209, 216, 220, 223, 225, 227, 239, 242, 244, 246, 257–8, 260, 265, 267, 273–4, 277, 279, 285, 296
historicism, 7, 51, 55, 225
historiography, 3–5, 55–71, 214, 239
history, 3, 5–20, 28, 31–2, 37, 41, 44, 48, 55–6, 58–60, 62–73, 118, 123, 132, 136–7, 145, 147, 156, 158, 187, 202–4, 210, 215, 221, 223–6, 228, 236–8, 247–8, 257, 267, 278–9, 286–94, 296, 299, 302–4, 331, 333
  and demarcation, 39–41
  and discourse theory, 2–53
  and empiricism, 118–41
  and historiography of medicine, 55–78
  and medicine, 78–112
  and millenarianism, 78–116
  and rationalism, 118–42
Hitler, Adolf, 253
Hobbes, Thomas, 217, 268
Hobhouse, Stephen, 97
Hodges, publishing house, 298
Hoffmann, Friedrich, 61, 121–2, 126, 133, 257
Hogarth, William, 181
Holland, 39, 189
Holland, Richard, *Practical Observations on the Small Pox*, 130
Holub, Miroslav, 5, 38
homeopathy, 28
Homer, 4, 49
Hone, J. M., 170
Hoofnail, J., *New Practical Improvements and Observations or Some of the Experiments and Considerations Touching Colors*, 132
Hooke, Robert, 281–2
  and *Micrographia*, 301
Horace, 267
Horstius, Jacobus, 184
Howard, John, 181
humanism, varieties of, 18–19, 33, 50, 56, 58

humanists, 28–9, 32–5, 50, 202
humanities, 3, 26, 55, 213, 219, 236, 241–2, 246, 253
humanity, 2
Hume, David, 80, 93, 132, 137, 193
  and *Treatise of Human Nature*, 193
Hunter, Richard, 60, 121, 128, 131
Hunter, William, 135, 184–5
Huntingdon, Countess of, 80, 95
Hurlock, Joseph, *A Practical Treatise upon Dentition*, 130
Husserl, Edmund, 248, 260
Hutton, James, 210
Huxham, J., *Observations de aere*, 133
Huxley, T. H., 220–2, 238, 247
Huygens, Christiaan, 281
Huysmans, Joris Karl, French symbolist novelist and decadent, 254
hydrostatics, 273, 278, 282
Hygeia, 153
hypertension, 12
Hyppolite, Jean, 257
hysteria, 31, 47, 177

iatrochemists, 121
iatromathematics, 82–3, 87, 121, 124–6, 130, 133
iatromechanics, 121, 124, 128, 133
Ibsen, Henrik, *Peer Gynt*, 3
iconography, 13, 17
idealism, 256, 258–9
ideologies, 27, 55, 118, 145, 202, 213, 240, 255, 260, 262, 264–70, 325
illness, 2, 9, 11–12, 19, 36, 38, 41–2, 44, 46–7, 86–7, 89, 94, 96, 104, 169, 176
imagination and theories of imagination, 5–19, 33, 36, 38–40, 44–5, 48, 51, 57, 78, 80, 82, 99, 103, 167, 176, 178–9, 182–6, 194, 202, 207, 217, 220, 224, 226–9, 236, 239, 247, 254–5, 257, 284, 291, 294–8, 300, 302, 332–3, 336
immortality, 106
Impressionism, 3–4, 224
Indiana University, 246
Indians, 192
  Narragansett, 146
Industrial Revolution, 243
infirmity, 45
influence models, 5–6
Ingram, Dale, 128, 131
Innys, William, 298–9
insanity, 9, 30, 137
intemperance, 45
internalists, 7–8, 14–15
Ireland, 148, 151–2, 161, 168, 302
  and Irish, 4, 145–6, 154, 161

Irvine, University of California at, 58
Isis, 213
Italy, 166, 336

Jacob, Margaret, C., 63, 269
  and *The Newtonians and the English Revolution*, 61–2
Jago, Richard, 283
Jamaica, 150, 190
James, Alice, 43
James, Robert, 124, 192, 279
  and *A Medical Dictionary*, 130, 279
James, Thomas, 271
Jarcho, Dr Saul, 151–2
Jerusalem, New, 86–7, 89, 243–4, 327
Jessup, T. E., 147
Jews, 102
Job, 17
Johns Hopkins University, 203
Johnson, Joseph, 301
Johnson, R. F., 300, 303–4
Johnson, Samuel, 36, 39–40, 42, 72, 80, 219, 289, 291–2
  and *Life of Boerhaave*, 34
  and *Rasselas*, 3, 240
Jones, R. F., 123, 226
Jonson, Ben, 6
Josselin, Ralph, 43
Joyce, James, 6
  and *Ulysses*, 3
Joyce, Jeremiah, *Scientific Diaglogues . . . in Six Volumes for Children*, 298
Jung, Carl, 18, 32, 43, 209, 223, 255, 257
Jurin, James, 157, 160

Kabbala, 89
Kafka, Franz, 38
Kant, Immanuel, 241, 245, 254–5, 260
Kaufman, Paul, 291, 303
Keats, John, 3, 5, 30–1, 33, 38, 42, 51
  and 'Cold Pastoral', 237
  and 'Ode on a Grecian Urn', 237
Keill, James, 82–3
Keill, John, 283
Keir, P., *Enquiry into the Nature and Virtues of the Medicinal Waters of Bristol*, 133
Keith, James, 103
Keller, Evelyn Fox, 244
Kenrick, William, 288–9
Kepler, Johannes, 216
Keuls, Eva, 244
Keynes, Geoffrey, 39, 120
Kilkenny College, 149
King, Lester, 63, 119, 122
  and *The Philosophy of Medicine: the Early Eighteenth Century*, 61–2

  and *The Road to Medical Enlightenment*, 61–2
Kinneir, D. Bayne, *Essay on the Doctrine of the Animal Spirits*, 133
Kircher, Anthanasius, *Mundus Subterraneus*, 304
Knapp, Professor Lewis, 185
Knight, J. Baker, 330
Knight, Thomas, 159–60
  and *Reflections on Catholicons or Universal Medicines*, 153
Koestler, Arthur, 238, 260
Korshin, Paul, 118
Kuhn, Thomas S., xi, 224, 238, 258–61
  and *Structure of Scientific Revolutions*, 258

Lacan, Jacques, 51, 258
Laing, R. D., 7
Lallemand, Claude François, physician and medical author, 68
Lamotte, Charles, author of *An Essay upon the state and condition of physicians among the antients* (1728), 124
Lancashire, 186
Landa, Louis, 145
language, 4, 8, 10, 18–19, 32, 34–5, 37–8, 40–1, 46, 49, 51, 56, 67, 203, 214–22, 226–9, 240, 256, 270, 289–90, 295
  and English, 3–4, 39–40, 44, 46, 61, 63, 65–6, 72, 78, 83, 85, 91–4, 97–103, 120, 125–6, 134, 136, 145, 157, 161, 165–6, 177, 185–6, 192, 206, 214, 216, 218–19, 221, 225, 237, 239–40, 242, 253, 256–60, 264–70, 272, 275, 284–5, 287–8, 290, 295–6, 299–300, 304, 325–38
  and French, 15–16, 57–8, 60, 65, 68, 72, 86, 88, 103, 145, 153, 185, 189, 214, 221, 223–5, 227, 240, 246, 254, 256, 258, 260, 265, 280, 288, 296, 304
  and Latin, 73, 84, 153, 214
  and Spanish, 66, 83, 126
Last, Sarah, 188
Latour, Bruno, xi
Laudan, Laurens, 132
Lautréamont, Comte de (Isidore Ducasse), hallucinatory symbolist writer, 254
Lavoisier, Antoine, 217
Law, William, 97–8, 103
  and *Serious Call to a Devout and Holy Life*, 89
Lawrence, D. H., 6
Lead, Jane, 82
Leake, James, 91
Leavis, F. R., 237–8, 240, 245–6
LeClerc, Jean, 85, 124

Index

Lecourt, Dominique, 258–61
Ledran, H. F., *Practical Treatise on Gunshot Wounds*, 130
Leeuwenhoek, Antony van, 121, 282
Leibniz's 'imaginary cells', x
Lemery, M. L., 192
Lenin, Vladimir Ilyich, 258
Leonardo da Vinci, 222
leprosy, 47
Lesser, F. C., 286
Lessius, Leonardus (1554–1623), author on health and long life, 91, 95
see also Cornaro, Luigi
leukaemia, 8, 37
Levine, Joseph M., 63
and *Dr Woodward's Shield: History, Science and Satire in Augustan England*, 61
Lévis-Strauss, Claude, xiii
Levy, Hyman, 239
Leyden, 83, 183, 189
University of, 124
Lhuyd, Edward, 288
Lichfield, 39
life, 14
Lindeboom, G. A., 60
linguistics, 224
Linnaeus, 135, 289–92, 304
Linnell, John, 147
Lintot, publishing house, 298–9
Lisbon, 169
literature (as discourse), 2–20, 26–51, 55–71, 82, 124, 137, 147, 176, 182, 202–10, 213–29, 236–48, 254–62, 264–305
literature and medicine, 2–20, 26–51, 55–71, 82, 124, 137, 176, 182, 224, 240, 257, 279
and George Berkeley, 145–72
and the case history, 10–11
and catharsis, 35–9
and the discourses of, 49–51
and the early English novel, 180–93
and the future, 17–19
and historiography, 3–5, 55–71
and the image of the physician and patient, 12–13
and interdisciplinary discussion, 5
and internalist–externalist problem, 7–8
and literature and science (as converging discourse), 2–20, 38, 47, 62, 147, 202–10, 213–29, 236–48, 254–62, 264–305
and metaphor, 2
and a methodology of inter-relationship, 29
and patients as authors, 41–9

and the physician as cultural hero and anti-hero, 16–17
and the physician as humanist, 32–5
and the physician as writer, 13–16
and the problem of demarcation, 39–41
and the problem of influence, 5–6
and publishing controversies, 145–72
and science fiction, 11–12
as emerging discipline, 2–75
in history, 4
in relation to theory, 2–75 *passim*
literature and science, 2–20, 38, 47, 62, 147, 202–10, 213–29, 236–48, 254–62, 264–305
and American theory, 224–8
and authors of, 305–9
and Gaston Bachelard, 253–61
and contemporary theory, 226–37
and discourse theory, 2–53, 202–50
and discourses of, 213–29
and empiricism, 223–31
and the Enlightenment, 215–23
and evidence, 224–37
and Foucault, 226–52 *passim*
and French theory, 223–36
and glossalalia, 215–16
and Grub Street, 325–40
and hermeneutics, 223–6
and historicism, 221–30
and the history of literature and science, 214–21
and the history of science, 221–37
and the imagination, 294–8
and the influence model, 206–7
and the influence problem, 206–7
and literalism, 227–9
and literature and medicine (as discipline), 2–20, 26–51, 55–71, 82, 124, 137, 176, 182, 224, 240, 257, 279
and mathematics, 225–7
and metaphor theory, 2, 8, 17, 31–2, 40–3, 48, 50, 50–6, 65–7, 227–9, 244, 255, 295
and metaphoric displacement, 226–8
and national literatures, 221–2
and natural philosophy, 270–94
and New Criticism, 15, 51, 205, 209, 223, 256
and Marjorie Hope Nicolson, 205–6
and the nineteenth century, 221–7
and philology, 203–4
and pluralism, 209–11
and positivism, x–xiii, 221–38 *passim*
and practice, 2–53
and readers, 264–305
and Romanticism, 222–8

and science fiction, 207–8
and the Scriblerians, 325–40
and C. P. Snow, 223, 237–8, 240, 243, 246–7
and specialisation, 216–20
and structuralism, 208–9
and theory, 2–53, 204–5
and traditional criticism, 203–4
and the 'two cultures', 216–33, 237–8, 240, 243–7
and universal languages, 214–16
and the Victorians, 220–3
Liverpool, 150
Lobb, Theophilus, *Practical Treatise of Painful Distempers*, 133
  and *Rational Methods of Curing Fevers*, 126
  and *A Treatise on Dissolvements of the Stone*, 133
Locke, John, 80, 89, 119, 123, 132, 137, 170, 215–19, 243, 271, 284–5, 296, 327
  and *An Essay Concerning Human Understanding*, 137, 193, 216–17
Lodge, Thomas, 30
London, 47, 83, 85–9, 131, 146, 151–2, 161, 163, 165, 181–3, 185, 189–92, 239, 268, 271, 276, 291, 299–300, 302, 328, 330, 332, 336, 338
  and Royal College of Physicians of, 126
  School of Economics, 118
love, 10, 57, 80, 87, 95, 100–1, 183
Lovejoy, Arthur O., 145, 204–5, 209, 221
Luce, A. A., 147–8, 151
Lucretius, 135
  and *De Rerum Natura*, 3
Lutherans, 45
Lydgate, Tertius, 3, 17
Lyell, Sir Charles, 220
lymphatics, 69
Lyotard, Jean-François, xi
Lyttelton, George, 1st Lord of, 79

Macalpine, Ida, 60
Machiavelli, Niccolò, 217
Mack, Maynard, 295
Maclaurin, Colin, 283
McLuhan, Marshall, 65
McSparran, the Revd James, author of *America Dissected* (1771), 146
Madden Richard, *Phantasmata or Illusions*, 9, 31
madness, 14, 55, 137
Mahomet (Mohammed), 335
maladies, 9, 11
Malebranche, Nicolas, 265
  and *Search after Truth*, 280
Mallarmé, Stéphane, 295

Manchester, 131
Mandeville, Bernard, 30, 90, 189
  and *Treatise of Hysterick and Hypochondriac Passions*, 129
mania, 47
Mann, Horace, Walpole's correspondent, 164, 224, 336–7
Mann, Thomas, 2, 18, 38, 47
  and *Magic Mountain*, 3–5
manners, 49
Mannheim, Karl, 268
Manning, Richard, *The Symptoms, Nature, Causes and Cure of the Febricula*, 130
Manningham, Richard 182
  and *Obstetrical Works in Latin*, 133
Mapp, Sarah, 13
Marcets, Jane, 297
Marion, Elie, 86
Marten, John, *Treatise of the Gout*, 132
Martin, Benjamin, 293–4, 337
Martin, George, *Essays Medical and Philosophical*, 133
Martyn, John, 269
Marx, Karl, 15, 26, 48, 51, 103, 225, 239, 257–9, 261
Mason, Simon, 131
materialism, 104, 258
mathematics, 81–2, 87, 94, 102, 124–5, 131, 137, 213, 215–17, 221, 225, 239, 270–1, 273–5, 279, 285, 325–6, 338
Maubray, Jon, *The Female Physician, Containing all the Diseases Incident to that Sex, in Virgins, Wives, and Widows*, 180
Mauclerc, John Henry, 184–5
Mauquest de al Motte, Guillaume (1655–1737), author of medical and surgical works and prolific commentator on midwifery, 130
Mayow, John, 123
  and *Tractatus Quinque*, 123
Mazzeo, Joseph, 206
Mead, Margaret, xiii
Mead, Dr Richard, 82, 126–7
mechanism, 105, 278
Meckel, J. F., 67–9
Medawar, Sir Peter, 238, 260
media, 6
medicine, 2–20, 26–51, 55–71, 78–106, 118–38, 145, 148–9, 156–8, 164, 167–9, 176–8, 182, 194, 224, 240, 257, 279
  contributing factors of, 134–8
  see also literature and medicine
melancholy, 4, 66, 90, 92–4
Melville, Herman, 46

Memoirs of Martinus Scriblerus, 329–30
Merquior, J. G., commentator on Foucault, 55
Merton, Robert, 207
metaphors and metaphor theory, 2, 8, 17, 31–2, 40–3, 48, 50, 55–6, 65, 67, 227–9, 244, 255, 295
and dualists, 227–9
metaphysics, 50, 101, 123, 125, 136, 163, 217, 254, 278
Methodism, 46
and Methodists, 45, 80
methodology, 28, 55, 69, 120, 208–9, 213, 228, 254, 275
Metternich, Baron, 89
Meyer, G. D., 280–1
Michelangelo, 222
Middle Ages, 45, 222, 296
Midwinter, David, 279
Mill, John Stuart, 134, 220
Millar, Andrew, 292, 298
Millar, John, 129–30
and *Observations on the Prevailing Diseases in Great Britain*, 129
millenarianism, 78–106
and medicine, 81–94
see also Cheyne, Dr George
Milton, 4, 203, 206, 217, 243–4, 326
Minerva, 332
modernism, 38, 256
Molière, Jean Baptiste Poquelin, 18
and *Le médecin malgré lui*, 4
Molyneux, Samuel, 181
monism, 241
Monro, Dr Thomas, 184–5, 308
Montagu, Elizabeth, 44–6
Montagu, Lady Mary Wortley, 166, 189, 335
Moore, Cecil, 93, 103
morality, 48
Morgagni, Giovanni Battista, 128
Morgan, John, 133
Morley, John, 220
mortality, 37
Muller, John, 274
Murray, John, publishing house, 298
myocarditis, 37
mysticism, 82, 96–8, 100, 105
mythology, 6, 57, 73, 210, 255–6
myth, 43

Namier, Sir Lewis (British historian), 120
Napier, 43
Napoleon, I, 298
Nash, Beau, 98
Nature, 87, 128, 136, 150–1, 155, 191, 214, 217–18, 227, 237, 254, 256, 269–70, 294, 303
Nelson, Gilbert, 91
nerves, 69, 94, 96, 134
Nesse, Christopher, 327
neurophysiology, 2, 78
Neve, Richard *Apopiroscopy: a Faithful History of Observations and Experiments*, 278
Newbery, John, 293, 297
Newcastle, Margaret Holles Cavendish, Duchess of, 280
Newcastle, Thomas Pelham-Holles, Duke of, 151, 296
Newcastle upon Tyne, Northumberland, 131
New Critics, the, 15, 51, 205, 209, 223, 256
New Jerusalem, the, see Jerusalem
Newport, 146
Newton, Sir Isaac, 40, 62, 78, 82–8, 91, 93–5, 97–8, 100–2, 104–5, 119, 124, 126–7, 135, 137, 154, 170, 182, 206–7, 215–17, 219, 223, 237, 242–5, 265, 269, 272, 274, 281–90, 298, 303–4, 325
and *Principia*, 217, 273, 282–4
New York, 151–2, 236
University, 206
Nicea, Council of, 86
Nicholson, William, 333
and *Introduction to Natural Philosophy*, 301
Nicolson, Marjorie Hope, 145, 176, 202–3, 205–10, 221, 238, 280, 293–5, 328
and *Newton Demands the Muse*, 282
see also literature and medicine; literature and science; Lovejoy, Arthur O.
Nietzsche, Friedrich, 254
Nihell, James, *New Observations Concerning the Pulse*, 130
Nile, 152
nineteenth century, the, 3, 8–9, 15, 17, 46, 48, 67, 134, 137, 146, 220, 241, 266, 279
Norfolk, Duke of, 296
Norway, 158, 162
Nourse, John, 291, 298
Novalis (F. von Hardenberg), 257

obesity, 81
Oedipus, 36
Oliver, Dr William, 101
Ophelia, 257
organicism, 105
Orrery, John Boyle, 5th Earl of, admirer of Swift, 91
Osborne, Thomas, 299

354

Oughtred, William, 272
Ould, Fielding, *Treatise of Midwifery*, 130, 186
Ovid, 57–8
Oxford, Earl of, *see* Harley, Robert

pain, 2, 16, 42–5. 48–9
painting, 9, 17, 73, 135, 219, 264
Paley, William, 286
pantheism, 105
Paracelsus, 91
Paré, Ambroise, 185
Paris, 42, 68, 161, 257, 259, 272, 334
Paris, John, *Philosophy in Sport made Science in Earnest: being an Attempt to Illustrate the First Principles of Natural Philosophy*, 279
Parker, E., 334–5
Parkinson, James, 134
Parkinson, Thomas, 192
Parnell, Thomas, 325
Parry-Jones, W., 60
  and *An Essay on the First Principles of Natural Philosophy*, 284
Parson, James, *A Mechanical Enquiry . . . into Hermaphrodites*, 126, 188
passions, 6, 36
pathology, 68
Pavlov, Ivan Petrovich, 35
Payne, Tom, 290
Pêcheux, Michel, 265
Peking, 328
Pemberton, Henry, 283
Pennant, Thomas, *British Zoology*, 293
Pepys, Samuel, 43
Perceval, Thomas, *A Narrative of the Treatment Experienced by a Gentleman, During a State of Mental Derangement*, 8–10
Percival, Sir John, 165
Perrault, Charles, 296
Pessoa, Alberto, 5
Peterborough, 131
Peterborough, Charles Mordaunt, 3rd Earl of, friend of Pope, 164
Peterfreund, Stuart, 242, 247–8
Petty, William, 123
phenomenology, 41
Philadelphians, 82–3, 103
Philips, Ambrose, *Free-Thinker*, 281
philologists, 204–5, 207
philosophy, 41, 44–5, 50, 62, 64, 72, 78, 80, 87–8, 96, 100, 119–20, 123, 136–7, 147, 150, 152, 160, 163, 166, 169–70, 193, 202, 208, 215, 218, 221, 224–7, 236, 253–5, 257, 259, 261, 264–94,

277–8, 280–2, 285–8, 290, 293, 300–2, 327
  and philosophers, 13, 63, 86, 118, 127, 122–3, 127, 145–6, 150, 179, 181, 192, 194, 202–3, 216–17, 227, 242, 244, 255, 257, 269, 280, 338
Philpotts's Library, 302
Phoebus, 153
physicians, *see* literature and medicine
physics, 3, 15, 101, 124–5, 137, 163, 215, 221, 253, 273, 275, 284, 287
physiology, 2, 4, 36, 80, 85, 87, 94–5, 99, 124–5, 136, 159, 194, 260
Pierce, C. S., 261
pineapples, 176–94
Pire, François, 257
Pitcairne, Archibald, 61, 81–3, 101–2, 105, 124
pity, 35–6
plagues, 48, 153, 162
Plath, Sylvia, 18, 37
Plato, 38, 90, 103–5, 135, 215, 217, 245
Plautus, 179
Pleshcheyev, Aleksyei Nikolaevich, political exile and friend of Chekhov, 46
Plomer, H. R., 298
Pluche, Noël Antoine (1688–1761), author of *Spectacle de la nature*, 286
Plumb, Professor J. H., 279
pluralism, 41, 58, 298–10, 225–6
  see also literature and science
Plutarch, 135
pneumatics, 278
pneumonia, 47
poetry, 6, 40, 51, 80–1, 148, 154–6, 158, 165–6, 169–71, 181–2, 187, 191–2, 204, 216–17, 219, 223, 228, 256, 258, 261, 282–3, 287, 290, 325, 332, 338
  and poets, 4–5, 13–14, 29, 36, 40, 80, 187, 191, 202, 208, 217, 237, 243, 255, 284, 293, 295
Poiret, Pierre, disciple of Mme Jeanne Guyon, 96, 98
polemicists, 73
politics, 15–16, 46–7, 72, 80, 123, 216, 262, 299, 327
Pope, Alexander, 39–40, 42–3, 46, 72, 78–80, 98, 102, 138, 145, 157, 159, 164–6, 168, 181–2, 190–1, 204, 206, 210, 219, 247, 267, 270, 275, 277, 280, 283, 295, 298–9, 303–5, 325–6, 328–36
  and *An Essay on Man*, 92, 101, 287
Popper, Karl, 227, 237–8, 260
pornography, 44
Porterfield, William, 80
Portsmouth Dockyard, 275

positivism, 220–1, 255–6, 258
  see also literature and science
Posnett, Macaulay, 221
postmodernism, 55, 213, 256
Pouchet, Felix, 31
Poulet, Georges, 67, 257, 261
Pound, Ezra, 253
pregnancy, 176–94
Price, de Solla, 267, 269
pride, 45
Priestley, Joseph, 101, 119, 217, 301
Princeton, 145, 148, 208
Prior, Thomas, 148, 150, 158–60, 166
  and *An Authentic Narrative of the Success of Tar-Water, in curing a great Number and Variety of Distempers*, 149, 153, 155
Protestantism, 269, 327
Proust, Marcel, 4, 18, 30–1, 42, 47
  and *Remembrance of Things Past*, 3–4
psyche, 36
psychiatry, 10, 15, 46, 51
  and psychiatrists, 11–12
psychology, 2, 6, 10, 34, 36, 46, 69, 90, 210, 221, 226–7, 245
psychotherapy, 34
puberty, 57
Pulteney, Richard, 290
  and *General View of the Writings of Linnaeus*, 290
purging, 36
Puritans, 9, 72, 265
Pyrmont, town in Hanover, Germany, 66
Pythagoras, 135
  and Pythagoreanism, 103–4
  see also Cheyne, Dr George

Quietism, 102–4
Quincy, Dr John, 133, 137
  and *London Pharmacopoeia*, 125
Quixote, Don, 79

race, 145
Ramsay, Dr James, 303
Ranelagh, park in Chelsea, London, 151
rationalism, 118–38, 227–8
Ray, John, 215, 218, 286–8
Read, Richard, *Secrets of Art and Nature . . . being the Summe and Sustance of Natural Philosophy*, 278
reason, 51, 127, 158, 183, 185, 245, 266
  Age of, 93, 101
  mechanical, 125–6
Récamier, Joseph Claude Anthelme (1774–1852), 68
reductionism, 50
Reeve, Thomas, 156

regeneration, 36
Reid, A., 132, 137, 160
  and *A Letter to Dr Hales Concerning the Nature of Tar*, 159
  and *Essay on the Intellectual Powers of Man*, 137
religion, 46, 80, 82, 89, 96–7, 99, 102–4, 123, 128, 216, 266–7, 278–9
Remonstrant-Arminian Seminary, Amsterdam, 85
  see also LeClerc, Jean
Renaissance, 4, 13, 28, 48–9, 206, 222, 248, 261, 288, 291, 296, 304
  and image of the 'Renaissance Man', x, xiv
Rendle-Short, John, medical historian, 63
Restoration, 9, 30, 39, 104, 123–4, 188, 206, 265–8, 270–1, 278–80, 282, 288, 293, 296, 300, 303, 327, 333, 337
Revelation, Book of, 94
revivalism, 55
Rhode Island, 145–6, 149
Richards, I. A., 256, 295
Richardson, Richard, 292
Richardson, Samuel, 47–8, 72, 79, 91, 94, 96–9, 104, 281
  and *Clarissa*, 100
Richmond, Sir Herbert, 274
Richmond, Surrey, 189, 237
Ricoeur, Paul, 58, 223, 238
'Risorius', author of *Remarks on the Bishop of Cloyne's Book Entitled Sins*, 160–1
Riverius, François de La Calmette (fl. 1684), 128
Rivington, Joseph, bookseller and printer, 298
Roach, Richard, 92
  and *The Great Crisis*, 89
  and *The Imperial Standard of Messiah Triumphant*, 89
Robinson, Nicholas, 121, 126, 298
Roger, Jacques, 145, 286–8
Rohault, Jacques (1620–75), author of works on natural philosophy, *Cartesian Physics*, 281
  and *System of Natural Philosophy*, 285
Rolandi, quack doctor, dispenser of the *aqua benedicta*, 66
the Romans, 56–8, 327
Romanticism, 3, 5–6, 33, 40, 45, 100, 105, 213, 220, 255–6
  and post-Romanticism, 3, 5, 47
Rome, 61
Rorty, Richard, 238, 255, 259, 261
Rose, John, 189
Rossi, M. M., 170

Rousseau, Jean-Jacques, 9, 100, 257, 297, 333
Rowning, John, *System of Natural Philosophy*, 301
Roxburgh, Earl of, 82
Roxburgh House, Aberdeenshire, centre of millenarian activity, 82–3
Royal Military Academy at Woolwich, 275
Ruskin, John, 247
  and *Political Economy of Art*, 9
Russia, 156
Ryswick, Treaty of, 82

St André, Nathaniel, 181–2
Sandham, Robert, 274
Sandhurst, 275
Santa Cruz, University of California at, 58
Sartre, Jean-Paul, 95, 257
Saussure, Ferdinand de, 261
Sauvage, François Boissier de (1706–67), prolific medical writer, 128
Saxony, 156
Schacterle, Lance, 238, 246–7
Schaffer, Simon, 265, 282
Schenkius, Johann, 184
Schiller, Johann von, 31
schizophrenics, 42
Schnitzler, Arthur, 31, 42
  and *Professor Bernhardi*, 4
Schofield, R., 103
science, 2–20, 26, 28, 38, 47, 55–6, 62, 67, 69, 72–3, 89, 101–2, 105, 119, 131, 135–7, 145, 147, 158, 161, 182, 192, 194, 202–10, 213–29, 236–48, 254–62, 264–305, 325
  authors, 303–9
  fiction, 207–8
  New Science, 120, 124, 206, 277
science fiction, 11–12
scientists, 119, 122, 125, 181, 202–10, 215, 239, 243, 262, 300
Scotland, 83, 89, 135, 178, 302
  and Scottish, 45, 80–3, 88, 132, 178, 284
  and Common-sense School, 132
Scott, Henry, 190–1
Scriptures, Holy, 73, 79, 94, 129
sculpture, 17, 73, 264
scurvy, 92, 98, 150, 154
secularism, 102, 327
Selzer, Richard, surgeon and author, 37–8, 47
semiotics, 49, 225
Senegal, 130
sensuality, 87
Serle, John, servant and gardener of Alexander Pope, 190

seventeenth century, the 6–7, 13, 42–3, 49, 62, 82, 119, 121, 123, 126–7, 130, 132, 179, 192, 206, 218, 227–8, 239, 248, 267, 271, 284, 300
sex, 14, 55, 186, 257, 281
sexuality, 69
  and discourses of sexuality, xiv
Shackleton, Richard, 164
Shaffer, Peter, 18
  Equus, 3, 7
Shaftesbury. Anthony Ashley Cooper, 4th Earl of, 103
Shakespeare, William, 4, 39–40, 42, 79, 179, 203, 326, 338
  and *Measure for Measure*, 179
shamanism, 28
Sharp, S., *A Treatise on the Operations of Surgery*, 133
Shaw, Peter, 4, 30, 51, 126, 276
  and *The Doctor's Dilemma*, 4, 46
Shelley, Percy Bysshe, 210, 254–5
Sherard, William, 292
Sherburn, George, Pope scholar, 333
Sherrington, Sir Charles Scott, *Man of His Nature*, 31
Short, T., *Mineral Waters of Cumberland*, 133
Siberia, 46
signs, 44, 46, 49, 214
sixteenth century, the, 4, 13, 300
Sloane, Sir Hans, 182
Slusser, George, 246
smallpox, 149–50, 152–3
Smart, Christopher, 169–70
Smellie, William, 178, 185, 194
  and *Treatise on the Theory and Practice of Midwifery*, 178, 188
Smith, character in the anonymous *Siris in the Shades* (1744), 162–3
Smollett, Tobias, 5, 7, 17, 30, 51, 98–9, 179, 184–8, 190–4, 292
  and *Ferdinand Count Fathom*, 160
  and *Humphry Clinker*, 4, 45
  and *The Life and Adventures of Sir Launcelot Greaves*, 7
  and *Peregrine Pickle*, 176–94
Snow, C. P., 223, 237–8, 240, 243, 246–7
  and *Two Cultures*, 222
societies, learned, 15, 58, 80, 95, 145, 240, 242
  Manchester Mathematical Society, 277
  Northampton Mathematical Society, 277
  Royal, 83–4, 135, 157–9, 182, 184–7, 265, 267–70, 280, 300, 328, 331, 337
  Spalding Gentleman's Society, 277
  Spitalfields Mathematical Society, 277

Index

socinianism, 333
sociology, 19, 57, 221, 237, 259
  and sociologists, 48, 79, 207
Solzhenitsyn, Alexander, 18
  and *Cancer Ward*, 4, 29
Somerville, Villiam, 80
Sontag, Susan, 19, 42
  and *Illness as Metaphor*, 65
Sorbonne, 257
Soubiran, André, 31
  and *Bedlam*, 13
soul, 17, 89
South Carolina, 146, 152
Spalding, Helen, 239
Spartan ephebes, 55
speech, 50
Spence, Joseph, *Anecdotes*, 159, 165
Spenser, Edmund, 217
Spilsbury, Francis, 192
spirits, 93
spleen, 93–4
Sprat, Bishop Thomas, 267
  and *History of the Royal Society* (1667), 218
  see also societies, learned, Royal
Stahl, Georg Ernst, 61, 121
Stahlians, 121
Stanhope, Philip, 4th Earl of Chesterfield, 80, 150, 155
Starobinski, Jean, 6, 261
Steele, Sir Richard, 55, 277, 328, 332, 336
Stephens, Joanna, 119
stereotypes, 2, 43
Sterne, Laurence, 4, 194, 303, 336–7
  and *Tristram Shandy*, 7, 176, 304, 336
Stevens, George Alexander, *Dramatic History of Master Edward*, 187
Stillingfleet, Benjamin, 290
Stirling, James, 276
Stoicism, neo-, 104
Stone, Lawrence, 227
  and *The Family, Sex and Marriage in England*, 66
Strahan, William, 279
Strawberry Hill, 336–7
Strother, Edward, 126
structuralism, 58, 65, 67, 208–10, 213
Stukeley, William, *Treatise of the Gout*, 133
supernaturalism, 326
superstition, 28
Swift, Johnathan, 4, 79–80, 125, 138, 145, 157, 164, 240, 247, 268, 270, 289, 295, 303–4, 325, 329, 333, 336
  and *Gulliver's Travels*, 3, 205, 297
  and *Tale of a Tub*, 4, 136
Swinburne, Algernon Charles, 257

Sydenham, Dr Thomas, 30, 61, 92, 119–25, 154, 168, 179

Talbot, Miss Catherine, 167, 283
tar water,
  and Batavia, 151
  and the Dutch, 151
Teddington, Surrey, 159
Tellende, Henry, 189
Temkin, Oswei, *The Falling Sickness*, 60
Tenison, Thomas, 269
Tennyson, Lord Alfred, 206
terror, 36
Thackeray, Arnold, 9, 207
Thebes, 57
Theiler, Max, *The Virus*, 31
theology, 101–2, 285–6
theology, physico-, 90, 94
Theophrastus, 135
theory, 26–51, 56, 64, 82, 120, 123, 125–6, 132, 134, 137, 145, 181, 184, 186, 202–3, 213, 225–9, 240, 246, 248, 253–6, 261, 264–5, 282, 291, 297, 330
  see also literature and medicine; literature and science
therapy, 2
Thomas, Lewis, *The New England Journal of Medicine*, 37
Thompson, E. P., *The Poverty of Theory*, 26
Thomson, James, 287
  and 'Summer', 191
Thomson, Dr Thomas, 166
  and *Discourse Concerning the Present State of Physick in Europe*, 133
  and *Syllabus Pointing Out Every Part of the Human System*, 133
Tipper, John, 280–1
Tiresias, 13
Tofts, Mary, 180–2, 184, 187
Tolstoy, Leo, 47
  and *The Death of Ivan Illych*, 4–5
Tom Jones, novel by Fielding, 160, 193
  see also Fielding, Henry
Toplitz, spa at, 66
Toulmin, Stephen, 238
traditionalism, 120, 124, 136, 203–5
tragedy, 36
Trautmann, Joanne, 36
Trembley, Abraham (1710–84), Swiss scientist, 128
Trimmer, Sarah, 297
Trinity, the, 154, 163–4, 170, 272
Trinity College, Dublin, 159
  see also Berkeley, George
Trollope, Anthony, 164

and *Barchester Towers*, 9
Tropham, Edward, *Letters from Edinburgh written in years 1774, and 1775*, 178
Tryon, Thomas, 91
tuberculosis, 65
Tunbridge Wells, 66
Turkey, 335
Turner, Daniel, 126, 133, 178, 183–6
Tussaud, Madame, 337
Tuveson, Ernest, 206
twentieth century, the, 8, 15, 33, 43, 49, 62, 98, 205, 221, 255, 260, 262
Twickenham, 190
'two-cultures', the, 216–24, 237–8, 240, 243, 246–7
  see also Snow, C. P.; literature and medicine; literature and science
Tyers, Thomas, 80
Tyndall, John, 247

Unitarians, 227–8
United States, 8
Updike, John, 146
utilitarianism, 28
utopias, 11–2

valetudinarians, 98
vapours, 93–4
Vartanian, Aram, 206
vegetarians, 91–2
veins, 69
Velikovsky, Immanuel, 208, 329
Vesalius, Andreas, 30, 222
Vichy, France, the spa at, 66
Vienna, 42
Vietnam, 258
Viets, Dr Henry, medical historian, 103
Virgil, 30, 80
Virginia, University of, 206
vitalism, 105, 128
Voltaire, François Marie Arouet de, 243

Wadlington, Katherine, 334–5
Wainwright, Jeremiah, *Mechanical Account of the Non-Naturals*, 126
Wales, Prince of, 181
Walker, George, 301
Wallis, John, *Treatise of Algebra*, 272
Wallis, Peter, 282
Wallis, Ruth, 282
Walpole, Horace, 164, 336–7
Walton, Christopher, 97
Warburton, William, 219, 270
Ward, Joshua, 119
  and 'Pill and Drop', 149, 161, 166, 168
Warren, Austin, 256

Warren, Samuel, *Passages from the Diary of a Late Physician*, 31
Warrington, Lancashire, 273
Warton, Joseph, 39, 165
Waterlands, Daniel, 273
Waterton, Charles, *Essays on Natural History*, 288
Watkins, J. W. N., 118
Watson, Richard, 272
Watts, Isaac, 297
Watts, Thomas, 275
Weber, Sam, xi
Wellek, René, 256
Wertenbaker, Lael T., *Death of a Man*, 18
Wesley, John, 45–6, 80, 119, 156, 163
  and *Primitive Physick*, 156
West Indies, 151, 189, 192
Westminster Hospital, 181
Westminster School, 271
Wetzler, Germany, 69
  see also Meckel, J. F.
Wharton, Thomas, 164
Whewell, William, 134, 243, 247
Whigs, 40, 102, 267, 296
Whiston, William, 86, 88–9, 92, 103, 182, 275–8, 284–5, 325–38
  and *The Cause of the Deluge Demonstrated*, 329
  and *The Copernicus*, 278
  and *Essay on the Revelation of Saint John*, 87
  and *New Theory of the Earth*, 327
White, Gilbert, 286
  and *Natural History of Selborne*, 287, 293
Whitefield, George, 80
Whitman, Walt, 46
Whytt, Robert, 121, 133
Wild, John, 147
Williams, Sir Charles Hanbury, 156, 166
Williams, Raymond, 223
Williams, Reeve, 271
Williams, William Carlos, 5, 12, 17–18, 31, 38, 47
Willis, Thomas, 32, 61, 123, 134, 282
  and *The Experienced Midwife*, 179
Winchester School, 270
Wiseman, Richard, 123
witchcraft, 28
Withering, William, 290
Wolff, Casper Friedrich, 128
women, 28
Wood, William, *Mechanical Essay Upon the Heart*, 126
Woodcock, John, 246–7
Woodward, Dr John, 61, 63, 127
Woolf, Virginia, 26

Index

Woolwich, 275
  see also Royal Military College
Wordsworth, Christopher, 272–3
Wordsworth, William, 47, 105, 219, 244, 247, 283
  and *Lyrical Ballads*, 202, 220, 247
World War II, 56, 204, 221, 225
Worster, Benjamin, 275

Wright, Thomas, 30

Yale University, 206
Yeats, William Butler, 170
York, Yorkshire, 131, 166, 292

Zola, Émile, *Lourdes*, 4
zoology, 278